APPLICATIONS OF REMOTE SENSING IN AGRICULTURE

Proceedings of Previous Easter Schools in Agricultural Science, published by Butterworths, London

*SOIL ZOOLOGY   Edited by D. K. McL. Kevan (1955)
*THE GROWTH OF LEAVES   Edited by F. L. Milthorpe (1956)
*CONTROL OF PLANT ENVIRONMENT   Edited by J. P. Hudson (1957)
*NUTRITION OF THE LEGUMES   Edited by E. G. Hallsworth (1958)
*THE MEASUREMENT OF GRASSLAND PRODUCTIVITY   Edited by J. D. Ivins (1959)
*DIGESTIVE PHYSIOLOGY AND NUTRITION OF THE RUMINANT   Edited by D. Lewis (1960)
*NUTRITION OF PIGS AND POULTRY   Edited by J. T. Morgan and D. Lewis (1961)
*ANTIBIOTICS IN AGRICULTURE   Edited by M. Woodbine (1962)
*THE GROWTH OF THE POTATO   Edited by J. D. Ivins and F. L. Milthorpe (1963)
*EXPERIMENTAL PEDOLOGY   Edited by E. G. Hallsworth and D. V. Crawford (1964)
*THE GROWTH OF CEREALS AND GRASSES   Edited by F. L. Milthorpe and J. D. Ivins (1965)
*REPRODUCTION IN THE FEMALE ANIMAL   Edited by G. E. Lamming and E. C. Amoroso (1967)
*GROWTH AND DEVELOPMENT OF MAMMALS   Edited by G. A. Lodge and G. E. Lamming (1968)
*ROOT GROWTH   Edited by W. J. Whittington (1968)
*PROTEINS AS HUMAN FOOD   Edited by R. A. Lawrie (1970)
*LACTATION   Edited by I. R. Falconer (1971)
*PIG PRODUCTION   Edited by D. J. A. Cole (1972)
*SEED ECOLOGY   Edited by W. Heydecker (1973)
 HEAT LOSS FROM ANIMALS AND MAN: ASSESSMENT AND CONTROL   Edited by J. L. Monteith and L. E. Mount (1974)
*MEAT   Edited by D. J. A. Cole and R. A. Lawrie (1975)
*PRINCIPLES OF CATTLE PRODUCTION   Edited by Henry Swan and W. H. Broster (1976)
*LIGHT AND PLANT DEVELOPMENT   Edited by H. Smith (1976)
*PLANT PROTEINS   Edited by G. Norton (1977)
*ANTIBIOTICS AND ANTIBIOSIS IN AGRICULTURE   Edited by M. Woodbine (1977)
*CONTROL OF OVULATION   Edited by D. B. Crighton, N. B. Haynes, G. R. Foxcroft and G. E. Lamming (1978)
*POLYSACCHARIDES IN FOOD   Edited by J. M. V. Blanshard and J. R. Mitchell (1979)
*SEED PRODUCTION   Edited by P. D. Hebblethwaite (1980)
 PROTEIN DEPOSITION IN ANIMALS   Edited by P. J. Buttery and D. B. Lindsay (1980)
 PHYSIOLOGICAL PROCESSES LIMITING PLANT PRODUCTIVITY   Edited by C. Johnson (1981)
 ENVIRONMENTAL ASPECTS OF HOUSING FOR ANIMAL PRODUCTION   Edited by J. A. Clark (1981)
*EFFECTS OF GASEOUS AIR POLLUTION IN AGRICULTURE AND HORTICULTURE
     Edited by M. H. Unsworth and D. P. Ormrod (1982)
 CHEMICAL MANIPULATION OF CROP GROWTH AND DEVELOPMENT   Edited by J. S. McLaren (1982)
 CONTROL OF PIG REPRODUCTION   Edited by D. J. A. Cole and G. R. Foxcroft (1982)
 SHEEP PRODUCTION   Edited by W. Haresign (1983)
*UPGRADING WASTE FOR FEEDS AND FOOD   Edited by D. A. Ledward, A. J. Taylor and R. A. Lawrie (1983)
 FATS IN ANIMAL NUTRITION   Edited by J. Wiseman (1984)
 IMMUNOLOGICAL ASPECTS OF REPRODUCTION IN MAMMALS   Edited by D. B. Crighton (1984)
 ETHYLENE AND PLANT DEVELOPMENT   Edited by J. A. Roberts and G. A. Tucker (1985)
 THE PEA CROP   Edited by P. D. Hebblethwaite, M. C. Heath and T. C. K. Dawkins (1985)
*PLANT TISSUE CULTURE AND ITS AGRICULTURAL APPLICATIONS   Edited by Lindsey A. Withers and P. G. Alderson (1986)
 CONTROL AND MANIPULATION OF ANIMAL GROWTH   Edited by P. J. Buttery, N. B. Haynes and D. B. Lindsay (1986)
 COMPUTER APPLICATIONS IN AGRICULTURAL ENVIRONMENTS   Edited by J. A. Clark, K. Gregson and R. A. Saffell (1986)
 MANIPULATION OF FLOWERING   Edited by J. G. Atherton (1987)
 FOOD STRUCTURE – ITS CREATION AND EVALUATION   Edited by J. R. Mitchell and J. M. V. Blanshard (1988)
 NUTRITION AND LACTATION IN THE DAIRY COW   Edited by P. C. Garnsworthy (1988)
 MANIPULATION OF FRUITING   Edited by C. J. Wright (1989)
 FEEDSTUFF EVALUATION   Edited by J. Wiseman and D. J. A. Cole (1990)
 GENETIC ENGINEERING OF CROP PLANTS   Edited by G. W. Lycett and D. Grierson (1990)

These titles are now out of print but are available in microfiche editions

Applications of Remote Sensing in Agriculture

M. D. STEVEN
Lecturer in Geography, University of Nottingham

J. A. CLARK
Senior Lecturer in Environmental Science, University of Nottingham School of Agriculture

BUTTERWORTHS
London   Boston   Singapore   Sydney   Toronto   Wellington

 PART OF REED INTERNATIONAL P.L.C.

All rights reserved. No part of this publication may be reproduced in any material form (including photocopying or storing it in any medium by electronic means and whether or not transiently or incidentally to some other use of this publication) without the written permission of the copyright owner except in accordance with the provisions of the Copyright, Designs and Patents Act 1988 or under the terms of a licence issued by the Copyright Licensing Agency Ltd, 33–34 Alfred Place, London, England WC1E 7DP. Applications for the copyright owner's written permission to reproduce any part of this publication should be addressed to the Publishers.

Warning: The doing of an unauthorised act in relation to a copyright work may result in both a civil claim for damages and criminal prosecution.

This book is sold subject to the Standard Conditions of Sale of Net Books and may not be re-sold in the UK below the net price given by the Publishers in their current price list.

First published 1990

© **Contributors 1990**

British Library Cataloguing in Publication Data

Clark, J. A. (Jeremy Austin), *1938–*
  Applications of remote sensing in agriculture.
  1. Great Britain. Agriculture
  I. Title   II. Steven, M. D.
  338.16

  ISBN 0-408-04767-4

Library of Congress Cataloging-in-Publication Data

Applications of remote sensing in agriculture/[edited by] J.A.
  Clark, M.D. Steven
      p.   cm.
    Includes bibliographical references
    ISBN 0-408-04767-4 :
    1. Agriculture–Remote sensing. I. Clark, J.A. (Jeremy Austin),
  1939– . II. Steven, M.D.
  S494.5.R4A66   1990
  630′.28–dc20   90-1778

Composition by Genesis Typesetting, Laser Quay, Rochester, Kent
Printed in Great Britain at the University Press, Cambridge

PREFACE

This book represents the proceedings of the 48th Easter School in Agricultural Science held at the University of Nottingham from 3rd -7th April 1989. The meeting attracted 146 delegates from over 22 countries and contributions to this volume come from 9 countries. The editors have divided the contributions on thematic lines according to the agricultural problems addressed, but within each theme a variety of technical approaches are described by different authors. The volume as a whole presents a review of the achievements of remote sensing in agriculture, establishes the state of the art and gives pointers to future developments.

At an international gathering of this nature there is naturally a diversity of approach, but there appears a surprising unity of purpose reflected in the recurrent theme of estimating the production and predicting the yields of crops. Indeed, this concern has an ancient history, for in the book of Genesis, chapter 41, we find the story of Joseph (he of the coat of many colours) interpreting the dreams of the Pharaoh to predict seven years of plenty followed by seven years of famine. The story serves as a reminder that the real needs of agricultural and economic planning are for long-term forecasts: Few would rely on remote sensing techniques to predict yields as far as 14 years ahead, despite the considerable scientific progress reported in this volume!

The characteristic spectral signature of vegetation in the visible and near-infrared bands was recognised early in the history of remote sensing and applications of remote sensing in agriculture were envisaged in the original specifications of Landsat. It is perhaps surprising, then, that satellite remote sensing techniques have found few "real-world" applications in agriculture, despite some early successes such as the LACIE programme. A large part of the difficulty in application has been due to logistics. Satellite systems that have the resolution to recognize individual fields are constrained to repeat cycles that are too long to characterize the growing season of a typical annual crop, especially when loss of data under cloud cover is considered.

A number of approaches to resolve this problem are reported in this volume. The French SPOT satellite can direct its sensors towards specified targets that are off the immediate sub-satellite track. This system can thus monitor a specified area on several occasions within its 26-day repeat cycle, but with the constraint that selection of a particular target for frequent monitoring is made at the expense of neighbouring potential targets. A second approach uses low-resolution satellite data to monitor regional phenological change or associated meteorological factors. A third approach is to use microwaves, either passively, using natural emissions, or actively, with radar. Microwaves are unaffected by cloud cover but their interactions with vegetation and soil are complex and the match between the information content of microwaves and the requirements of agricultural users is as yet unclear. A fourth approach is to liberate the remote sensing techniques from the constraints of a satellite platform. Aerial photography is still the best option for investigation of many agricultural problems at scales smaller than the field. And, in the future, 'ultra-low altitude' platforms such as tractors may allow the use of the full range of monitoring techniques to solve the down-to-earth problems of farmers, rather than those of governments or commodity brokers.

M.D. Steven
J.A. Clark

ACKNOWLEDGEMENTS

We would like to extend our thanks to the many people who helped to make the conference a success: To Tim Malthus and Mark Danson who provided invaluable help in the organization and administration of the meeting; To Brenda Champion and Patricia Steven who ran the registration desk and sorted out many of the "real-world" problems of participants; and to John Corrie who managed the visual aids. We also thank the team at Butterworths, particularly Patricia Horwood and Sharon Cooper, for their help in producing this volume, and John Burgess, who tackled the typesetting 'in-house', overcoming his own limited experience, and the complexities of the text program provided on the University Computer.

The Editors and the University of Nottingham School of Agriculture are grateful to the following organizations for their generous financial assistance towards the expenses of the School:

U.K. Overseas Development Administration,
Campbell Scientific Ltd.,
The Royal Society,
Eurimage,
Logica Space and Defence Systems Ltd.,
The Remote Sensing Society.

CONTENTS

Preface		v
Acknowledgements		vii
Contents		ix

I. PRINCIPLES

1 SENSORS, PLATFORMS AND APPLICATIONS; ACQUIRING
  AND MANAGING REMOTELY SENSED DATA   3
  J.A. Allan, *School of Oriental and African Studies, University of London, Russell Square, London WC1H 0XG, U.K.*

2 OPTICAL PROPERTIES OF VEGETATION CANOPIES   19
  G. Guyot, *INRA Bioclimatologie, BP 91, 84143 Montfavet Cedex, France*

3 FACTORS AFFECTING THE RADIATIVE TEMPERATURE
  OF A VEGETATIVE CANOPY   45
  P. Boissard, *INRA Bioclimatologie, 78850 Thiverval-Grignon, France*,
  G. Guyot, *INRA Bioclimatologie, Domaine St. Paul B. P. 91, 84143 Montfavet Cedex, France*, and R.D. Jackson, *USDA, ARS, U.S. Water Conservation Laboratory, Phoenix, Arizona 85040, U.S.A.*

II. CLIMATE AND SOIL

4 DISCRIMINATION AND MONITORING OF SOILS   75
  R. Evans, *Department of Geography, University of Cambridge, Downing Street, Cambridge CB2 3EN, U.K.*

5 ESTIMATION OF RAINFALL USING GEOSTATIONARY
  SATELLITE DATA   97
  J.R. Milford and G. Dugdale, *Department of Meteorology, University of Reading, U.K.*

6 APPLICATION OF REMOTE SENSING AND
  GEOGRAPHICAL INFORMATION SYSTEMS IN WATER
  MANAGEMENT   111
  G.J.A. Nieuwenhuis, J.W. Miltenburg and H.A.M. Thunnissen, *The Winand Staring Centre for Integrated Land, Soil and Water Research, P.O. Box 125, 6700 AC Wageningen, The Netherlands.*

III. LAND CLASSIFICATION AND CROP INVENTORIES

7   THEORETICAL PROBLEMS IN IMAGE CLASSIFICATION   127
P.M. Mather, *Department of Geography, University of Nottingham, Nottingham, NG7 2RD, U.K.*

8   ESTIMATING PRODUCTION OF WINTER WHEAT BY REMOTE SENSING AND UNIFIED GROUND NETWORK. I. SYSTEM VERIFICATION   137
Liu Guoxiang and Zheng Dawei, *Beijing Municipal Academy of Agricultural and Forestry Sciences, Beijing, China*

9   ESTIMATING PRODUCTION OF WINTER WHEAT BY REMOTE SENSING AND UNIFIED GROUND NETWORK. II. NATIONWIDE ESTIMATION OF WHEAT YIELDS   149
Li Yuzhu, *Academy of Meteorological Science, Beijing 100081, China.*

10   CROP INVENTORY STUDIES USING LANDSAT DATA ON A LARGE AREA IN HUNGARY   159
G. Csornai, O. Dalia, J. Farkasfalvy and G. Nádor, *FÖMI Remote Sensing Centre, H-1149 Budapest, Bosnyák tér 5, Hungary*

IV. PRODUCTIVITY

11   HIGH TEMPORAL FREQUENCY REMOTE SENSING OF PRIMARY PRODUCTION USING NOAA AVHRR   169
S.D. Prince, *Geography Department, Room 1113, Lefrak Hall, University of Maryland, College Park, Maryland 20742-8225, U.S.A.*

12   ESTIMATING GRASSLAND BIOMASS USING REMOTELY SENSED DATA   185
E.T. Kanemasu, T.H. Demetriades-Shah and H. Su, *Evapotranspiration Laboratory, Waters Annex, Department of Agronomy, Kansas State University, Manhattan, Kansas, 66502, U.S.A.* and A.R.G. Lang, *CSIRO Center for Environmental Mechanics, GPO Box 821, Canberra, ACT 2601, Australia.*

13   REMOTE SENSING TO PREDICT THE YIELD OF SUGAR BEET IN ENGLAND   201
K.W. Jaggard and C.J.A. Clark, *AFRC Institute of Arable Crops Research, Broom's Barn, Higham, Bury St. Edmunds, Suffolk IP28 6NP, U.K.*

V. STRESS

14 HIGH-SPECTRAL RESOLUTION INDICES FOR CROP STRESS   209
M.D. Steven, T.J. Malthus, T.H. Demetriades-Shah, F.M. Danson and J.A. Clark, *University of Nottingham, U.K.*

15 THE IDENTIFICATION OF CROP DISEASE AND STRESS BY AERIAL PHOTOGRAPHY   229
R.H. Blakeman, *Aerial Photography Unit, ADAS, Block B, Government Buildings, Brooklands Avenue, Cambridge, CB2 2DR, U.K.*

16 ESTIMATION OF PLANT WATER STATUS FROM CANOPY TEMPERATURE: AN ANALYSIS OF THE INVERSE PROBLEM   255
G.S. Campbell, *Department of Agronomy and Soils, Washington State University, Pullman, Washington, U.S.A.* and J.M. Norman, *Department of Soil Science, University of Wisconsin, Madison, Wisconsin, U.S.A.*

17 A SIMPLIFIED ALGORITHM FOR THE EVALUATION OF FROST-AFFECTED CITRUS   273
M.A. Gilabert, D. Seggara and J. Meliá, *Departament de Termodinámica, Facultat de Física, Universitat de Valencia, 46100-Burjassot, Valencia, Spain*

VI. NEW TECHNIQUES

18 APPLICATIONS OF CHLOROPHYLL FLUORESCENCE IN STRESS PHYSIOLOGY AND REMOTE SENSING   287
Hartmut K. Lichtenthaler, *Botanisches Institut (Plant Physiology), University of Karlsruhe, Kaiserstrasse 12, D-7500 Karlsruhe, F.R.G.*

19 APPLICATIONS OF RADAR IN AGRICULTURE   307
M.G. Holmes, *Botany School, University of Cambridge, Downing Street, Cambridge CB2 3EA, U.K.*

20 MICROWAVE RADIOMETRY FOR MONITORING AGRICULTURAL CROPS   331
G. Luzi, S. Paloscia and P. Pampaloni, *Centre for Microwave Remote Sensing, Viale Galileo, 32 - 50125 Firenze, Italy*

21	ON THE USES OF COMBINED OPTICAL AND ACTIVE-MICROWAVE IMAGE DATA FOR AGRICULTURAL APPLICATIONS J.F. Paris, *Department of Geography, California State University, Fresno, CA., 93740-0069, U.S.A.*	355
VII.	OPPORTUNITIES, PROGRESS AND PROSPECTS	
22	REMOTE SENSING IN AGRICULTURE: FROM RESEARCH TO APPLICATIONS C. King, *Remote Sensing Department,* and J. Meyer-Roux, *Project Manager, Institute for Remote Sensing Applications, CEC Joint Research Centre, 21020 Ispra, Italy*	377
23	REMOTE SENSING IN AGRICULTURE: PROGRESS AND PROSPECTS J.L. Monteith, *International Crops Research Institute for the Semi-Arid Tropics, Patancheru P.O., Andhra Pradesh 502 324, India*	397
	LIST OF POSTER PRESENTATIONS	403
	LIST OF PARTICIPANTS	405
	INDEX	413

I.
PRINCIPLES

1

SENSORS, PLATFORMS AND APPLICATIONS; ACQUIRING AND MANAGING REMOTELY SENSED DATA

J.A. ALLAN

School of Oriental and African Studies,
University of London, Russell Square, London WC1H 0XG, U.K.

Introduction

Current and future sensors will be discussed in terms of the evolution of the suite of technologies which contribute to remote sensing and of the requirements of agricultural applications. The initial treatment will be historical, in order to provide a perspective on the significant and enduring place of some platforms and sensors, such as the aircraft and the camera. The main emphasis will be given to the experience gained in the past two decades with multi-spectral scanners and the 'push-broom' devices deployed from space in the civil domain since the early 1980s. The limited operational use of microwave sensing in agricultural applications will be commented upon.

Future platforms and sensors will be discussed in terms of the alternative strategies of synergism and free flying platforms. The commitment of the USA and ESA to Space Station and the Columbus Project indicate an emphasis on the monitoring of the Earth's surface at a small scale for global science, an explicit goal of NASA. Agricultural applications will be served by the 1990's NASA and ESA programmes, but it is argued that it is likely to be other free-flying systems which provide the operational data for agricultural monitoring and management in future.

Some general features of agricultural applications

Agricultural applications have many features in common with other land applications, for example the relevance of the large area covering capacity and the great usefulness of multi-date and multi-spectral data. But agricultural applications differ from the other major land application areas in a number of ways. Agricultural applications are characterized by a number of phenological, land management and economic features, which together ensure that remote sensing has and will continue to play an increasingly significant role in monitoring agricultural tracts, especially at the national level. The community interested in the remote sensing of agricultural areas is large and diverse, and includes scientists, members of national and international government institutions (eg the EEC,

ESA, ASEAN etc), national agencies (eg, USAID, ODA), international agencies (FAO, World Bank, IFAD) as well as consulting organisations and many large and small public companies concerned with resource and environmental management.

Agricultural applications also differ from many major application groupings in that there is often an especially large body of complementary, and often alternative, data with which remotely sensed data have to be integrated or for which they substitute. Often the complementarity between the various data sources is obvious, but on occasions the remotely sensed data appear to compete with land surface data otherwise derived. Without any attempt to prepare a comprehensive list of agricultural applications, it has been easy to list ten major scientific applications and over thirty renewable and non-renewable resource survey and management applications, ranging from land surface climate monitoring to studies of wild-life. While the main role of agricultural applications is, and will continue to be, at the national level, there will be an increasing role for them at the international level (*Table 1.1*).

Although the list of current and potential applications is long and the outcomes significant, not all the potential users of data recorded by remote sensing systems assimilate such information into operational systems with equal facility. The scientific community in the fields of land evaluation and agro-ecological studies is usually only deterred from using such data by their cost. The management community is often also inhibited by cost, but in many cases it is the unfamiliarity of the data which prevents their deployment. Professionals using existing systems are often unwilling to see the advantages of new systems, fearing that existing jobs might be replaced by the comprehensive coverage of satellite observation.

Trends in technological development

Earth observation from airborne platforms has a one hundred and thirty year history, although the majority of the innovation and development has taken place in the last twenty years. The developments of the next two decades are likely to be characterized by rates of growth similar to those of the recent past, despite the problems of funding space platforms and observing systems. While changes have been rapid there are a number of enduring features of Earth observing technologies.

Amongst platforms the aircraft is the most important example, although by now the volume of observation from satellites has massively overtaken air-photo archival materials in terms of coverage if not of scale. The number of platforms in orbit is large and increasing. The Soviet Union has launched more than four times as many missions as the United States, but many of these were of short duration for surveillance purposes. Other nations have the capacity to construct and launch platforms, including France, Japan, and China. Launch services are being provided by the two major participants in the development of space as well as by France and China. A useful survey of the civil and military programmes is available (Warwick, 1985), see *Table 1.2*.

Of sensors, the camera has a classic place but, though present on many space projects as a large format camera, it has been mainly replaced for routine observation by other sensors: the multi-spectral scanners; charge-coupled linear and array devices (CCDs); thermal sensors; microwave sounders and scanners; spectrometers and lidars. Global positioning systems are also becoming commonplace, greatly enhancing the geometric potential of all observing systems. The most important development in sensors will be

TABLE 1.1. Agricultural applications of remote sensing.

Scientific applications of relevance in agriculture

    Crop science
        including all types of
            vegetation indices
        Stress detection
    Agro-climatology
    Agro-ecology
    'Global change'

    Hydrology
        Surface run-off
        Groundwater
    Land surface climatology
    Soil science
    Terrestrial ecosystems
    Vegetation science,
        including forest science

Management applications of remote sensing in agriculture

    Agriculture
        Agro-ecology
        Crop damage
        Crop condition
        Crop inventory - local, regional,
            national and global
        Crop classification
        Livestock inventory
        Yield estimation
        Yield prediction
    Degraded land studies
    Environmental impact studies
        - of agricultural development
        - of urban development
    GIS - contributions to regional
        agricultural information systems
    Irrigated lands
        Inventory
        Monitoring extent
        Crop condition
        Water utilisation
        Regulation of water use

    Land evaluation
    Land use
    Land degradation
        Soil erosion
        Soil salinity
    Land drainage
    Land classification
    Land reclamation
        Agricultural
        Urban
    Pollution
        of soil
        of vegetation
    Rangeland management
        Rangeland condition
        Rangeland classification
        Range production
    Soil survey

TABLE 1.2. The technological history of land applications. The data provided by such technologies had relatively little impact on agriculture until the 1930s when air photographs began to be used for soil survey, for example in the advisory services of the USDA. Both before and since the input of remotely sensed data into topographic mapping was of some importance in farm planning and in rural engineering activities connected with soil conservation, land drainage and irrigation.

	Platforms	Sensors	Data storage	User technology and software
1860	Balloon	Camera Eye/brain	Photograph	Cartographic techniques and the eye/brain
1960	Aircraft Satellites	Camera, LFC and MS MSS scanners for most of e.m. spectrum Radiometers	Photograph Digital tape/disc	Cartographic techniques and the eye/brain Automated digital cartography (ADC) Computer
1985	Aircraft Satellites	Camera, LFC and MS Super MSS and CCD devices Microwave sensors and altimeters Spectrometers Lidars GPS NB. The eye/brain is still useful	Photograph Digital tape/disc Optical	Cartographic techniques and the eye/brain ADC, GIS Microcomputer networked Super-computer
2000	Aircraft Satellites Space stations	Camera - LFC ATSR, SAR, HIRIS, HMMR, MODIS type instruments GPS	Photograph Digital tape/disc	Cartographic Techniques and the eye/brain ADC, GIS A computer beyond superlatives

in the deployment of microwave sensors for all-weather monitoring of the land surface.

Data storage was originally achieved on the photograph, which was read and interpreted by the amazing and apparently unsurpassable eye-brain combination. The eye-brain element of the original user's equipment has also earned an enduring place in the systems of all users of remotely sensed imagery and is still useful. With the improvement of publicly available spatial resolution, to ten metres or better for imagery from space since 1986, the eye-brain has, if anything, increased again in importance through the unique capacities of its acuity and associated intelligence, as yet not remotely emulated by artificial systems. The eye-brain has had other advantages. It proved especially flexible at the interface between photograph and cartographer, and the very large cartographic community has not always been very friendly to the arrival of new digital methods of capturing, storing, selecting and, finally, displaying spatial information. They have even resisted the application of digital techniques to traditional cartographic applications, never mind to data from new Earth observing systems. Undoubtedly, the main technology of the recent past, which has transformed the potential of Earth observing systems of all kinds, is the computer. While the development of platforms and sensors has progressed steadily, it is the computer and its peripherals (especially disk storage and now optical disks) which have very rapidly transformed the storage of, and access to, the massive data sets being assembled daily and yearly from space. More important even than the enabling role of new supercomputers, as well as of networked microcomputers, has been the impact of these new data handling technologies in the area of data integration. The registration of conventionally acquired spatial data with Earth observations is now possible, will become commonplace, and will provide a completely new information environment for scientists and managers of agricultural resources worldwide.

Information needs for agricultural applications

Since 1972, when satellite data became publicly available for Earth observation from the Landsat platform, both scientific and land managing professionals for the first time had data with the coverage and frequency which approximated to their needs. Before 1972 the only spatial records available were map series at scales ranging from 1:250000 to 1:10000, not always of the tracts of interest, which were updated rarely and never more frequently than once in five years. Earth scientists had been told for over a century that the surface cover of interest to them was not worth mapping, because such phenomena were not sufficiently valuable to be mapped, mainly because they changed so rapidly. Of earth scientists, only the geologist and the soil surveyor are undemanding with respect to the temporal resolution of land surface observations, but even some of the professionals in these communities require high temporal resolution, namely those interested in hydrogeology and the soil scientists concerned with soil moisture. Changes in soil moisture lag only a little those of the atmosphere, and the hydrogeologist has need of land surface data of high frequency to model infiltration rates and evaporation. These applications are amongst a wide range of subjects which are directly concerned with the consequences of changes in the climate near the surface. Such changes are so rapid that monitoring them is extremely demanding of the temporal resolution of all types of observing system, whether deployed remotely or on the ground. Those involved in such applications, and all scientists and managers interested in agriculture, require data which are relevant to annual changes and more often to the much more rapid land surface changes determined by the phenology of vegetation and crops (Allan, 1984; 1987; Guyot and Verbrugghe, 1981; 1984; 1986; 1988). The Landsat programme allowed a potentially huge community of scientists and land resource managers to observe, monitor and manage soil, vegetation, river catchments, land use and a myriad of other activities. It also established the role of satellite data in global monitoring in the International Land Surface Climatology Programme (ISLSCP, 1986).

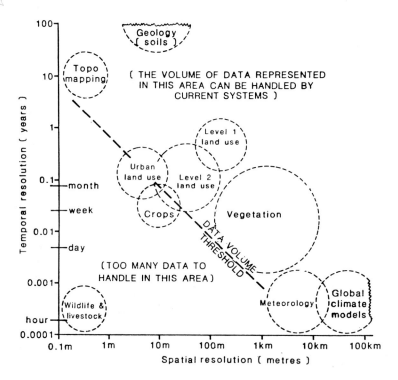

FIGURE 1.1. The requirements of agricultural applications with respect to spatial and temporal resolution of remote sensing systems compared with those of other users of remotely sensed data

It is evident from comparisons of the resolution requirements of particular applications, that agricultural applications were never able to be addressed by traditional mapping systems, as is evident in *Figure 1.1*. The frequency of data acquisition of between five and ten years, which had satisfied the topographic mapping community, was irrelevant to all environmental scientists except some geologists. At the same time, the global coverage provided by the meteorological satellite sensing systems, for example the orbiting AVHRR sensors on the NOAA series, or the information from the geostationary Meteosat, were useful in frequency but irrelevant for most agricultural applications in terms of spatial resolution. The organisation of agricultural space varies from continent to continent, and within any one country it varies according to terrain and climatic circumstances. In very broken terrain, and generally speaking in tracts which have been used for agriculture for a long time, the fragmentation of agricultural parcels is great and parcel sizes are small. Thus, in the long used terraces of South and East Asia the parcels sizes are rarely larger than one hectare, while in Western Europe almost all parcels are over one hectare and the majority of agricultural production takes place on agricultural parcels of more than three hectares. In contrast in North America and in the Soviet Union the parcels sizes are often large and in the Soviet Union can be many square kilometres in extent. In the Southern Hemisphere parcel sizes range from massive units, kilometres in extent in Argentina and Australia, to very fragmented and impermanent parcels in Africa. Small parcel sizes are also usual in South Asia and South-East Asia. However, even in the U.S.A. the most common parcel size was about 10 acres, compared to about 50 acres in Canada.

The size of parcels is an important factor for the use of remote sensing from the air and from space in monitoring crop area and condition. It is obvious that it is not the individual plant which is detectable, but the cultivation of areas, planted with the same species and sufficiently large to enable the resolving power of the observing system to be relevant. In other words, the agricultural parcels must be at least four times the nominal area of the recording pixel, to be sure that there is no effect from the mixed pixels at the parcel boundaries. Taking into account the point spread function associated with scanning sensors, it is preferable to have parcels which are ten times the area of the recording pixel. Fortunately, the farmer works very hard to make whole parcels appear the same from above: farmers are seeking the consistency of production demanded by the market, and this pressure, along with cultivation and crop protection technologies, conspire to affect an outcome in terms of the crop cover which makes it possible to discriminate crops even though individual plants would not be separable. For applications of remote sensing carried out on the ground from vehicles such as tractors, the size of the parcel is not a problem. However, it is important to stress in such studies and monitoring that the footprint being recorded must be of a size sufficient to represent the crop or soil cover in the parcel. It is generally agreed that a footprint of about a metre diameter is required to represent correctly a crop such as wheat, while a footprint of about two metres diameter is required for a row crop such as maize.

The need for all-weather systems

Because cloud cover interferes with the acquisition of timely and comprehensive information in the visible and near-IR, it is very important that the micro-wave bands be exploited for the inventory and monitoring of agricultural crops. Sensors recording in the visible and infrared wavelengths have a poor potential in very extensive regions which are of great economic and political significance. The temperate latitudes are beset by cloud-cover problems, so that the probabilities of acquiring sequential data in the United Kingdom, for example, at particular seasons are low and generally under ten per cent. The probability of acquiring good quality data from even a single specified pass can be as low as 25 per cent in some parts of the United Kingdom.

Although the promise of microwave data in agricultural studies has been loudly proclaimed, to date radar has only been deployed to overcome the problem of cloud-cover at the national scale in equatorial latitudes in, for example, Brazil and Nigeria (Radam, 1981; Allan, 1980). The main purpose in both cases was the survey of forest resources but the survey of other land cover types, including agricultural use, was also shown to be possible. In future there will be satellite coverage from such platforms as the

ESA ERS-1, which will provide thirty metre (nominal) spatial resolution microwave data, which will be useful for the vast majority of the agricultural environment world-wide.

To date radar data has not proved to be ideal for discriminating crop cover, mainly because radar returns are sensitive to roughness and moisture, rather than to the multi-spectral properties which have proved to be so useful in the visible and the infrared regions of the electro-magnetic spectrum. The potential of the discriminating power of multi-spectral radar sensors in crop and vegetation has yet to be thoroughly investigated, but it is not too extravagant to claim that it is upon the successful development of multi-spectral radar imaging systems that the medium and long-term future of operational remote sensing in agriculture will depend. Meanwhile, there is a great deal of scientific research in train into the microwave signatures of canopies, which will yield useful operational procedures for monitoring crops and vegetation (Askne, 1987; Harris and Hobbs, 1984; Holmes, Luzi et al. and Paris, these proceedings). Successful deployment will depend as much on developments in the ground segment, in the further evolution of real-time data processing capacities, as on the design and construction of new radars. Contracts have been placed in this regard for the reception and processing of the ERS-1 data, which is imminent.

The inescapable place of sampling: the agricultural case

Figure 1.1 not only indicates the relationship of spatial and temporal resolution to agricultural applications, it also points up the importance of sampling. Agricultural applications need adequate temporal resolution, in other words the frequency of the temporal sample must be appropriate for the particular application. Since some of the critical and indicative changes in crop phenology take place over a period of as little as four days the ideal potential temporal resolution should be less than four days. In other words, there should be the option of acquiring data every three days even if this frequency is only needed once during the growing season.

Some environmental applications, such as meteorology, where there are even more demanding temporal resolution requirements, can cope with the associated data volume problem by relaxing the spatial resolution of their data sets. Meteorological data have spatial resolution cells of between one and ten kilometres, and global climate modellers are prepared to handle data on a raster with 200 kilometre pixels. We have seen already that agriculture cannot relax the spatial resolution to less than 100 metre square pixels globally, and for much of the most productive agriculture of the world it will be necessary to deploy sensors with five metre resolution.

With such high spatial resolution the burden of data processing will be unacceptable even with new generations of computers. Happily, the development of spatial information systems which can merge the information from the relatively static environment of parcel boundaries with the raster data from crop monitoring satellite-borne sensors will make it possible to use only a fraction of the total information. Such processing can be post-acquisition, but in due course it will be possible to build systems which are pre-programmed to acquire sampled data from agricultural parcels so that only one, or at least a very small number, of high spatial resolution pixels (say five metre resolution) are acquired for each parcel. It has been shown that it is possible to discriminate individual crops and parcels reliably with data sets which have been reduced by factors of 100 times (Allan, 1986; Shearn 1987). It behoves those involved in the application of remote sensing in agriculture to offer procedures by which data reduction can be achieved, and also to encourage the integration of remotely sensed data and other types of agricultural data. In addition to the advantage of massive data reduction, such integration will enable the concentration of attention on administrative and management priorities as well on crop area and condition. Such integration will also enable the new data acquisition technologies to match rural applications. Sampling will make possible the economic acquisition of data on cropping which can be applied on the various scales in which the agricultural environment is organised; namely the agricultural parcel, the farm management unit and the administrative divisions of local government.

Applications at various scales

Remotely sensed data have played a role in a wide variety of agricultural applications at a very wide range of scales, from the micro-level to global surveys of internationally significant crops such as wheat. A good example of the local level survey was the demonstration that irrigated farming, and particularly the unregulated use of water by particular farmers, could be policed by remote sensing. The example comes from California where remotely sensed data was shown to be able to detect land being irrigated. When these data were registered with farm and field boundaries, and then with data on the parcels in which irrigation had been authorised, the parcels where unauthorised water was being applied were very clear. The same big-brotherly role has been deployed to detect and regulate many crops, such as tobacco, which attract the interest of tax officials. A similar surveillance role was threatened in EEC countries in an attempt to regulate the production of such crops as olives, where it was recognised that subsidy arrangements were being abused and where it was difficult to police such abuses at the countrywide scale. In this case it would have been necessary to deploy aerial photographic systems to detect the very varied habitats of the olive. Sometimes, as in this last case, even the threat of the use of remote sensing in such a regulatory role has been enough to change behaviour, and it is a matter which should be recognised by remote sensing practitioners that once the management goal of changing behaviour has been achieved it is often no longer necessary to deploy the remote sensing system!

At a different scale, a good example of how satellite remote sensing can be used effectively to gain access to otherwise absent, or prohibitively expensive, data was a study in Libya, where the purpose was to provide data on water use in a region where it had become clear that groundwater was being very heavily over-used. Here a combination of double sampling procedures enabled a sound estimate to be made of the irrigated land in this very diverse partially irrigated region. The biases of the satellite data were anticipated as well as the precision of the ground survey. The combination of the two sources of information via procedures of double-sampling and regression, enabled better estimates of irrigated land to be derived than had been available previously. The cost of the survey at $US 120000, with half going to the satellite data acquisition and other data processing and the other half to fieldwork, was modest for the 110000 km^2 extent of the study area (Wall, 1982; Latham, Allan and McLachlan, 1981). A similar approach was used to estimate agricultural and other socio-economic features of the Yemen Arab Republic (Schock, 1982).

At the global level, there was an early attempt to test the ability of the global coverage of Landsat to monitor crops of international economic and political significance, such as wheat. The Large Area Crop Inventory Experiment (LACIE) (MacDonald, 1976; NASA, 1978) of the 1976-78 period established that remotely sensed data could be used as one of a number of inputs to models of regional and national production. The results were as good as those being used by the USDA in the United States itself and for other areas of the world, such as the Soviet Union, it was deduced that the satellite coverage provided an improved prediction of crop area, and therefore of production (NASA, 1978).

Another very important development in regional and global monitoring is the ARTEMIS system at FAO in Rome. ARTEMIS stands for the Automated Real Time Environmental Monitoring Information System (Hielkema, 1988). Using Meteosat and AVHRR data, it provides rainfall prediction based on cloud-top temperatures and greenness/vegetation index maps which have proved very useful in the field for those trying to locate locust breeding grounds. The USGS is also distributing such information in hard copy form, which is flown regularly to the Sahel countries (FAO, 1989). A very important technique which should at least be mentioned, is that of low level aerial survey. There are a number of approaches which differ according to whether systematic or random sampling is the basis of the sample frame (Watson, 1981). The techniques were developed in East Africa but are widely applied throughout the continent and also in North America.

Earth observation and holistic information systems - the future

Global systems

Since 1980 it has become become increasingly clear that Earth observing systems have a special place in spatial information systems, at the local as well as at the global scale. We are at the beginning of a revolution in the handling and use of geographical information systems (GIS). Earth scientists, and especially the community concerned with the evaluation of land resources, have long used laboriously assembled cartographic overlays of the very wide range of land surface features - amongst them topography, soil qualities, drainage, vegetation and land use etc - to study and manage particular tracts of land. The GIS provides the means to assemble and interact with an even more comprehensive body of conventionally acquired as well as remotely observed data. Not only environmental data can be integrated, but also a wide range of socio-economic and administrative data.

At the global level, the ability to bring together data observed over the already well mapped parts of the land surface, together with coverage from tracts which have never been recorded and never will be by any other means, is stimulating the US National Aeronautical and Space Administration (NASA), the major world agency promoting the use of Earth observation, to turn its whole applications attention to global science (NASA, 1987).

It should be recognised that NASA has turned its back on land management applications, for example agriculture and forestry, and in the past year has produced a very large volume of analysis and reports, as well as extremely well produced glossy explanatory leaflets aimed at persuading the US voter, and especially US legislators, that the next exciting goal for Earth observation is the provision of data on the interactions among the Earth's atmosphere, land and biosphere, with a view to explaining Earth dynamics, Earth evolution and global change. The goal of the studies will be to gain an understanding of the entire Earth system on a global scale, and the challenge will be the development of a predictive capability with respect to natural changes and those which derive from human intervention at the land surface. The observing programme in the United States will involve the creation of a comprehensive Earth observing remote sensing system which will be based on US, ESA as well as Canadian and Japanese contributions, and which will from 1995, or soon thereafter, provide a global information system. Recent discussions between US and Soviet scientists are likely to enable both Western and Soviet data to be used in these new spatial recording and predictive systems.

The significance for the agriculturalists of the world of this emphasis on global science will be in the improvement in our understanding of the global climate and ecological systems which have an impact on the agricultural sectors of all countries. The greater understanding derived from these programmes will also be of importance to international agencies attempting to predict agricultural potential, and will enable them to intervene in a timely fashion when agricultural and grazing sectors, impaired by natural and man-induced pressures, cannot cope with demand. The negative significance for agriculturalists is that on-going research at the local and national scale into agriculture itself, and the development of new operational monitoring systems, will depend on other agencies than NASA. In this regard, it is relevant to mention some recent initiatives by the European Economic Commission (EEC), which has funded research on yield prediction on the basis of remotely sensed AVHRR data, which has been found to be very useful for environmental monitoring in Africa (Justice, 1986).

Geographical information systems at the local and national scales

Even more important for agriculture than the integration of data at the global level is the explosion of such integration at the local and the national levels (Ripple, 1987). Some countries are already at an advanced stage, notably in the United States, Canada and Australia. In such countries the areas to be covered are huge, and the use of remote sensing for the survey, and especially for the evaluation, of resources has been commonplace for over thirty years, and their capacity to absorb the new technologies of data handling is advanced. Such institutional environments are ideal for the installation and development of geographical information systems, which include remotely sensed data.

All practitioners in resources surveys must grasp the potential of the systems and be prepared to become familiar with the various cartographic conventions of the many professionals who contribute data to rural geographical information systems (Burrough, 1987). The powerful multiplication which results from the integration of climatic, terrain, environmental, agronomic, economic, social, institutional and management data, makes available for scientist and manager alike a new and very powerful tool with which to optimise agricultural systems, taking into account the goals of improving production and productivity as well as those of environmental protection (Dale and MacLaughlin, 1988). The availability of constantly improving desk top computing power enables the merging of processed raster imagery with the many layered data sets in geographical information packages, specially designed for land evaluation and rural management (Rossiter, 1988).

The first phase of data integration during the period 1975 to 1985 was not very comfortable, as there were two articulate groups advocating mutually exclusive procedures. The traditional mapping community was certain that the vector system of spatial data management would solve all problems and those that it would not solve were not worth recognising. This argument was particularly attractive to the vector group as their approach meant that they need make no change in the scope of their mapping activities, since the vector approach was wholly compatible with their conventional system of coordinated geometric control and presentation of detail. They were interested in linear and point data and not at all concerned with providing data on the spaces between the lines. As far as agriculturalists were concerned, such maps would provide the parcel boundaries but no data on the cropping and management of the parcels.

Agriculturalists are only a little concerned with linear and point data. On the other hand they are very greatly concerned with information on the variation in environmental and agricultural variables within parcels, and much prefer the raster data provided by remote sensing systems. The problem of integrating the two types of data impeded the development of satisfactory geographical information systems, but happily the enlightened in both communities are beginning to recognise that the computer is a very flexible device which can provide information storage which accommodates the needs of those handling spatial data where the intensity of activity varies spatially from tract to tract. The solution to this problem has come through the recognition that quad-tree procedures and tesselar arithmetic provide ideal systems, so that cells which need high spatial resolution can be nested and readily accessed within a background where lower resolution data are adequate (Mason and Townshend, 1988; Rhind and Green, 1988; Raper and Green, 1989).

Future systems

In some ways ESA's European Remote Sensing Satellite I (ERS I) provides the next important step in expanding the scope of remote sensing for agricultural applications. Its launch in 1991 will also represent the first step in the global Earth Observation System. It will be some years later before the US Space Station begins to be assembled in space, probably by 1995. The European Polar Platform (Columbus) element (BNSC, 1986; ESA, 1986) of the overall observing programme will be one of two orbiting elements; the other will be constructed and managed by NASA (NASA, 1984). Both platforms are expected to be operational by the second half of the 1990s. The philosophy behind these large platforms, which are variously quoted as being able to carry payloads of between two and three tons, is that the benefits of synergism would be a major advantage. Through their capacity in payload and power both the US and the European platforms would be able to carry five or more sensors, enabling data acquisition in the visible, the infrared and the microwave channels. The Canadians are also associated with the programme, and have offered a maintenance service as their major contribution. The final configurations have not been decided, but it seems likely that there will be some or all of the following sensors on board:

HIRIS High Resolution Imaging Spectrometer, a multichannel spectrometer ($c.$ 250 channels). Platform at 834 km nominal altitude, GIFOV 30 m, swath 30 km, spectral range 0.4-2.5 μm. Average spectral sample width for 0.4-1.0 μm will be 9.4 nm and for the 1.0-2.5 μm range it will be 11.7 nm. The sensor will be especially useful in geological surveys but is also likely to be useful in discriminating canopy signatures. However, 'the data rate precludes it as an instrument for repetitive mapping' (NASA 1986, MODIS section).

MODIS Moderate Resolution Imaging Spectrometer:
: MODIS T. 17 spectral bands of 10 nm width in the 0.4-1.0 μm range. The spectrometer employs a scanning mirror system, collecting optics and a 64x64 element silicon detector array with a + 60° rotation giving global cover every two days. It will provide data fore and aft of the satellite, which will be useful in minimising specular reflectance of surface radiance and in examining the bi-directional reflectance distribution function.
MODIS N. non-pointing. 25 channels in the spectral range 0.4-12.0 μm, 500 m nadir resolution, swath 1513 km, bandwidth mainly 20 nm. Also, thermal channels for cloud and atmospheric studies with 7 channels, Instantaneous Field Of View (IFOV) 1000 m.

LASA Lidar Atmospheric Sounder and Altimeter - this instrument has no likely relevance for agricultural applications.

HRMMR High Resolution Multi-frequency Microwave Radiometer - relevant for soil moisture and hydrology studies but with a very large footprint and only relevant in some regions for agricultural applications.

SAR A Synthetic Aperture Radar with ten frequencies, multi-polarised and multi-frequency - (L,C and X band). For vegetation canopy morphology studies, also soil and soil moisture and arid lands; will offer synergism with HIRIS having the same spatial resolution of 30m.

LAWS Laser Atmospheric Wind Sounder. This instrument has no likely relevance for agricultural applications.

Altimeter System - advanced microwave altimeter.

For agriculturalists, the promise of HIRIS is considerable, although those of us who have struggled to make sense of spectral signatures from complex canopies using Landsat MSS and Thematic Mapper data judge that expectations that generalised signals derived from complex cover will yield distinctive spectra may prove to be over-optimistic. Meanwhile, the AVIRIS sensor, the airborne precursor of HIRIS, is proving troublesome at the development phase. Data from the MODIS sensor is likely to be of very great importance for the monitoring of vegetation condition generally, and for estimating crop areas and predicting crop yields. Its temporal resolution will ensure that it provides coverage of vast areas comprehensively, through the filtering procedures developed by NASA in their AVHRR experiments (Justice, 1986). For reasons of payload and power economy MODIS T may not fly.

Meanwhile the French SPOT programme is scheduled to be extended to a second platform with the same sensors by the early 1990s, and there are indications that a third platform will also be launched, which will have an additional large area, medium resolution sensor on board by the late 1990s (CNES, 1988). There will be no shortage of data, although many will still find the cost high despite the real low cost per square kilometre which all types of remote sensing systems achieve (Allan, 1984). News that funding for Landsat 6 has been approved is very reassuring following a tense period of lobbying which was necessary after existing and future Landsat programmes had been threatened with termination by the end of March 1989.

References

ALLAN, J.A. (1980). The Nigerian Radar (NIRAD) survey of forest resources: an application of side-looking airborne radar in Nigeria. *Remote Sensing Quarterly*, **2**, 36-44

ALLAN, J.A. (1984). The role and future of remote sensing. In *Satellite remote sensing: review and preview*. Remote Sensing Society, Reading

ALLAN, J.A. (1986). How few data do we need: some radical thoughts on renewable natural resources surveys? In *Resources, development and environmental management*. Balkema, Rotterdam with ITC, Wageningen

ALLAN, J.A. (1987). Strategies for the intelligent classification of crops: sampling ephemeral data from permanent parcels. In *Image processing in remote sensing*. Remote Sensing Society, Nottingham

ASKNE, J. (1987). International symposium on microwave signatures in remote sensing. *International Journal of Remote Sensing, Special Issue*, **8**, 1577-1723

BURROUGH, P. (1987). *Principles of geographical information systems for land assessment studies*. Oxford Science Publications, Clarendon Press, Oxford

BNSC (British National Space Centre). (1986). Land science and land applications, Vol V of Columbus/Space Station, UK Utilization Study, 1985/86, BNSC, Department of Trade and Industry, London. See also Polar Platform Data Instrument Sheets, BNSC/DTI. London

CNES (1988). *SPOT 1 image utilization assessment results*. Cepaudes-Editions, Toulouse

DALE, P. and MacLAUGHLIN, J.D. (1988). *Land Information Management*. Clarendon Press, Oxford

EUROPEAN SPACE AGENCY (ESA) (1986). *Earth observation requirements of the Polar Orbiting Platform - elements of the International Space Station*. ESA, Paris

FOOD AND AGRICULTURAL ORGANISATION (FAO) (1989). Report on the workshop on desert locusts monitoring and management held in FAO Rome in November 1988. FAO, Rome

GUYOT, G. and VERBRUGGHE, M. (1981). *Spectral signatures of objects in remote sensing*. Proceedings of the 1st Colloquium of Working Group 3 of Commission VII of ISPRS, Avignon. ESA, Paris

GUYOT, G. and VERBRUGGHE, M. (1984). *Spectral signatures of objects in remote sensing,*. Proceedings of the 2nd Colloquium of Working Group 3 of Commission VII of ISPRS, Bordeaux. ESA, Paris

GUYOT, G. and VERBRUGGHE, M. (1986). *Spectral signatures of objects in remote sensing*. Proceedings of the 3rd Colloquium of Working Group 3 of Commission VII of ISPRS, Les Arcs. ESA, Paris

GUYOT, G. and VERBRUGGHE, M. (1988). *Spectral signatures of objects in remote sensing*. Proceedings of the 4th Colloquium of Working Group 3 of Commission VII of ISPRS, Aussois. ESA, Paris

HIELKEMA, J.J. (1988). Use of satellite remote sensing for desert locust survey forecasting at FAO. In *FAO, Report on desert locust research - defining future research priorities*. FAO, Rome

HARRIS, R. and HOBBS, A. (1984). Progress in radar remote sensing of vegetation and hydrology. Department of Geography, University of Durham. Part of Final Report for Natural Environment Research Council (NERC), Study of microwave remote sensing techniques for land surface and hydrological applications. NERC Contract F60/66/08, Swindon

ISLSCP (International Land Surface Climatology Project) (1986). Parameterisation of land-surface climatology: use of satellite data in climate studies: first results of ISLSCP, ESA/Estec, Noordwijk, ESA.SP.248

JUSTICE, C. (1986). Monitoring the grasslands of semi-arid Africa using NOAA (National Oceanographic and Atmospheric Adminstration) AVHRR (Advanced Very High Resolution Radiometer) data. *International Journal of Remote Sensing - Special Issue*, 1383-1609

LATHAM, J.S., ALLAN, J.A. and McLACHLAN, K.S. (1981). *Monitoring the changing areal extent of irrigated lands of the Gefara Plain, Libya.* Secretariat of Agricultural Reclamation and Land Development, Tripoli, Libya

MacDONALD, R.B. (1976). *Large Area Crop Inventory Experiment (LACIE)*. Report to the Pecora Memorial Symposium, Sioux Falls, South Dakota

MASON, D.C. and TOWNSHEND, J.R.G. (1988). Research related to geographical information systems at the Natural Environment Research Council's Unit for Thematic Information Systems. *International Journal of Geographical Information Systems*, 2, 121-141

NASA, (1978) LACIE: briefing materials for technical presentation, Johnson Space Flight Centre, Report JSC-14857

NASA, (1984). Earth observing system, NASA Technical Memorandum 86129, Vol I NASA, Washington DC

NASA (1986). Report on the EOS Data Panel, Parts a-h, (see especially Vol IIb on MODIS) NASA Technical Memorandum 87777, NASA, Washington DC

NASA. (1987). Earth observing system, NASA Technical Memorandum 86129, Vol II Science and Mission requirements. NASA, Washington DC

RADAM, (1981). Projecto Radambrasil, Levantamento de recurrios naturais, Ministero das Minas e Energia, Rio de Janeiro

RAPER, J. and GREEN, N.P.A. (1989). Geographical Information System Tutor, GIST, software demonstrating GIS concepts and practice. Birkbeck College, University of London

RHIND, D.W. and GREEN, N.P.A. (1988). Design of geographical information system for a heterogeneous scientific community. *International Journal of Geographical Information Systems*, 2, 171-189

RIPPLE, W.J. (1987). *GIS for resource management: a compendium.* ASPRS/ACSM, Falls Church, Virginia

ROSSITER, D. (1988). ALES (Automated Land Evaluation System): a microcomputer program to assist in land evaluation, Department of Agronomy, Cornell University, Ithaca, NY. A paper presented at the Symposium on land qualities in space and time, August 1988, Wageningen

SCHOCK, R. (1982). *Land use studies and crop acreage estimates from aerial photography and satellite imagery: a case study of Turbah, Yemen Arab Republic.* Geography Department, University of Zurich

SHEARN, V.J. (1987). Data volume reduction techniques from land cover classification. Unpublished MSc Thesis, University College London, London

WALL, S. (1982). A Landsat-based inventory procedure for the estimation of irrigated land in arid areas. In *Proceedings of the ERIM thematic conference on remote sensing in arid and semi-arid lands, Cairo.* Environmental Research Institute of Michigan, Ann Arbor, Michigan

WARWICK, G. (1985). International satellite directory. *Flight International*, January 1985, 29-55

WATSON, R.M. (1981). Down-market remote sensing. In *Matching remote sensing technologies and their applications.* Remote Sensing Society, Reading

2
OPTICAL PROPERTIES OF VEGETATION CANOPIES

G. GUYOT

INRA Bioclimatologie, BP 91, 84143 Montfavet Cedex, France

Introduction

Remote sensing data relative to vegetation canopies can only be interpreted if the mechanisms of interaction of electromagnetic radiation with plants and the underlying soil are known. This short review is concentrated on the visible, near and middle-infrared domains.

The optical properties of vegetation canopies are not static. They not only vary with time as a function of plant status but also as a function of external (solar elevation, orientation and inclination of the view axis, atmospheric conditions ...) and internal factors (soil colour, row orientation, canopy geometry). These different points will be considered briefly.

Optical properties of the components of a vegetation canopy

Leaves

The optical properties of crop or forest canopies depend mainly on the optical properties of leaves and the underlying soil, but in some cases they are also affected by the optical properties of other parts of the plants, such as the bark on tree branches, flowers, fruit, etc. As an example, *Figure 2.1* displays reflectance and transmittance spectra of wheat leaves from the visible to middle-infrared (Guyot, 1984). All of the reflectance spectra of plant leaves (low crops or forest trees including conifers) have the same shape. Differences just appear in the magnitude of the reflectance (Guyot, Guyon and Riom, 1989; Baldy, Guyot and Merelle, 1981). As is shown in *Figure 2.1*, three spectral domains can be considered according to the different leaf optical properties (Gausman and Allen, 1973; Knipling, 1970; Guyot, 1984).

THE VISIBLE DOMAIN (400-700 NM):

In this domain leaf reflectance is low (less than 15%) and leaf transmittance is very low. The main part of incident radiation is absorbed by leaf pigments such as chlorophyll,

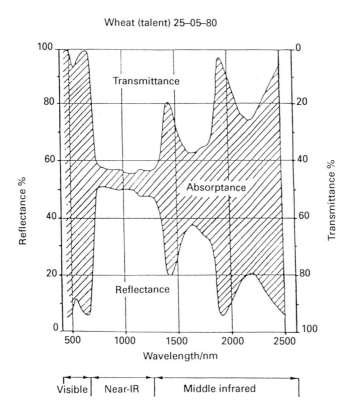

FIGURE 2.1. Reflectance and transmittance spectra of wheat leaves from visible to middle-infrared. The hatched part corresponds to leaf absorptance (After Guyot, 1984)

xanthophyll, carotenoids and anthocyanins. The main pigments affecting leaf absorption are chlorophylls a and b (65% of pigments of higher plant leaves) which exhibit two absorption bands centered in the blue and in the red. For this reason leaves have a maximum reflectance at 550 nm in the yellow-green region.

THE NEAR-INFRARED DOMAIN (700-1300 NM):

In this spectral domain leaf pigments and the cellulose of cell walls are transparent, so leaf absorptance is very low (less than 10%) and incoming radiation is either reflected or transmitted. Reflectance reaches about 50% on the "infrared plateau", this level depending on the anatomical structure of the leaves. Its level increases with the number of cell layers (Gausman, Allen, Cardenas and Richardson, 1970) and the size of the cells, the orientation of their walls and the heterogeneity of their content (Gausman, 1974; Grant, 1987).

THE MIDDLE-INFRARED DOMAIN (1300-2500 NM):

In this domain leaf optical properties are mainly affected by their water content (Allen, Gausman, Richardson and Thomas, 1969; Tucker and Garratt, 1977). Beyond 1300 nm strong water absorption bands at 1450, 1950 and 2500 nm produce leaf reflectance minima. But, between these bands, water absorption still exists and affects leaf optical properties. The levels of the two relative maxima therefore vary according to leaf water content (Thematic Mapper channels TM5 and TM7 are centered on these two maxima).

MECHANISMS OF LIGHT DIFFUSION BY LEAVES

The theory of Willstatter and Stoll (1928) has been used for a long period to explain light diffusion by leaves. From this theory, light diffusion is mainly due to total reflections which occur at air-cell interfaces when the incidence angle of the light is equal to or greater than the total reflection angle. As the refractive index of the cells is about 1.5 the angle of total reflection is 41.8°. In this case the diffusion of the light should be mainly due to the spongy mesophyll (*Figure 2.2*) where the orientations of the cell walls are randomly distributed and where there are a large number of air-cell interfaces (Gausman and Allen, 1973). The role played by the palisade parenchyma would be of little importance.

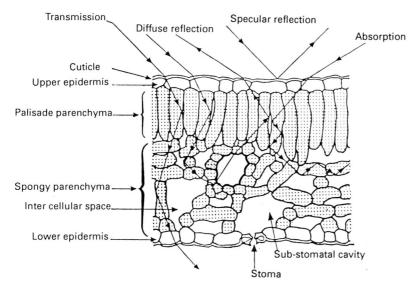

FIGURE 2.2. Schematic representation of the interaction of incoming radiation with leaf tissues. The cross-section corresponds to a dicotyledon leaf (*Eleborus niger*) (adapted from Lichtenthaler and Pfister, 1978)

The real mechanism of light diffusion is, in fact, more complex, as has been shown by Kumar and Silva (1973) and Sinclair, Schreiber and Hoffer, (1973). Light is also reflected at cell walls when the incidence angle is inferior to the total reflection angle, but to a lesser extent. Moreover, light is also scattered by the heterogeneities of cell contents. Any change in the refractive index induces light refraction (cell wall - protoplasm, chloroplast - cell wall). These different phenomena play a significant role in light diffusion.

In order to give a more accurate and exhaustive description of leaf optical properties, physical models have recently been developed (Jacquemoud, Baret and Guyot, 1989). They introduce the specific optical properties of the different tissues of the leaf.

Directional properties of leaves

In the visible, near and middle-infrared domains the contribution of the cuticle to leaf reflectance was often neglected in the past. The main factor considered was the scattering of incident light by the internal tissues of the leaf.

If we consider an incident flux, normal to the leaf surface, the specular reflection of the cuticle can be neglected and some authors have considered the leaves as Lambertian diffusers (Fuchs, Stanhill and Waanders, 1972; Suits, 1972a, 1972b). However, for large incidence angles, specular reflection can play an important role. For this reason leaf reflectance depends on the incidence angle of the incoming radiation, and its angular variation is due to the combination of diffuse and specular reflection (Breece and Holmes, 1971). Recent work by Grant (Grant, 1985; 1987; Grant, Daughtry and Vanderbilt, 1987a; 1987b) shows that the diffuse reflectance is spectrally modified and depends on the internal structure of the leaf and its pigments and water content, whereas the specular component, which is reflected by the cuticle, is not spectrally affected but, on the other hand, is partially polarized.

The amount of incident light which is specularly reflected is practically the same in the visible and near-infrared. As the diffuse component of reflected radiation is low in the visible and high in the near-infrared, the shape of reflection polar distributions will vary with the wavelength (*Figure 2.3*). In the near-infrared the leaves can be considered as Lambertian diffusers, whereas in the visible they exhibit a strong directional effect.

Factors affecting leaf optical properties

ANATOMICAL STRUCTURE OF THE LEAVES

Near-infrared reflectance is strongly affected by leaf anatomical structure. It depends on the number of cell layers, the size of the cells and the relative thickness of the spongy mesophyll. Thus, the leaves of dicotyledons have a higher reflectance than those of monocotyledons having the same thickness because their spongy mesophyll is more developed (Hoffer and Johanssen, 1969; Sinclair, Hoffer and Schreiber, 1971)

In the same manner, the leaves of drought adapted plants, like olive trees can have a very high near-infrared reflectance (decreasing with the cell size and increasing with the number of cell layers, Baldy et al., 1981). The leaves also present an asymmetry in their optical properties: the reflectance of the lower side is generally higher than the reflectance of the upper side because of the greater chloroplast density in the palisade parenchyma (Gausman, Allen, Cardenas and Richardson, 1970; Gausman and Allen, 1973). Moreover, the presence of hairs, on the lower side of some leaves or on both sides of pubescent leaves, increases the reflectance in the visible and middle-infrared but has little effect on the near-infrared reflectance (Gausman et al., 1978; Baldy et al., 1981). The hairs are made of cellulose and are dry. For this reason they are white in the visible, transparent in the near-infrared and also have a high middle-infrared reflectance.

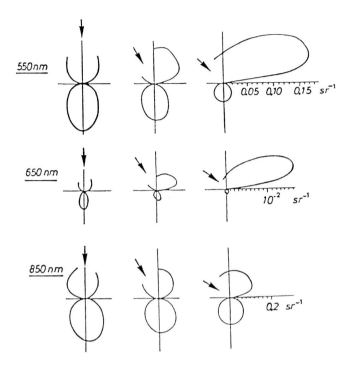

FIGURE 2.3. Polar plots of reflected radiant intensity for a soybean leaf drawn for different wavelengths and different incidence angles (adapted from Breece and Holmes, 1971)

LEAF AGE

Leaf optical properties change significantly only during the juvenile stages and senescence. During the major part of their life the leaves of annual plants or deciduous trees have practically constant optical properties (Gausman, Allen, Escobar, Rodriguez and Cardenas, 1971; Guyot, 1980; Guyot, Dupont, Joannes, Prieur, Huet and Podaire, 1985). However, the optical properties of conifer needles (fir or spruce) evolve as a function of the time (from year to year) with increasing chlorophyll content (Lichtenthaler and Rinderle, 1988).

Figure 2.4 displays the evolution of the reflectance of wheat leaves during senescence. The disappearance of chlorophylls and their replacement by brown pigments, resulting from chlorophyll degradation, produces increases of the yellow-green and red reflectances. In the near-infrared, the reflectance only evolves when the leaf dries and when its internal structure is changed. In the middle-infrared the increase of the leaf reflectance is connected with its drying, but we must note that the decrease of the water content begins relatively late, when the leaf is yellow.

LEAF WATER CONTENT

Leaf water content has not only a direct effect on leaf optical properties in the middle-infrared but also an indirect effect on the visible and near-infrared reflectances because

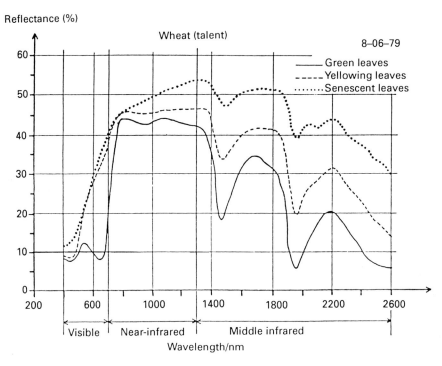

FIGURE 2.4. Reflectance spectra of wheat leaves during senescence (after Guyot, 1984)

it affects the cell turgor (Myers, 1970). Thus, a decrease of the leaf water content induces an increasing reflectance in the whole spectrum. However, this is larger in the middle-infrared than in the other parts of the spectrum. But these results were obtained in laboratory conditions: Thomas, Namken, Oerther and Brown, (1971) and Guyot et al., (1985) have shown that, in natural conditions, it is necessary to have an extremely severe water stress to affect the leaf optical properties. Usually, water stress cannot be detected when only the optical properties of the leaves are measured.

MINERAL DEFICIENCIES

Mineral deficiencies mainly affect chlorophyll content (Al Abbas, Barr, Hall, Crane and Baumgardner, 1974; Horler, Barber and Barringer, 1980; Horler, Dockray and Barber, 1983) and possibly the anatomical structure of the leaves, depending on their severity (Thomas, Myers, Heilman and Wiegand, 1966: Gausman et al., 1978). Chlorosis (iron deficiency) and nitrogen deficiency are the more common phenomena. Chlorosis mainly affects leaf reflectance in the visible (Moreau, Boissard and Bonhomme, 1981; Oester, 1981; Sandwald, 1981). Nitrogen deficiency changes the whole reflectance spectrum. The visible reflectance is increased (due to decreasing chlorophyll content) and the near and middle-infrared reflectances are decreased (due to decreasing number of cell layers, Thomas and Oerther, 1972).

Pest and disease attacks

Pest and disease attacks can :

- change the leaf pigment content (yellowing). In this case the leaf optical properties are only affected in the visible (Moreau et al., 1981; Oester, 1981; Sandwald, 1981);

- induce necroses. The reflectances of necrosed parts can be compared with the reflectance of senescent leaves (*Figure 2.4*);

- produce other pigments which can increase or decrease the reflectance in different parts of the spectrum (Keegan, Schleter, Hall and Haas, 1956; Baldy et al., 1981);

- modify the transpiration rate of leaves without changing their optical properties. In this case the radiative temperature of leaves is changed and the effect of disease or pest attacks can be detected in the thermal infrared.

Optical properties of other plant parts

For annual crops, flowers can significantly change the reflectance during a certain time. This is the case, for example, for flowers of rape or sunflower which exhibit a strong yellow reflectance. In spring the flowers of herbaceous plants of meadows can also change their reflectances (Girard, 1986).

In forest the optical properties of trees depend not only on the optical properties of their leaves but also on those of bark and cones (for conifers). Bark reflectance is quite different from leaf reflectance: it progressively increases from the visible to the middle-infrared (Guyot et al., 1989). The visible reflectance, which does not exhibit chlorophyll absorption bands, is higher than the leaf reflectance. In the near-infrared bark reflectance is lower than leaf reflectance but it is higher in the middle-infrared.

Optical properties of soil

Figure 2.5 displays an example of reflectance spectra of a bare soil for different moisture contents (Bowers and Hanks, 1965). It shows that the reflectance increases progressively from the visible to the middle-infrared, where water absorption bands appear as in leaf reflectance spectra. *Figure 2.5* also shows that the soil moisture status affects the whole reflectance spectrum. For any wavelength, the reflectance decreases when the soil moisture content increases.

Soil reflectance also depends on its mineral composition but, because of its complexity, it is difficult to identify the characteristic spectral bands of a given mineral. They generally overlap and, practically, only the soil iron content can be determined from spectral measurements, because iron exhibits strong and wide spectral bands in the visible and near-infrared (Myers, Bauer, Gausman, Hart, Heiman, MacDonald, Park, Ryerson Schmugge and Westin, 1983; Coleman and Montgomery, 1987). The soil organic matter content, when higher than 2%, can significantly reduce the soil reflectance in the visible and near-infrared. It can also mask the spectral bands of different minerals (Bowers and Hanks, 1965).

The soil reflectance in the whole spectral domain also depends on the size of its particles. For a given soil type the smaller are the particles the higher is the reflectance (Piech and Walker, 1974). This is due to the fact that the small particles tend to form

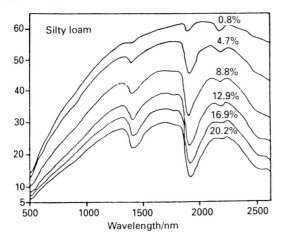

FIGURE 2.5. Reflectance spectra of a silty loam soil for different moisture contents (after Bowers and Hanks, 1965)

a plane surface with a small number of asperities, whereas the larger particles tend to form a surface presenting micro-asperities, which play the role of light traps (Myers et al., 1983).

At a macroscopic scale clods or surface irregularities affect the distribution of shadowed and illuminated surfaces. Thus, when the soil roughness increases its reflectance decreases in the whole spectral domain (King, 1986; Epiphanio and Vitorello, 1984), because the soil elements do not transmit the incident light and have the same effect for any wavelength. The variation of soil reflectance due to the surface roughness can be very large, and in some cases it can be larger than the variation caused by the growth of a plant canopy (Huete, Jackson and Post, 1985).

Surface irregularities also introduce a variation of the reflectance as a function of the slope of the soil surface and of its orientation (Cierniewski, 1987). This also affects the variation of the directional reflectance factor as a function of the inclination and orientation of the view axis. The reflectance factor is a maximum when the view axis is parallel to the sunbeams (no shadow in the field of view) (Guyot, Jacquin, Malet and Thouy, 1978; Eaton and Dirmhirn, 1979; Cooper and Smith, 1985). Thus, a bare soil is not a Lambertian reflector.

Reflectance spectra of plant canopies

The reflectance spectra of plant canopies are a combination of the reflectance spectra of the plants and underlying soil. When a plant canopy grows the soil contribution progressively decreases: the bare soil reflectance spectrum is replaced by the plant reflectance spectrum. Thus, as shown in *Figure 2.6*, during the growth of the plants the visible and middle-infrared reflectances decrease and the near-infrared reflectance increases. During senescence the reverse phenomenon is observed (Ahlrichs and Bauer, 1983; Baret and Guyot, 1986; Guyot, 1989).

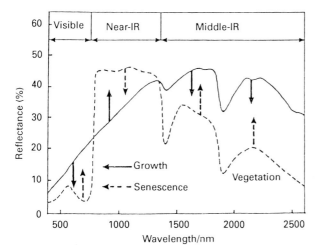

FIGURE 2.6. Schematic representation of the changes in plant canopy reflectance during growth and senescence (after Guyot, 1989)

When the green biomass increases during the growing season, the canopy reflectance can reach saturation levels; *Figure 2.7* shows this phenomenon. The variations of red and near-infrared reflectances are represented as a function of the leaf area index (LAI, total leaf area per unit ground area). In the visible and middle-infrared the saturation level is reached for LAI around 3, and in the near-infrared for LAI around 5 or 6. But we must note that the value of the LAI corresponding to the saturation level also depends on the plant geometry. Higher LAI values are needed to reach this level with plants having erect leaves than with plants having horizontal leaves.

The reflectance spectrum of a forest stand depends on the reflectance spectra of trees and underlying soil, which can be partially or totally covered by low vegetation. If the density of tree crowns is relatively low the effect of the soil and of the low vegetation can be dominant and mask the effect of the trees. This is, for example, the case of the Les Landes forest in the South-West of France (Guyon-Haeck, 1985; Heois, 1979) or of sub-arctic forest in Alaska (Dean, Kodana and Wendler, 1986).

Effects of external factors on the reflectance of plant canopies

Size of the viewed area

The variability of the spectral response of a plant canopy (forest or low vegetation) depends on the size of the viewed area. It is necessary to view a representative sample of the studied area in order to minimize the effects of local heterogeneities. *Figure 2.8* displays an example of the effect of the size of the viewed area on the variation coefficient of the reflectance of a wheat field. The measurements were performed at different levels

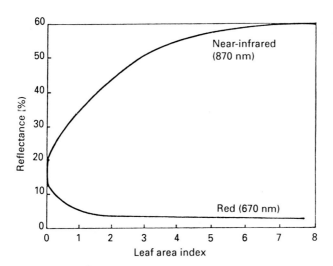

FIGURE 2.7. Variation of the reflectance of a plant canopy as a function of the leaf area index. Figure drawn from Verhoef and Bunnik's (1981) data for a plant canopy, the leaf inclination of which is 40° measured from the stem (after Guyot, 1989)

above the ground and centered on the same points. The variation coefficient decreases with the increasing size of the viewed sample and asymptotically reaches a constant level which corresponds to the characteristic variability of the whole canopy (Guyot, 1980). Guyon-Haeck (1985) has observed the same phenomenon on pine forest in the South-West of France (Les Landes). The increasing of pixel size is accompanied by a decrease in the local variability. In this case, the asymptotic level is reached with pixels of about 80 × 80 m (16 SPOT pixels).

Solar elevation

Sunbeams penetrate deeper into a plant canopy as they are less inclined. Thus, the proportion of shadowed and illuminated surfaces in the area viewed by a radiometer changes with the sun elevation. This effect can be analyzed at two different time scales: the day and the year. It also depends on the latitude.

ANNUAL EFFECTS OF SOLAR ELEVATION

When the characteristics of a natural surface do not change strongly during the year (coniferous forest, bare soil, rocks ...) the annual course of their spectral response is mainly due to changes in solar elevation. This effect can be corrected by using models (Suits, 1972a; Richardson, 1981; Cavayas and Teillet, 1986; Li and Strahler, 1985; Otterman and Weiss, 1984). For annual vegetation, the evolution of bi-directional reflectance through the year is not only influenced by the solar elevation but also by changes in canopy geometry, and it is difficult to separate these two effects.

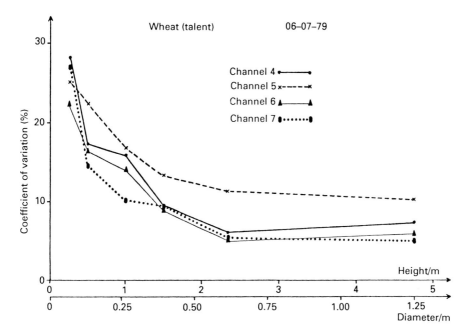

FIGURE 2.8. Effect of the size of the viewed area (or of the height of measurement) on the variation coefficient of canopy reflectance of wheat for the 4 Landsat MSS channels (after Guyot, 1980)

DIURNAL EFFECTS OF SOLAR ELEVATION

At the scale of the day it is possible to monitor the effect of solar elevation with field measurements (Guyot, 1980; Eaton and Dirmhirn, 1979; Kimes, 1983) or model simulations (Schnetzler, 1981; Kirchner, Schnetzler and Smith, 1981; Verhoef and Bunnik, 1981; Kimes, 1986). These different studies show that the bi-directional reflectance increases in the visible and decreases in the near-infrared with increasing solar elevation.

As we have seen, the leaf transmittance is very low in the visible. When the sun is near the horizon shadows are predominant within a plant canopy and the reflectance reaches its lowest values. On the contrary, in the near infrared, the leaves scatter about 50% of incident radiation and transmit about the same amount. When the sun is near the horizon, the upper layers of the canopy are brilliantly illuminated and the canopy reflectance reaches its maximum value.

These explanations are only valid for continuous plant canopies having a complete ground cover. Coniferous forests (spruce, fir ...) present some differences. Each tree has a conical shape and the crowns are often not adjacent (particularly in boreal forests). As the crowns are dense enough to prevent transmission of any incident radiation, the reflectance of such forests increases with solar elevation from the visible to the middle infrared. Models considering trees as cones casting shadows on a contrasting background allow simulation of the effect of solar elevation which agree quite well with experimental data (Li and Strahler, 1985; Otterman and Weiss, 1984).

Zenith view angle

Vegetation canopies are not perfect Lambertian diffusers. Their spectral radiance varies as a function of the inclination and orientation of the view axis (Guyot *et al.*, 1978; Guyot, Malet and Baret, 1980; Eaton and Dirmhirn, 1979; Coulson, 1966; Egbert and Ulaby, 1972; Suits, 1972b; Bunnik, 1978; Kriebel, 1978; Kimes, 1983; 1986; Otterman, 1985; Li and Strahler, 1985; Shibayama and Wiegand, 1985; Shibayama, Wiegand and Richardson, 1986; Bartlett, Johnson, Hardisky and Klemas, 1986). For low crops and forest with a continuous cover, the angular variation of the reflectance depends on the spectral domain.

Figure 2.9 displays polar diagrams of the reflectance of a wheat canopy in two vertical planes, parallel to and perpendicular to the direction of the solar beams (Guyot *et al.*, 1978). The reflectance measured with nadir viewing is taken as a reference for each spectral band (Landsat MSS 4, 5, 6, 7). All of the off-nadir measurements are presented as ratios of the nadir measurement. In this case it can be easily shown that the reflection locus, corresponding to a Lambertian surface, is a half circle with radius equal to unity (Guyot *et al.*, 1980).

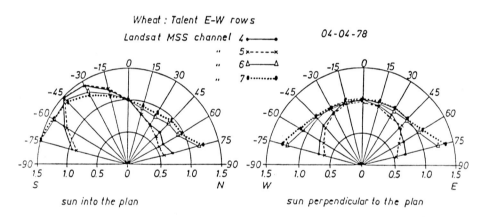

FIGURE 2.9. Polar plots of reflectance of a wheat canopy in two vertical planes. The reflectance measured with nadir viewing is taken as a reference for each spectral band and all the reflectance data are represented by their relative values (after Guyot *et al.*, 1978)

In the vertical plane, perpendicular to the Sun's direction, reflectance dependence is symmetrical. The curves corresponding to visible (MSS channels 4 and 5) and near-infrared (MSS channels 6 and 7) are clearly separated. In the visible Kimes (1983) has shown that reflectance decreases for small off-nadir viewing angles and increases for large off-nadir viewing angles. In the near-infrared the reflectance increases monotonically with the inclination of the view axis. In the vertical plane, parallel to the sun's direction, we clearly observe an assymetry. The reflectance strongly increases when the measurements are performed in the same direction as the incoming radiation (hot spot). On the contrary, it decreases when the measurements are performed in the opposite direction. The amplitude of this effect is larger in the visible than in the near-infrared for low crops or continuous forest canopies. These effects can easily be explained. The radiation flux reaching a radiometer is composed of elementary fluxes coming from the different components of the canopy, which are situated at different levels above the ground. The measured signal depends on the proportion of highly illuminated and shadowed elements in the field of view. In the visible, leaves absorb most of the incident radiation, and consequently shadows are dark. Moreover, the soil has generally a higher reflectance than the vegetation. When the inclination of the view axis increases, the proportion of soil decreases and the proportion of shadowed elements increases in the field of view, yielding a decrease in the reflectance. For continuous canopies, for large inclinations of the view axis the upper layers play a predominant role, and as they are brilliantly illuminated the canopy reflectance increases again.

In the near-infrared leaves reflect about 50% of incoming radiation and transmit about 45%. Thus, shadows are less dark than in the visible and, as the soil has a lower reflectance than the plants, the canopy reflectance increases with the inclination of the view axis.

Nebulosity

Two factors can be considered : the cloud cover and the density of atmospheric aerosols.

CLOUD COVER

Cloud cover not only changes the irradiance level, for a given sun elevation, but also changes the proportion of direct and diffuse radiation reaching the Earth's surface. When diffuse radiation is predominant, the plants are equally illuminated on all sides and the dark shadows disappear. Thus, for nadir viewing, the radiance of a plant canopy will decrease with the irradiance but its reflectance will increase. This phenomenon has been illustrated by Lord, Desjardins, Dube and Brach (1985a) on low crops. They have noted a 10% increase in reflectance during overcast days as compared to clear conditions.

As the upper layer of a plant canopy is illuminated without shadows under overcast conditions, off-nadir reflectance can be strongly increased as compared to clear conditions (Guyot, 1980). Satellite data cannot be acquired with such irradiance conditions, but ground based radiometric measurements can be performed, and some large errors can be introduced if the view axis of the radiometer is not vertical.

ATMOSPHERIC AEROSOLS

Atmospheric aerosols modify the optical path radiance between ground and satellite. This effect increases when the wavelength becomes shorter. The extent of spectral modifications due to atmospheric aerosols can be determined with model simulations (Schnetzler, 1981; Kirchner et al., 1981; Holben, Kimes and Fraser, 1986). The case of off-nadir viewing has been particularly studied. As an example, *Figure 2.10* shows the variation of the reflectance of a plant canopy (fescue) as a function of the off-nadir view angle. This effect is determined both on ground level and outside the atmosphere. This figure shows a large variation of the reflectance for typical NOAA-7 AVHRR viewing and illumination geometry and for various aerosol content of the atmosphere. However, the work of Holben et al. (1986) shows that the normalized difference (ND) can be used to minimize the effect of viewing geometry (*Figure 2.11*), but they concluded that the interpretation of satellite data requires knowledge of the angular dependance of reflectance for the viewed surfaces.

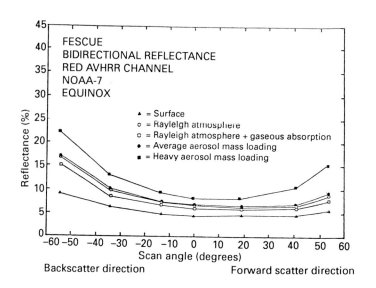
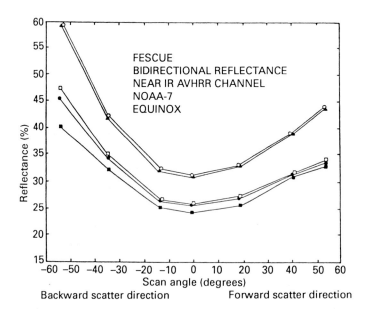

FIGURE 2.10. Response of reflectance of a fescue canopy as a function of off-nadir view angle in the red and near-infrared bands of NOAA AVHRR, on ground level and outside the atmosphere and for different atmospheric conditions (after Holben et al., 1986)

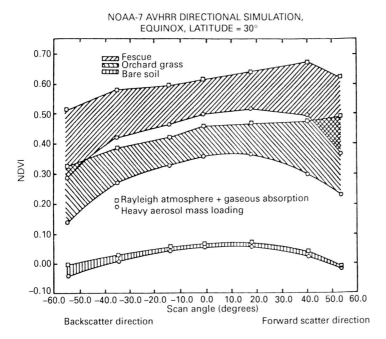

FIGURE 2.11. Variation of the Normalised Difference Vegetation Index **NDVI** as a function of the NOAA-AVHRR scanning angle, for different vegetation canopies and for two extreme atmospheric conditions (after Holben et al., 1986)

Wind speed

The wind modifies the geometry of a plant canopy: it agitates the leaves and changes their orientation and it can also lodge the plants. These effects expose to view a larger part of the stems and underlying soil and they introduce a supplementary variability of the reflectance. For example, for a barley crop the coefficient of variation of the reflectance can increase from 12% and 8% in red and near-infrared bands, for calm weather, up to 60% and 40%, respectively, for windy conditions (Lord, Desjardins and Dube, 1985). The mean level of the reflectance is also affected. It can either increase or decrease with the wind speed, depending on the type of crop (Guyot, 1980; Emori, Yasuda, Fujimoto, Yamamoto and Isaka, 1978; Wright, 1986).

Influence of internal factors

Effect of soil cover on canopy reflectance

The reflectance of plant canopies depends on the percentage of ground cover and on the optical properties of the soil background. *Figure 2.12* presents, as an example, results of measurements performed by Vanderbilt, Kollenkark, Biehl, Robinson, Bauer and Ranson, (1981) on a soybean canopy having a 60% soil cover. Measurements were performed with three different backgrounds : natural soil and black and white painted boards. More detailed studies were performed by Huete (Huete, 1986; 1987; 1988; Huete et al., 1985), who has tried to separate the soil contribution to the reflectance of a plant canopy. His analysis and model simulations (Huete, 1988) show that the effect of the soil can be seen up to a LAI of around 3, which corresponds to almost complete cover of the soil. For LAI greater than 3 a saturation is observed. In the near-infrared the saturation occurs for LAI greater than 5. In the middle-infrared the background effect is also seen for relatively large LAI values. In this domain the bi-directional reflectance of a plant canopy can be strongly affected by surface soil moisture (Baret, Guyot, Bégué, Maurel and Podaire, 1988).

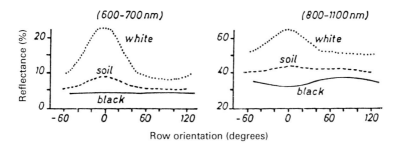

FIGURE 2.12. Effect of the crop row orientation and of the soil colour on the reflectance of a soybean canopy. The sun azimuth is taken as a reference (after Vanderbilt et al., 1981)

Effect of crop row orientation

As shown in *Figure 2.12*, the crop row orientation with respect to the sun azimuth can have a large effect on the reflectance. The effect of row orientation is larger in the visible than in the near infrared because of the stronger contrast between soil and leaf reflectance in this domain. The effect of row orientation also depends on the proportion of ground cover. Vanderbilt et al., (1981) have shown that the maximum effect is observed when the ground cover ranges between 40% and 60%. This experimental result is confirmed by model simulation performed by Goel and Grier (1986) on forest canopies.

Effect of canopy geometry

The most important factor that acts on the optical properties of a plant canopy is its geometrical structure. The relationship between the geometrical structure of a plant canopy and its optical properties has been studied experimentally and theoretically. Model simulation enables to test the effect of each geometrical parameter independently (Suits, 1972a; Verhoef, 1984; 1985; Kimes, 1984; Goel, 1988).

One of the main factors characterizing the geometry of a plant canopy is the leaf inclination angle (Verhoef and Bunnik, 1981). As in a plant canopy all of the leaves do not have the same inclination, it is necessary to not only consider their mean inclination but also their distribution (measured by leaf inclination distribution function, LIDF). The role played by the LIDF has been studied in detail by Verhoef and Bunnik (1981), and *Figures 2.13* and *2.14* display results of their model simulations plotted in the form of feature space diagrams for nadir observation. For each canopy geometry green reflectance is plotted versus red reflectance in *Figure 2.13*; and near-infrared versus red is shown in *Figure 2.14*. The LAI varies between 0 (bare soil) and 10. In both figures points with equal LAI are connected by straight lines surrounding the area of possible leaf inclination distribution. From these figures the following can be concluded:

- the position of a point in the feature space is mainly determined by the LAI and average leaf inclination angle;

- the dispersion due to differences in canopy geometry is much greater in the red - near-infrared subspace than in the red - green subspace;

- the dispersion due to differences in LIDF increases with LAI.

These points have been confirmed by experimental results obtained by Kimes (1983), Pinter, Jackson, Ezra and Gausman (1985), and Ranson, Daughtry, Biehl and Bauer, (1985).

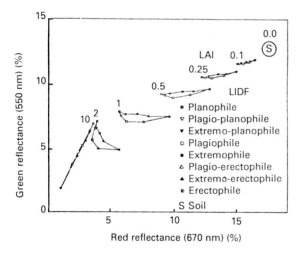

FIGURE 2.13. Feature space diagram red - green for different canopy geometries under vertical view (after Verhoef and Bunnik, 1981)

In forest, supplementary factors must be considered, particularly the shape and the size of the tree crowns (Hughes, Evans, Burns and Hill, 1986; Walsh, 1980; Guyon-Haeck, 1985). The shape of the statistical distribution of the reflectance will vary with these factors.

Effect of physiological activity of the plants

The physiological activity of the plants can affect the reflectance of a vegetative canopy if it changes the LIDF. This is, for example, the case of several stressors which do not affect the optical properties of the leaves (eg, water stress and some diseases or pest attacks affecting the root system). Moreover, for certain plant canopies the LIDF changes

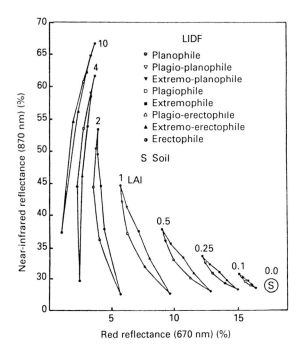

FIGURE 2.14. Feature space diagram red - near-infrared for different canopy geometries under vertical view (after Verhoef and Bunnik, 1981)

during the day as a function of the sun azimuth (eg, sunflower, soybean, cotton) and the reflectance can be affected as a consequence (Kimes and Kirchner, 1983).

Conclusions

This short review shows the complexity of the optical properties of plant canopies, which depend on numerous factors.

For discriminating different plant canopies with single or multi-date satellite data it is not necessary to use sophisticated correction models. Discrimination is generally based on the contrast existing between the different surfaces, but this approach is only valid in the case of nadir viewing. When satellite data having a large field of view are considered it is necessary to correct for the angular variation both of atmospheric effects and the optical properties of a given target on the signal received. Without these corrections the same object will have different spectral signatures in different parts of the same image. The correction of the angular variation of the optical properties of crop canopies is difficult because it varies from one crop to another: we cannot apply a standard correction to a given image, and research is needed on this.

However, in agricultural remote sensing, we are not only interested in discriminating crops but also in determining their standing biomass and monitoring their evolution with time. In order to reach these objectives it is necessary to have quantitative data which can be compared from date to date. For this purpose atmospheric correction models must be used.

References

AHLRICHS, J.S. and BAUER, M.E. (1983). Relation of agronomic and multispectral reflectance characteristics of spring wheat canopies. *Agronomy Journal*, **75**, 987-993

AL-ABBAS, A.H., BARR, R., HALL, J.D., CRANE, F.L. and BAUMGARDNER, M.F. (1974). Spectra of normal and nutrient deficient maize leaves. *Agronomy Journal*, **66**, 16-20

ALLEN, W.A., GAUSMAN, H.W., RICHARDSON, A.J. and THOMAS, J.P., (1969). Interaction of isotropic light with a compact plant leaf. *Journal Optical Society of America*, **59**, 1376-1379

BALDY, Ch., GUYOT, G. and MERELLE, F. (1981). Contribution a l'étude des propriétés optiques des feuilles d'olivier (Olea europea L.). *Comptes Rendues Academie des Sciences, Paris*. **T.193**, Série III, 253-256

BARET, F. and GUYOT, G. (1986). Suivi de la maturation de couverts de blé par radiométrie dans les domaines visible et proche-infrarouge. *Agronomie*, **6**, 509-516

BARET, F., GUYOT, G., BEGUE, A., MAUREL, P. and PODAIRE, A. (1988). Complementarity of middle infrared with visible and near infrared reflectance for monitoring wheat canopies. *Remote Sensing of Environment*, **26**, 213-225

BARTLETT, D.S., JOHNSON, R.W., HARDISKY, M.A. and KLEMAS, V. (1986). Assessing impacts of off-nadir observation on remote sensing of vegetation: use of Suits model. *International Journal of Remote Sensing*, **7**, 247-264

BOWERS, S.S. and HANKS, R.J. (1965). Reflection of radiant energy from soils. *Soil Science*, **100**, 130-138

BREECE, H.T. and HOLMES, R.H. (1971). Bidirectional scattering characteristics of healthy green soybean and corn leaves *in vivo*. *Applied Optics*, **10**, 119-127

BUNNIK, N.J.J. (1978). The multispectral reflectance of shortwave radiation by agricultural crops in relation with their morphological and optical properties. Mededelingen Landbouwhogeschool, Wageningen, Netherlands

CAVAYAS, F. and TEILLET, P.M. (1986). Geometric model simulations of conifer canopy reflectance. In *Proceedings 3rd International Colloquium Spectral Signatures of Objects in Remote Sensing*. Les Arcs (France). 16-20 Dec. 1985. ESA SP-247. European Space Agency, Paris

CIERNIEWSKI, J. (1987). A model for soil surface roughness influence on the spectral response of bare soils in the visible and near-infrared range. *Remote Sensing of Environment*, **23**, 97-115

COLEMAN, T.L. and MONTGOMERY, O.L. (1987). Soil moisture, organic matter and iron content effect on the spectral characteristics of selected vertisols and affisols in Alabama. *Photogrammetric Engineering and Remote Sensing*, **53**, 1659-1663

COOPER, K.D. and SMITH, J.A. (1985). A Monte Carlo reflectance model for soil surfaces with three-dimensional structure. *IEEE Transactions on Geoscience and Remote Sensing*, **GE-23**, 668-673

COULSON, K.K. (1966). Effect of reflection properties of natural surfaces in aerial reconnaissance. *Applied Optics*, **5**, 905-917

DEAN, K.G., KODANA, Y. and WENDLER, G. (1986). Comparison of leaf and canopy reflectance of subarctic forests. *Photogrammetric Engineering and Remote Sensing*, **52**, 809-811

EATON, F.D. and DIRMHIRN, I. (1979). Reflected irradiance indicatrices of natural surfaces and their effect on albedo. *Applied Optics*, **18**, 994-1008

EGBERT, D.D. and ULABY, F.T. (1972). Effect of angles on reflectivity. *Photogrammetric Engineering*, **38**, 556-564

EMORI, Y., YASUDA, Y., FUJIMOTO, M., YAMAMOTO, H. and ISAKA, I. (1978). Data acquisition and data base of spectral signature in Japan. In *Proceedings International Symposium on Remote Sensing for Observation and Inventory of Earth Resources and the Endangered Environment*, Freiburg *International Archives Photogrammetry*, **22**, 543-566

EPIPHANIO, J.C.N. and VITORELLO, I. (1984). Inter-relationships between view angles (azimuth) and surface moisture and roughness conditions in field-measured radiometer reflectances of an oxisol. In *CR IIe Colloque International Signatures Spectrales d'Objets en Télédétection*. Bordeaux (France), 12-16 Sept. 1983. INRA Publishers. *Les Colloques de l'INRA*, **23**, 185-192

FUCHS, M., STANHILL, G. and WAANDERS, A.G. (1972). Diurnal variations of the visible and near infrared reflectance of a wheat crop. *Israel Journal of Agricultural Research*, **22**, 63-75

GAUSMAN, H.W. (1974). Leaf reflectance of near infrared. *Photogrammetric Engineering*, **40**, 57-62

GAUSMAN, H.W. and ALLEN, W.A. (1973). Optical parameters of leaves of 30 plant species. *Plant Physiology*, **52**, 57-62

GAUSMAN, H.W., ALLEN, W.A., CARDENAS, R. and RICHARDSON, A.J. (1970). Relation of light reflectance to histological and physical evaluation of cotton leaf maturity. *Applied Optics*, **9**, 545-552

GAUSMAN, H.W., ALLEN, W.A., ESCOBAR, D.E., RODRIGUEZ, R.R. and CARDENAS, R. (1971). Age effect of cotton leaves on light reflectance, transmittance and absorptance and on water content and thickness. *Agronomy Journal*, **63**, 465-468

GAUSMAN, H.W., ESCOBAR, D.E. and RODRIGUEZ, R.R. (1978). Effect of stress and pubescence on plant leaf and canopy reflectance. In *Proceedings International Symposium on Remote Sensing for Observation and Inventory of Earth Resources and the Endangered Environment*. Freiburg, F.R.G. 2-5 July, 1978. *International Archives of Photogrammetry*, **23**, 719-749

GIRARD, C.M. (1986). Spectral and botanical classification of grasslands: Aussois example. In *Proceedings International Symposium on Remote Sensing for Resources Development and Environmental Management*. Enschede. Ed by Balkema, A.A., Damen, M.C.J., Sicco Smit,

G. and Verstappen, T.Th. *International Archives Photogrammetry and Remote Sensing*, **26**, 209-272

GOEL, N.S. (1988). Models of vegetation canopy reflectance and their use in estimation of biophysical parameters from reflectance data. *Remote Sensing Reviews*, **4**, 1-222

GOEL, N.S. and GRIER, T. (1986). Estimation of canopy parameters for inhomogeneous vegetation canopies from reflectance data. *International Journal of Remote Sensing*, **7**, 665-681

GRANT, L. (1985). Polarized and non polarized components of leaf reflectance. Ph. D. Thesis. Purdue University, West Lafayette, Indiana, USA

GRANT, L. (1987). Diffuse and specular characteristics of leaf reflectance. *Remote Sensing of Environment*, **22**, 309-322

GRANT, L., DAUGHTRY, C.S.T. and VANDERBILT, V.C. (1987a). Polarized and non-polarized reflectances of *Coleus blumei*. *Environmental and Experimental Botany*, **27**, 139-145

GRANT, L., DAUGHTRY, C.S.T. and VANDERBILT, V.C. (1987b). Variations in the polarized leaf reflectance of Sorghum bicolor. *Remote Sensing of Environment*, **21**, 333-339

GUYON-HAECK, D. (1985). Intérêt du satellite de télédétection SPOT (simulations) pour la connaissance de la forêt. Etude de peuplements de pin maritime du massif forestier des Landes de Gascogne. Thèse Docteur Ingénieur. Université Paul Sabatier. Toulouse (Sciences). No. d'ordre 923

GUYOT, G. (1980). Analysis of factors acting on the variability of spectral signatures of natural surfaces. In *Proceedings International Symposium I.S.P.* Hamburg. *International Archives Photogrammetry*, **22**, 382-393

GUYOT, G. (1984). Caractérisation spectrale des couverts végétaux dans le visible et le proche infrarouge, application a la télédétection. *Bulletin Societé Francaise de Photogrammetrie et de Télédétection*, **95**, 5-22

GUYOT, G. (1989). *Les signatures spectrales des surface naturelles*. Télédétection Satellitaire No. 5. Paradigme, Caen (France)

GUYOT, G., DUPONT, O., JOANNES, H., PRIEUR, C., HUET, M. and PODAIRE, A. (1985). Investigation into the mid-IR spectral band best suited to monitoring vegetation water content. In *Proceedings 18th International ERIM Symposium*. Paris. Oct. 1984. Vol.2: 1049-1063. Environmental Research Institute of Michigan, Ann Arbor, Michigan

GUYOT, G., JACQUIN, C., MALET, P. and THOUY, G. (1978). Evolution des indicatrices de réflexion de cultures de céréales en fonction de leurs stades phénologiques. In *Proceedings International Symposium on Remote Sensing for Observation and Inventory of Earth Resources and Endangered Environment*. Freiburg, F.R.G., 2-5 July 1978. *International Archives of Photogrammetry*, **22**, 705-718

GUYOT, G., MALET, P. and BARET, F. (1980). Analyse des indicatrices de réflexion de l'orge et du blé, possibilité de la stéréoradiométrie. In *Proceedings International Symposium I.S.P.* Hamburg, F.R.G. *International Archives of Photogrammetry*, **22**, 372-381

GUYOT, G., GUYON, D. and RIOM, J. (1989). Factors affecting the spectral response of forest canopies: a review. *Geocarto International*, **3**, 3-18

HEOIS, B. (1979). Etude par télédétection de peuplements de pins maritimes dans le Sud-Ouest de la France. Mémoire de fin d'Etudes d'Ingénieur des Techniques Forestières. INRA Recherche Forestière Bordeaux, ENITEF, Domaine des Barres, Nogent-sur-Vernisson (France)

HOFFER, R.M. and JOHANSSEN, C.J. (1969). Ecological potentials in spectral signature analysis. In *Remote Sensing in Ecology*. Ed. by Johnson, P.L. University of Georgia Press, Athens, G.A. USA

HOLBEN, B.N., KIMES, D.S. and FRASER, R.S. (1986). Directional reflectance response in AVHRR red and near IR bands for three cover types and varying atmospheric conditions. *Remote Sensing of Environment*, **19**, 213-236.

HORLER, D.N.H., BARBER, J. and BARRINGER, A.R. (1980). Effect of heavy metals on the absorbance and reflectance spectra of plants. *International Journal of Remote Sensing*, **1**, 121-136

HORLER, D.N.H., DOCKRAY, M. and BARBER, J. (1983). The red edge of plant leaf reflectance. *International Journal of Remote Sensing*, **4**, 273-288.

HUETE, A.R. (1986). Reconstruction of vegetation spectra from soil plant canopies. In *Proceedings 3rd International Colloquium Spectral Signatures of Objects in Remote Sensing*. Les Arcs (France). 16-20 Dec. 1985. ESA SP-247, European Space Agency, Paris

HUETE, A.R. (1987). Soil dependent spectral response in a developing plant canopy. *Agronomy Journal*, **79**, 61-68

HUETE, A.R. (1988). Soil and atmosphere influences on the spectra of partial canopies. *Remote Sensing of Environment*, **25**, 89-106

HUETE, A.R., JACKSON, R.D. and POST, D.F. (1985). Spectral response of plant canopies with different soil backgrounds. *Remote Sensing of Environment*. **17**, 37-53

HUGHES, J.S., EVANS, D.L., BURNS, P.Y. and HILL, J.M. (1986). Identification of two southern pine species in high-resolution aerial MSS data. *Photogrametric Engineering and Remote Sensing*, **52**, 1175-1180

JACQUEMOUD, S., BARET, F. and GUYOT, G. (1989). Modelization of leaf optical properties for interpreting high spectral resolution reflectance measurements. In *Proceedings EARSeL Symposium*. Espoo, Finland

KEEGAN, H.J., SCHLETER, J.C., HALL, W.A. and HAAS, G.M. (1956). Spectrophotometric and calorimetric study of diseased and rust-resisting cereal crops. National Bureau of Standards Report 4591

KIMES, D.S. (1983). Dynamics of directional reflectance factor distributions for vegetation canopies. *Applied Optics*, **22**, 1364-1372

KIMES, D.S. (1984). Modeling the directional reflectance from complete homogeneous vegetation canopies with various leaf-orientation distributions. *Journal Optical Society of America*, **1**, 575-588

KIMES, D.S. (1986). Modelisation of the optical scattering behaviour of the vegetation canopies. In *Proceedings 3rd International Colloquium Spectral Signatures of Objects in Remote Sensing*. Les Arcs, France. 16-20 Dec. 1985. ESA-SP247. European Space Agency, Paris

KIMES, D.S. and KIRCHNER, J.A. (1983). Diurnal variations of vegetative canopy structure. *International Journal of Remote Sensing*, **4**, 257-271

KING, C. (1986). Les qualités spectrales des sols nus: analyse des spectres radiométriques acquis sur le terrain dans le Bassin Parisien. In *CR IIe Colloque International Signatures Spectrales d'Objets en Télédétection*. Bordeaux 12-16 Sept. 1983. *Les Colloques de l'INRA*, **23**, 253-264

KIRCHNER, J.A., SCHNETZLER, C.C. and SMITH, J.A. (1981). Simulated directional radiances of vegetation from satellite platforms. *International Journal of Remote Sensing*, **2**, 253-264

KNIPLING, E.B. (1970). Physical and physiological basis for the reflectance of visible and near infrared radiation from vegetation. *Remote Sensing of Environment*, **1**, 155-159

KRIEBEL, K.T. (1978). Measured spectral bidirectional reflection properties of four vegetated surfaces. *Applied Optics*, **17**, 253-259

KUMAR, R. and SILVA, L. (1973). Light ray tracing through a leaf cross section. *Applied Optics*, **12**, 2950-2954

LI, X. and STRAHLER, A.H. (1985). Geometric optical modeling of a conifer forest canopy. *IEEE Transactions on Geoscience and Remote Sensing*, **GE-23**, 705-721

LICHTENTHALER, H.K. and PFISTER, K. (1978). *Praktikum der photosynthese*. Quelle & Meyer Verlag, Heidelberg

LICHTENTHALER, H.K. and RINDERLE, U. (1988). Chlorophyll fluorescence spectra of leaves as induced by blue light and red laser light. In *Proceedings 4th International Colloquium Spectral Signatures of Objects in Remote Sensing*. Aussois, France. ESA-SP 287. European Space Agency, Paris

LORD, D., DESJARDINS, R.L. and DUBE, P.A. (1985). Influence of wind on crop canopy reflectance measurements. *Remote Sensing of Environment*, **18**, 113-123

LORD, D., DESJARDINS, R.L., DUBE, P.A. and BRACH, E.J. (1985). Variations of crop canopy spectral reflectance measurements under changing sky conditions *Photogrammetric Engineering and Remote Sensing*, **51**, 689-695

MOREAU, J.P., BOISSARD, P. and BONHOMME, R. (1981). Caractérisation des maladies type jaunisse grâce aux mesures de réflectance. In *CR Colloque International Signatures Spectrales d'Objets en Télédétection*. Avignon (France). 8-11 Sept. 1981. *Les Colloques de l'INRA*, **5**, 443-453

MYERS, V.I. (1970). Soil, water, plant relationships. In *Remote Sensing with spectral reference to agriculture and forestry*. National Academy of Science. Washington D.C.

MYERS, V.I., BAUER, M.E., GAUSMAN, H.W., HART, W.G., HEIMAN, J.L., MacDONALD, R.B., PARK, A.B., RYERSON, R.A., SCHMUGGE, T.J. and WESTIN, F.C. (1983). Remote sensing application in agriculture. In *Manual of Remote Sensing*, 2nd edition. American Society of Photogrammetry and Remote Sensing, Falls Church, Virginia

OESTER, B. (1981). Signatures spectrales des aiguilles de pins sylvestres. In *CR Colloque International Signatures Spectrales d'Objets en Télédétection*. Avignon, France. 8-11 Sept. 1981. *Les Colloques de l'INRA*, **5**, 191-199

OTTERMAN, J. (1985). Bidirectional and hemispheric reflectivities of a bright soil plane and a sparse dark canopy. *International Journal of Remote Sensing*, **6**, 897-902

OTTERMAN, J. and WEISS, G.H. (1984). Reflection from a field of randomly located vertical protrusions. *Applied Optics*, **23**, 1931-1936

PIECH, K.R. and WALKER, J.E., (1974). Interpretation of soils. *Photogrammetric Engineering*, **40**, 87-94

PINTER, P.J. Jr., JACKSON, R.D., EZRA, C.E. and GAUSMAN, H.W. (1985). Sun angle and canopy architecture effects on the spectral reflectance of 6 wheat cultivars. *International Journal of Remote Sensing*, **6**, 1813-1825

RANSON, K.J., DAUGHTRY, C.S.T., BIEHL, L.L. and BAUER, M.E. (1985). Sun view angle effects on reflectance factors of crop canopies. *Remote Sensing of Environment*, **18**, 147-161

RICHARDSON, A.J. (1981). Measurements of reflectance factors under daily and intermittent irradiation variation. *Applied Optics*, **20**, 3336-3340

SANDWALD, E.F. (1981). Laboratory determined spectral signatures of leaves healthy and rizomania diseased sugarbeets and disease interpretability from aerial IRC photographs. In *CR Colloque International Signatures Spectrales d'Objets en Télédétection*. Avignon, France. 8-11 Sept. 1981. *Les Colloques de l'INRA*, **5**, 201:208

SCHNETZLER, C.C. (1981). Effect of sun and sensor geometry, canopy structure and density, and atmospheric condition on the spectral response of vegetation, with particular emphasis on accross-track pointing. In *Proceedings International Colloquium Spectral Signatures of Objects in Remote Sensing*. Avignon, France. 8-11 Sept. *Les Colloques de l'INRA*, **5**, 509-520

SHIBAYAMA, M. and WIEGAND, C.L. (1985). View azimuth and zenith, and solar angle effects on wheat canopy reflectance. *Remote Sensing of Environment*, **18**, 91-103

SHIBAYAMA, M., WIEGAND, C.L. and RICHARDSON, A.J. (1986). Diurnal patterns of bidirectional vegetation indices for wheat canopies. *International Journal of Remote Sensing*, **7**, 233-246

SINCLAIR, T.R., HOFFER, R.M. and SCHREIBER, M.M. (1971). Reflectance and internal structure of leaves from several crops during a growing season. *Agronomy Journal*, **63**, 864-868

SINCLAIR, T.R., SCHREIBER, M.M. and HOFFER, R.M. (1973). Pathway of solar radiation through leaves. *Agronomy Journal*, **63**, 276-283

SUITS, G.H. (1972a). The calculation of the bidirectional reflectance of a vegetative canopy. *Remote Sensing of Environment*, **2**, 117-125

SUITS, G.H. (1972b). The cause of azimuthal variation in directional reflectance of vegetative canopies. *Remote Sensing of Environment*, **2**, 175-182

THOMAS, J.R., MYERS, V.I., HEILMAN, M.D. and WIEGAND, C.L. (1966). Factors affecting light reflectance of cotton. In *Proceedings 4th Symposium on Remote Sensing of Environment*. Environmental Research Institute of Michigan, Ann Arbor, Michigan

THOMAS, J.R., NAMKEN, L.N., OERTHER, G.F. and BROWN, R.G. (1971). Estimating leaf water content by reflectance measurements. *Agronomy Journal*, **63**, 845-847

THOMAS, J.R., and OERTHER, G.F. (1972). Estimating nitrogen content of sweet pepper leaves by reflectance measurements. *Agronomy Journal*, **64**, 11-13

TUCKER, C.J. and GARRATT, M.W. (1977). Leaf optical system modeled as a stochastic process. *Applied Optics*, **16**, 635-642

VANDERBILT, V.C., KOLLENKARK, J.C., BIEHL, L.L., ROBINSON, B.F., BAUER, M.E. and RANSON, K.J. (1981). Diurnal changes in reflectance factor due to sun-row direction interactions. In *Proceedings International Colloquium Spectral Signatures of Objects in Remote Sensing*. Avignon, France 8-11 Sept. 1981. *Les Colloques de l'INRA*, **5**, 499-508

VERHOEF, W. (1984). Light scattering by leaf layers with application to canopy reflectance modeling: the SAIL model. *Remote Sensing of Environment*, **16**, 125-141

VERHOEF, W. (1985). Earth observation modeling based on layer scattering matrices. *Remote Sensing of Environment*, **17**, 165-178

VERHOEF, W. and BUNNIK, N.J.J. (1981). Influence of crop geometry on multispectral reflectance determined by the use of canopy reflectance models. In *Proceedings International Colloquium Spectral Signatures of Objects in Remote Sensing*. Avignon, France. 8-11 Sept. 1981. *Les Colloques de l'INRA*, **5**, 273-290

WALSH, S.J. (1980). Coniferous tree species mapped using Landsat data. *Remote Sensing of Environment*, **9**, 11-26

WILLSTATTER, R. and STOLL, A. (1928). *Investigations on chlorophyll*. Science Press, London

WRIGHT, G.G. (1986). Some observations of the effect of wind turbulence on near infrared/red ratio. *International Journal of Remote Sensing*, **7**, 173-178

3

FACTORS AFFECTING THE RADIATIVE TEMPERATURE OF A VEGETATIVE CANOPY

P. BOISSARD
INRA Bioclimatologie, 78850 Thiverval-Grignon, France

G. GUYOT
INRA Bioclimatologie, Domaine St. Paul B. P. 91, 84143 Montfavet Cedex, France

and

R.D. JACKSON
USDA, ARS, U.S. Water Conservation Laboratory, Phoenix, Arizona 85040, U.S.A.

Introduction

In recent years, considerable progress has been made in the development of instruments that measure thermal infrared radiation. Scanners aboard several satellites, as well as aircraft, have thermal infrared channels, and different types of IR detecting instruments are being used in ground-based studies. These devices range from boom-mounted thermal imaging devices to small hand-held IR thermometers. The availability of these instruments has given rise to a number of research programs that use thermal data to obtain information about the growth and condition of plant canopies.

The data recorded by a radiometer represent an integration of a series of elementary fluxes of radiation that originate from various layers of leaves and soil, all of which may be at different temperatures. The question arises as to what is the significance of the composite signal measured by a radiometer? What is the relation of this signal to the condition of the plants?

The temperature distribution within a plant canopy is determined by both external environmental factors and by internal plant factors. In this report we will attempt to analyse a number of the factors on which the radiative temperature of a plant canopy depends. It should be noted that, although the factors are analysed separately, in reality they are interactive and highly dependent on each other.

External environmental factors

The aerial environment that surrounds plant canopies plays a major role in determining their radiative temperature. Furthermore, instrument characteristics and measurement techniques determine the amount and the quality of information that can be obtained.

Climatic conditions

Aerial factors that influence plant canopy temperature include the temperature and water vapour pressure of the air, net radiation and wind. These factors can be written in terms of an energy balance, from which their relative effect on plant canopy temperatures can be ascertained. The surface temperatures of different elements of a plant canopy depend on the instantaneous energy balance of each element. The enegy balance equation can be written as:

$$R_n + H + G + \lambda E = 0, \tag{3.1}$$

where R_n is the balance of radiative energy exchanges, H the sensible heat flux due to convection, G the heat flux due to conduction (its magnitude is generally considerably less that the other fluxes and will be considered negligible in the following equations) and λE the latent heat flux due to the evaporation of water (this flux is the energy equivalent of the evapotranspiration of the plant canopy). The sensible heat flux can be expressed as

$$H = -h(T_s - T_a) = -\rho c_p(T_s - T_a)/r_a \tag{3.2}$$

where T_s and T_a are the temperatures of the surface and the air, respectively, $h = \rho c_p/r_a$ is the coefficient of turbulent transfer for heat, ρc_p is the volumetric heat capacity of air and r_a is an aerodynamic resistance.

GENERAL RELATIONSHIPS

The net radiation R_n can be expressed in the form

$$R_n = (1 - \alpha)R_s + \epsilon R_l - \epsilon \sigma T_s^4 \tag{3.3}$$

where R_s is the global solar radiation, α the albedo, R_l the atmospheric thermal radiation, σ the Stefan-Boltzmann constant, ϵ the emissivity of the surface and T_s the absolute (Kelvin) temperature of the surface.

The energy balance equation can then be written as:

$$(1 - \alpha)R_s + \epsilon R_l - \epsilon \sigma T_s^4 - h(T_s - T_a) + G + \lambda E = 0 \tag{3.4}$$

To evaluate the influence of climatic factors on the surface temperature T_s, we make a number of approximations (Seguin, 1980a). Taking $\epsilon = 1$ and $G = 0$, and writing the thermal radiation term as the linear approximation:

$$\sigma T_s^4 - \sigma T_a^4 = 4\sigma T^3(T_s - T_a), \tag{3.5}$$

Substitution in the energy balance equation gives:

$$(1 - \alpha)R_s + R_l - \sigma T_a^4 = (h + 4\sigma T^3)(T_s - T_a) - \lambda E. \tag{3.6}$$

If one defines R_{nc} as the "net climatic radiation" (Seguin, 1980a) (defined by Monteith (1973) as the isothermal net radiation)

$$R_{nc} = (1 - \alpha)R_s + R_l - \sigma T_a^4, \tag{3.7}$$

then the temperature difference becomes:

$$T_s - T_a = (R_{nc} + \lambda E)/(h + 4\sigma T_a^3), \tag{3.8}$$

with $R_{nc} > 0$ and $\lambda E < 0$ for the daylight period.

This approximate relationship shows that, with all the other parameters being constant, the difference (T_s - T_a) varies linearly with R_{nc} and linearly in a direction opposite λE, but inversely with h and therefore with the wind velocity (h is proportional to the wind velocity), and inversely with the air temperature.

Resistance form of energy balance equation

Monteith (1973) coupled plant and climatic factors by including aerodynamic and canopy resistance terms in the energy balance. In addition to the aerodynamic resistance (r_a) we may define a canopy resistance (r_c) (to be discussed in more detail later), and write the latent heat flux as

$$\lambda E = \rho c_p (e_c^* - e_a)/[\gamma(r_a + r_c)]. \tag{3.9}$$

Combining *Equations 3.2* and *3.9* with *3.1* and rearranging results in:

$$T_c - T_a = \frac{r_a R_n}{\rho c_p} \cdot \frac{\gamma(1 + r_c/r_a)}{\Delta + \gamma(1 + r_c/r_a)} - \frac{(e_a^* - e_a)}{\Delta + \gamma(1 + r_c/r_a)}, \tag{3.10}$$

which relates the difference between the canopy and the air temperatures to the vapour pressure deficit of the air ($e_a^* - e_a$), the difference between the saturation vapour pressure at air temperature e_a^* and the vapour pressure in the air, e_a the net radiation (R_n) and the wind, the latter through the aerodynamic resistance. In the above equations γ is the psychrometer constant, e_c^* is the saturation vapour pressure at the canopy temperature and Δ is the slope of the saturation vapour pressure - temperature curve. *Equation 3.10* was given in a slightly different form by Monteith and Szeicz (1962). The upper limit of ($T_s - T_a$) can be obtained from *Equation 3.10* by allowing r_c to become infinitely large;

$$(T_s - T_a) = r_a R_n / \rho c_p, \tag{3.11}$$

and the lower limit is found by setting $r_c = 0$, to yield

$$T_c - T_a = \frac{r_a R_n}{\rho c_p} \cdot \frac{\gamma}{\Delta + \gamma} - \frac{e_a^* - e_a}{\Delta + \gamma} \tag{3.12}$$

Equations 3.10 and *3.12* show that the temperature difference decreases linearly as the vapour pressure deficit of the air increases, provided that the canopy resistance is sufficiently low to allow transpiration, and thus evaporative cooling of the leaves. Although r_c is not 0 for well-watered crops (Van Bavel and Ehrler, 1968), it can be quite low and canopy temperatures can be considerably less than air temperature (Jackson, 1982a). This relationship was demonstrated experimentally (Idso, Reginato, Reicosky and Hatfield, 1981) and is reproduced in *Figure 3.1*. The upper limit (*Equation 3.11*) varies only with the net radiation and the aerodynamic resistance, with the latter being somewhat difficult to evaluate (Hatfield, Perrier and Jackson, 1983).

Effect of wind velocity

For neutral conditions, the aerodynamic resistance is given by the expression

$$r_a = \{ln[(z - d)/z_o]\}^2/(k^2 u) \tag{3.13}$$

where z is a reference height, d is the zero plane displacement, z_o is the surface roughness height, k is von Karman's constant and u is the wind-speed. Neutral conditions imply that $T_s = T_a$. Under conditions of medium to low atmospheric humidity, $T_s - T_a$ may range from -10 to +5°C, depending on the water status of the crop (Jackson, Idso, Reginato and Pinter, 1981). Thus, stability corrections must be applied to *Equation 3.11* if actual conditions are to be approximated (Hatfield et al., 1983).

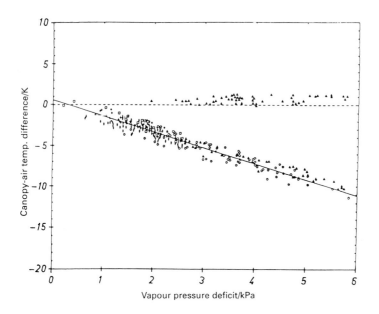

FIGURE 3.1. Canopy air temperature differences versus vapour pressure deficit for several well-watered plots of alfalfa ——, assumed to be transpiring at the potential rate, and one severely water-stressed plot - - -, for which all temperature differences were positive

Monteith (1963) showed that stability could be accounted for (at least approximately) by inclusion of the Richardson number. Expressing this number in terms of temperature differences and combining it with *Equation 3.13* results in the relation:

$$r_{ac} = r_a[1 - n(z - d)g(T_s - T_a)/(T_o ku^2)] \qquad (3.14)$$

where r_{ac} is the stability corrected aerodynamic resistance, n is an empirical constant (usually assigned a value of 5), g is the acceleration due to gravity and T_o the average temperature (usually taken as the air temperature). A detailed discussion of the development of and the assumptions underlying *Equations 3.13* and *3.14* was given by Monteith 1963; 1973).

A surface which is cooler than air will have a vertical transfer of sensible heat toward it. This effect partially counters the momentum-induced aerodynamic resistance, and depends upon windspeed, canopy roughness and the temperature gradient. Conversely, if the surface temperature is warmer than the air, the aerodynamic resistance will decrease, because of buoyancy effects. Values of r_a for neutral conditions (horizontal dashed lines), and stability corrected values (solid lines) at several windspeeds and temperature gradients for a rough canopy ($z_o = 0.13$ m), and for a smooth canopy ($z_o = 0.05$ m) are shown in *Figure 3.2*.

The value of T_o was taken as 303 K. The calculations show that, as the windspeed increases, the stability corrected r_{ac} decreases, regardless of whether the canopy is warmer or cooler than the air. Also, as the surface becomes smoother (smaller z_o) values of r_{ac} become larger, at the same value of windspeed and temperature difference.

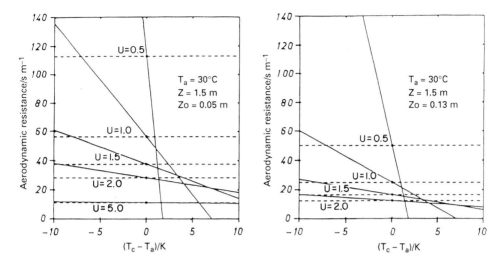

FIGURE 3.2. Stability corrected (solid lines) and uncorrected (dashed lines) aerodynamic resistances for surface roughnesses of 0.05 and 0.13 m, as a function of canopy-air temperature differences (Data from Hatfield et al., 1983)

Figure 3.2 and *Equation 3.10* show that an increase in wind velocity will cause the canopy temperature to approach that of air, regardless of whether the canopy was warmer or cooler than the air. Therefore, the scatter of values of the radiative temperature of a plant canopy should decrease as the wind velocity increases. This result is shown in *Figure 3.3*. The data correspond to temperatures of maize foliage grown at Avignon in the south-east of France. The radiative temperatures were measured using an infrared imaging radiometer mounted on a hydraulic boom and held 7 m above the canopy (Guyot and Chasseray, 1981). The surface viewed was about 4m × 4m and the resolution element was a square of 30 mm per side. Thermal resolution was about 0.1 K. *Figure 3.3* shows the contraction of the histogram of temperature frequencies when the wind velocity increased and the average temperature was higher than that of the air.

Measurements carried out in California on a 259 ha area planted to barley allowed the study of variability in radiation temperature as a function of the size of the resolution element (Millard, Goettelman and Le Roy, 1981). The results showed that the highest variability in surface temperatures was observed during calm periods with a dry soil sparsely covered with vegetation. Increasing the wind velocity, moistening the soil and the development of a plant canopy had the effect of reducing the variability of radiation temperatures.

SUN ELEVATION ANGLE

Solar elevation has a major influence on the apparent radiative temperature of a plant canopy. Two effects are manifested; the average canopy temperature increases with increased sun elevation because of the increase of energy received, and the temperature distribution within the canopy is a function of the angle of penetration of solar radiation.

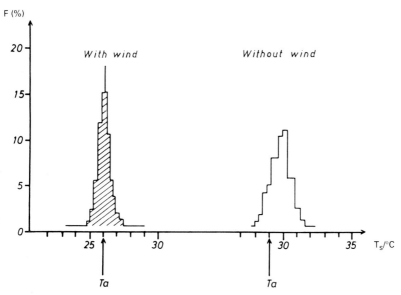

FIGURE 3.3. Histogram of radiative temperature frequencies (F) determined on a maize canopy at solar noon at Avignon-Montfavet for 2 different days, with wind and without wind. Numbers on the histogram represent measurement frequencies for intervals of temperature of 0.25 K, expressed as a percentage of the total number of measurements. T_a indicates the value of air temperature

Equation 3.8 shows that the temperature difference $T_s - T_a$ increases linearly with the net radiation, which in turn increases with increasing sun elevation.

The effect of sun elevation on the penetration and distribution of solar radiation and therefore on temperature is shown schematically in *Figure 3.4*. When the sun elevation is low some of the canopy and most of the soil are shaded. On the other hand, when the sun is high, the soil can be illuminated and, under these conditions, the dynamics of temperature distribution is much greater, and the interpretation of composite temperatures is much more difficult.

A geometric model was developed by Jackson, Reginato, Pinter and Idso (1979) to determine the effects of sun elevation and of view angle on row crops. In their model the rows are assumed to be solid rectangular sections. Four elements are taken into account; sunlit and shaded vegetation and sunlit and shaded soil. The proportion of the four different elements within the field of view of an instrument can be calculated as a function of sun elevation. The model of Jackson et al. (1979) was improved by Kimes and Kirchner (1983a). *Figure 3.5* shows part of their experimental results (obtained from measurements in Phoenix, Arizona). The four elements, sunlit and shaded vegetation and soil, are shown as a function of sun elevation. It can be seen that the vegetation temperatures in the shade and in the sun are not significantly different and that the temperatures of plants and shaded soil increase less rapidly than the temperatures of sunlit soil, as the sun elevation increases.

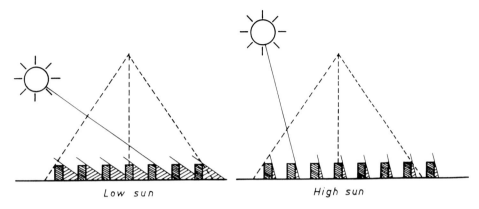

FIGURE 3.4. Diagram showing the effect of sun elevation on the distribution of shadows in a row crop

MEASUREMENT CONDITIONS

Since the radiative temperature of a plant canopy is a function of its geometry, it is apparent that the angle from which the canopy is viewed will have an effect on the measurement results. For example, if the view is from nadir, the probability that both soil and vegetation will be seen is much greater than if the view is from an oblique angle. In fact, some experimental procedures have been designed to use view angles from 60° to 70° from nadir in order to obtain a measure of the vegetation temperature by reducing the effects of soil background on the measured radiative temperatures.

Spatial variability in surface temperatures within a scene may be caused by differences in soil conditions. This variability may cause the temperatures to be dependent on the size of the surface viewed.

VIEW ZENITH ANGLE

A number of investigations have been dedicated to studying the effect of the view angle on the apparent radiative temperature of a plant canopy (Jackson et al. 1979; Kimes and Kirchner 1983b; Byrne, Begg, Fleming and Dunin, 1979; Kimes, Idso, Pinter, Jackson and Reginato, 1980a; 1980b; Jackson, 1981). At present it is well known that the spectral response from a plant canopy in the visible and near-infrared varies as a function of the view angle (Guyot, 1983). This variation is a result of the geometric structure of the canopy. This is also true in the thermal infrared. Although the various elements of a plant canopy can be considered isotropic radiators, radiation issuing from the canopy does not follow Lambert's law. Indeed the radiation flux detected by a radiometer originates from plant parts situated at different levels in the canopy and therefore they may be at different temperatures. The proportion of various canopy elements seen varies with the view angle, resulting in a temperature variability within the field of view of the radiometer. Some experimental data (Guyot and Chasseray, 1981; Boissard, Bertolini, Valery and Renard, 1981) and model simulations (Kimes et al., 1980a; Jackson 1981) allow us to evaluate the importance of these variations.

Two types of model have been perfected: geometric models for row crops (Jackson

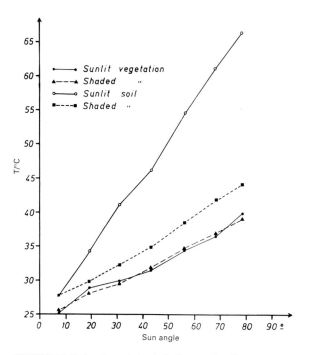

FIGURE 3.5. Sunlit and shaded plant and soil temperatures in a field of cotton as a function of sun angle. Data from measurements made by Kimes and Kirchner (1983a) in Phoenix, Arizona (25th June, 1981)

et al., 1979; Kimes and Kirchner, 1983a; Sobrino, Caselles and Becker, 1988) and statistical models for homogeneous crops (Kimes et al., 1980a; 1980b; Kimes, 1980; Smith, Ranson, Nguyen, Balick, Link, Fritschen, and Hutchinson, 1981; Prévost, 1985; McGuire, Smith, Balick and Hutchinson, 1989). For the row crops the geometric model is in good agreement with the experimental data and shows a variation in the radiative temperature with the view angle, as could be predicted by examining the diagrams in *Figure 3.4*. *Figure 3.6* gives as an example the results of measurements carried out on 23rd and 25th June 1981 for different sun elevations and for different zenith view angles. The variations are maximum when the sun is at the zenith. On 25th June a difference of 16.2°C was noted between the nadir view and a view at 80° from the nadir (Kimes and Kirchner, 1983a).

When the crop rows are not apparent, it is necessary to use another type of model to account for the view inclination effect. This case is included in the statistical model developed by (Kimes et al., 1980a; Kimes, 1980).

In the thermal infrared the emissivity ϵ of leaves is generally high, of the order of 0.97 (Gates, 1964). As a first approximation, it can be considered that only the leaves or other plant parts and the soil elements that are directly viewed by a radiometer contribute to the measured radiative temperature (an analysis of the possible error introduced by this approximation will be given later). The measurement then depends on two factors: the profile of the surface temperature of the various elements, and the visibility of layer Z in

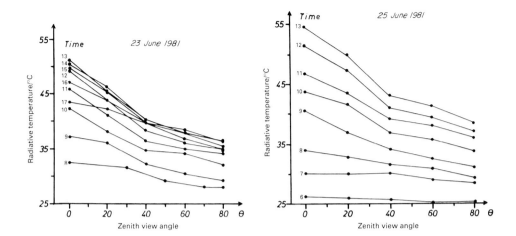

FIGURE 3.6. Effect of the view inclination and the hour of the day on the radiative temperatures measured above a field of cotton when the rows are oriented east-west (Re-drawn from Kimes and Kirchner (1983a), data from Phoenix, Arizona)

the direction (θ, ϕ), as shown in *Figure 3.7*.

The canopy is separated into layers of equal height (Kimes, 1980), and for each the probability ($\mathbf{P_{GAPi}}$) of the presence of a gap through the layer i, for a ray from a given direction, can be determined experimentally. Likewise the probability of contact ($\mathbf{P_{HITi}}$) can be determined. Then,

$$\mathbf{P_{GAPi}}(\theta,\phi) + \mathbf{P_{HITi}}(\theta,\phi) = 1 \tag{3.15}$$

At the first approximation the radiance of the canopy in the direction (θ,ϕ) (*Figure 3.7*) can be written,

$$\mathbf{L}(\theta, \phi) = \pi^{-1} \times \sum_{i=1}^{n}\{[\prod_{k=0}^{i-1} \mathbf{P_{GAPk}}(\theta,\phi)] \times \mathbf{P_{HITi}}(\theta, \phi)\epsilon_i\sigma T_i^4\} \tag{3.16}$$

where n is the number of layers, ϵ_i the emissivity of the layer i, T_i the average absolute temperature of layer i, and σ is the Stefan-Boltzman constant, as previously defined. Tests of the model on soybean (Kimes, 1980) and on wheat (Kimes et al, 1980a) showed good agreement between experimental and theoretical results.

VIEW AZIMUTH ANGLE

Measured radiative temperatures of vegetative canopies are affected by the geometrical relationship between solar azimuth and the radiometer view azimuth. Radiative temperatures of the sunlit side of a crop have been reported to be as much as 3°C warmer than those taken of the shaded side (Monteith and Szeicz, 1962; Fuchs, Kanemasu, Kerr and Tanner, 1967; Kimes, 1982).

Nielsen, Clawson and Blad (1984) measured radiative temperatures of soybeans at two locations in Nebraska. Maintaining a zenith angle of about 75°, they found that

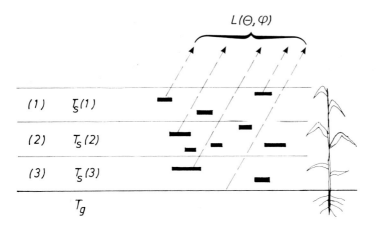

FIGURE 3.7. Modelling of layers of a plant canopy

the measured canopy temperatures decreased linearly as the difference between the solar azimuth and the radiometer view azimuth increased from 0 to about 110°, then from about 110 to 180° the temperature remained relatively constant at a value of about 0.3°C below the average of temperatures measured in the four cardinal directions.

SPATIAL VARIABILITY

The precision with which the radiative temperature of a plant canopy can be determined depends on the spatial variability within the field and on the size of the resolution element (pixel). Spatial variability of surface temperatures results from several soil and plant conditions. Variability in soil physical properties may cause unequal spatial distribution of soil water. The slope and aspect of the terrain alter the radiation balance, and soil chemical properties can cause unequal plant growth. The importance of evaluating spatial variability of surface temperatures becomes apparent when one considers the question of how many samples are required to obtain a representative temperature for the area of interest.

Hatfield, Millard and Goettleman (1982) used both ground-based and aircraft-based thermal measurements to study the spatial variability of surface temperatures. They found that the variability was larger for bare, dry, fields than for vegetation covered fields. To evaluate the variability within individual fields they measured surface temperatures at one metre intervals in north-south (340 m) and east-west (180 m) transects across a cultivated field. The data showed that the variability was considerable, but that random sampling would be as efficient as grid sampling to estimate the mean. These results were confirmed in a subsequent study in irrigated sorghum fields (Hatfield, Vauclin, Vieira and Bernard, 1984). Surface temperatures were measured at 1 m intervals along 85 m transects at a number of times during the growing season. Semi-variograms of the data indicated no spatial dependence of the measurements, hence the temperature variability was random. The number of samples needed to achieve a confidence limit of 1°C about the mean was 10 samples for fields having more than 40% available soil water and 20 samples when the soils were drier.

The spatial variability of soil properties led Aston and van Bavel (1972) to suggest that, for full canopies, various locations in a field may become water deficient before others, and the canopy temperature would consequently show a greater variability than when the plants were well watered. This suggestion was examined by Gardner, Blad and Watts (1981), who measured midday canopy temperatures of corn in Nebraska. They found standard deviations of 0.3°C in fully irrigated plots, and deviations as high as 4.2°C in non-irrigated plots. They concluded that canopy temperature variability may be a useful tool to signal the onset of water deficits.

EFFECT OF SIZE OF THE RESOLUTION ELEMENT

A study conducted by Millard et al., (1981) in California on a large, varied contour surface planted to barley allowed a test of the influence of pixel size on the precision of radiative temperature measurements using an airborne scanner. The calculations carried out on pixel sizes of 4, 16, 65, and 259 ha showed that the size of the surface seen was of relatively little importance. They found that the fraction of the pixel area having temperatures of ±1, ±2, ±3 or ±5°C of the mean was essentially the same regardless of pixel size. Therefore at the spatial resolution scale of satellites, it does not appear advantageous to have pixels of 4 ha rather than pixels of 259 ha to characterize the temperature of homogeneous plant canopies, on condition, however, that they are part of large homogeneous surfaces. The other alternative for measuring temperature is the use of high spatial resolution (of the order of a metre). This would permit a precise characterization of temperature variability. Millard et al., (1981) suggested that scales intermediate between several square metres and several hectares appear to have little practical interest.

Plant factors

Plant factors that influence the apparent radiative temperature of a canopy can be grouped into two classes; geometrical and physiological. Geometric factors include canopy height, width, density, areal spacing of plants, and the amount of soil background exposed. Physiological factors include transpiration (which provides a cooling mechanism), and leaf geometry changes such as leaf droop and curl, due to water deficits, and azimuthal and zenithal changes, due to heliotropism.

Canopy geometry

As we have seen, model studies have delineated the fundamental role played by geometric structure in determining the radiative temperature of plant canopies. In the following sections the amount of soil exposed, the orientation of the crop rows, and the height of the canopy, will be examined from an experimental point of view.

THE AMOUNT OF EXPOSED SOIL

Figure 3.8 depicts the variation in radiative temperatures of wheat canopies, as a function of the view angle, that occurred near solar noon (the time of maximum heating) and at sunrise (the time of minimum surface temperatures). The plots identified on the curves have considerably different amounts of soil cover, as is shown in *Table 3.1*.

Plot 5A was essentially bare, having plants only 8 cm tall. For this plot very little

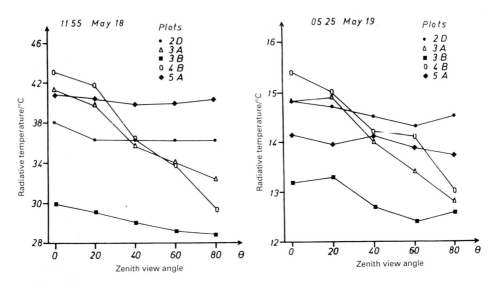

FIGURE 3.8. Effect of view angle on the radiative temperature of wheat canopies having different amounts of exposed soil in Phoenix, Arizona (Redrawn from Kimes et al., 1980a)

TABLE 3.1. Characteristics of the wheat canopies depicted in *Figure 3.8*. (Data from Kimes et al., 1980a)

Plot	2D	3A	3B	4B	5A
Wheat height (m)	1.04	0.66	0.79	0.53	0.08
Leaf Area Index	5.4	2.1	2.7	1.5	0.05
Green fraction	0.00	0.30	0.54	0.91	0.72

variation in apparent radiative temperature was measured as a function of view angle, either at dawn or midday.

The effect of changing the view angle was also barely noticeable during the day for the very dense and completely senesced canopy (2D) and for the dense and actively growing canopy (3B). The effect of view angle was highest on the partial canopies because the solar radiation readily penetrated the canopy to the soil, and when the zenithal view angle increased from nadir the amount of exposed soil within the field of view of the instrument decreased.

At dawn, the observed variations are in the same direction but the temperature ranges are much less. For both the dawn and midday data, the highest radiative temperature measured at a vertical angle was not obtained from the bare soil plot but from a rather dense senesced canopy. In this case the canopy and soil are heated by the sun, but losses by radiation and convection are reduced because of the presence of stalks and senesced leaves. Essentially the same phenomenon occurs at dawn; the soil underneath the vegetation cools less quickly than the bare soil and a partial canopy appears warmer than a dense canopy.

Measurements carried out in the Southeast of France (Guyot and Chasseray, 1981) on two canopies having different growth patterns showed that wheat having an erect growth pattern (*cv.* Talent) appeared systematically warmer than wheat having a spreading growth pattern (*cv.* Capitole). Both varieties were at essentially the same physiological state. The thermal variations observed from a vertical view were between 1 and 3°C throughout the growing season.

Figure 3.9 shows frequency histograms for radiative temperatures of two adjacent maize canopies obtained near solar noon (Guyot and Chasseray, 1981; Chasseray, 1981). One of the plots had rows orientated north-south, the other east-west. The measurements were made in a manner similar to those in *Figure 3.3* (using a thermal infrared imaging device with a vertical view). In *Figure 3.9* one may note that the slope of the temperature histograms changed considerably with time and as the amount of soil exposed decreased. On 4 June the maize plants were 0.25 m high, with bare soil being the predominant component within the field of view of the instrument. Also, the temperature distribution was asymmetrical with a maximal frequency for the high temperatures. For this experiment all temperatures were above the air temperatures (measured under a cover at 2 m).

On 16 June the maize plants were 0.50 m high and the shapes of the histograms were considerably different from the 4 June data. The plots with rows arranged east-west and north-south were distinctly different. For the plot whose rows were orientated east-west the lower temperatures were the most frequent. They correspond to sunlit and shaded maize plants. The large spread of higher temperatures correspond to bare soil. For the north-south plot the histogram is bi-modal. Since the measurements were carried out near solar noon the sun penetrated between the rows and the frequency of high temperatures was greater than in the east-west plot. By 16 July the plants were 1.5 m high (Leaf Area Index 4.1), the soil was almost completely covered, and the histograms for two plots were very similar and symmetric. The temperature distribution was also much less than when the soil was not completely covered. The histograms are also well centred on the air temperature (irrigated maize).

The amount of soil exposed may also be related to plant geometrical changes due to water stress. Some plants will orient their leaves normal to incident radiation, when well watered, thus maximising the interception of radiation. When stressed, the leaves are oriented parallel to the incident radiation. Under these conditions, the apparent ground cover may be 100% for well watered plants and less than 20% when stressed (Byrne *et al.*, 1979). The measurement of radiative temperatures near solar noon with a nadir view would therefore be considerably different for stressed and nonstressed conditions.

Row orientation

Figure 3.9 shows that the temperature distribution of a plant canopy depends on the amount of soil covered and on crop row orientation. A simulation made by Kimes and Kirchner (1983a), using a geometric model derived from that of Jackson *et al.*, (1979), allowed them to obtain the results presented in *Figure 3.10*. This shows the changes in apparent radiative temperatures for canopies of different densities, for a vertical view and a view inclined at an 80° angle from nadir. The simulation was carried out as a function of the orientation of the rows relative to the sun. *Figure 3.10* shows that there exists an optimum amount of soil cover (of the order of 30% in this case) for which the thermal variation is maximal. *Figure 3.10* also shows that, for the higher amounts of

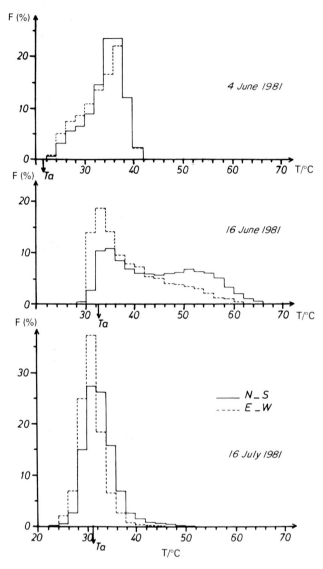

FIGURE 3.9. Change with time of temperature frequency histogram on two neighbouring plots of maize, one of which had rows orientated north-south and the other, east-west. Measurements were made near solar noon (Redrawn from Chasseray, 1981)

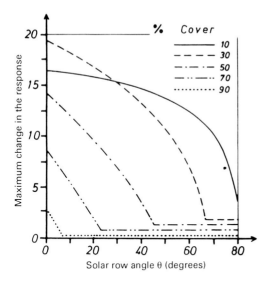

FIGURE 3.10. Maximum change in the response of a radiometer viewing a canopy from the vertical and an angle of 80°, as a function of crop row orientation relative to the sun. The curves correspond to plant canopies having different amounts of soil cover. Row spacing was 1 m (After Kimes and Kirchner, 1983a)

vegetative cover, the radiative temperature of the canopy is only slightly influenced by row orientation and instrument view angle.

CANOPY HEIGHT

Canopy height is another parameter of plant canopy structure that can influence the distribution of radiative temperatures. Thermographs, especially those taken at night, can often distinguish forests or other tall plants from low growing crops. However, during the day, with canopies that are totally covered and well-watered, it is sometimes impossible to distinguish a short crop (beets) from a tall crop (maize), or from a wooded area. An example of this is given by the histograms of temperature obtained for the Paris basin depicted in *Figure 3.11* (Perrier, Itier, Boissard, Goillot, Belluomo and Valery, 1980).

On the other hand, at night these three types of canopy can be clearly separated. The taller the plants, the warmer they appear. The explanation for this phenomenon is relatively simple. The radiative temperature of a homogeneous canopy is determined by temperatures taken of the different layers as seen by a radiometer. A low plant canopy will have a small vertical thermal gradient. On the other hand, a vertical thermal profile with higher temperatures near the soil and lower temperatures close to the tops of the plants will develop in a tall plant canopy. Therefore, a vertical view from above a tall but relatively open canopy will see a greater proportion of warm zones than above a low canopy. This conclusion was confirmed by measurements carried out by Boissard *et al.* (1981).

When a dense canopy consists of trees either randomly distributed or planted in straight lines the trees appear warmer on the night-time thermographs than the soil below

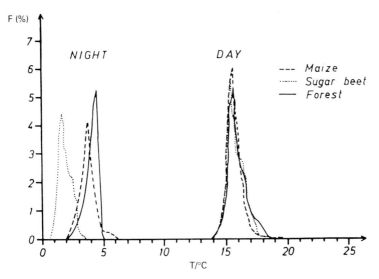

FIGURE 3.11. Histograms of thermographs obtained with a airborne scanner (Daedalus) at an altitude of 1700 m at 0100 h and at 1300 h above the Beauce plain at the end of September. (Redrawn from Perrier et al. (1980)

(Balick and Wilson, 1980; Goillot, 1976). Another phenomenon, however, is superimposed on the radiative effect described previously. During one windless night, temperatures increased with altitude in the layers of air near the soil (inversion profile). Therefore the tops of the trees were in air warmer than that near the soil. The measurements carried out by Balick and Wilson (1980) showed that, under such conditions, the crown of the trees had a temperature equal to that of air. It is for this reason that the trees appeared warmer on the night-time thermographs.

Plant physiological factors

As the energy balance equation shows, the radiative temperature of a plant canopy depends on evapotranspiration. Several studies have shown that the radiative temperature of the canopy can be used to determine evapotranspiration rates (Seguin, 1980a; 1980b; Hatfield et al., 1983; Seguin, Baelz, Monget and Petit, 1982; Seguin and Itier, 1983; Soer, 1980), and for water requirements (Jackson, Reginato, and Idso, 1977; Millard, Jackson, Goettleman, Reginato and Idso, 1978; Katerji, Itier and Ferreira, 1988).

As an example, *Figure 3.12* shows the results of measurements carried out in southeast France (Avignon-Montfavet) on a maize crop subjected to a water deficit. Irrigation was stopped on the evening of 5 July and the crop was not watered for a week. *Figure 3.12* shows that a progressive change of the histograms relative to the air temperature took place. As water became limiting, canopy temperature gradually became warmer than the air. Upon irrigation the canopy started to cool with respect to the air.

Figure 3.13 shows additional data from the same experiment, but here histograms are compared for temperatures determined on plots whose rows are oriented east-west and north-south. In this figure, as in *Figure 3.9* when the water deficit was slight or zero, the histograms for the two plots were very similar. However, as the water deficit increased, the histograms were separated clearly. The measurements were carried out at solar noon. At this time the sun penetrated to the ground between the N-S rows. The penetration of light was facilitated by changes of geometric structure due to the wilting of the leaves (Kimes and Kirchner, 1983b). For the canopy having east-west rows, there was always shade between the rows, regardless of the changes in geometric structure. Direct solar flux was absorbed at different levels in each of the canopies. When the water deficit was large, the leaves reduced their evaporation flux (according to the energy balance equation) resulting in an increase in surface temperature, which was different for each plot.

Figure 3.14 shows vertical air temperature profiles and surface temperatures for shaded and sunlit leaves and soil for the same experiment. Within the canopy having east-west rows, the maximum air temperature was observed at the top of the canopy, whereas for the north-south rows, air temperatures were warmer near the soil surface. The temperatures of both shaded and sunlit leaves for the east-west rows were below that of air, a result of evaporative cooling by transpiration. In the north-south case, the soil was much warmer than the rest of the canopy (dry surface, wind checked by vegetation).

The results obtained in the first experiment (*Figure 3.12*) indicated that the two maize canopies had the same water status (they were always treated identically). However, the different row orientation modified the temperature distribution at solar noon, causing the north-south canopy (because of the better penetration of solar radiation) to have evapotranspiration rates slightly higher than for the east-west canopy, although the inferred differential was small. A rough calculation showed that, at solar noon, the north-south canopy had an evapotranspiration rate approximately 100 W m^{-2} greater than the east-west canopy. The difference in evaporation flux was manifested only for a maximum of two hours around solar noon.

STOMATAL CONTROL

The physiological response of plants to a water deficit is to reduce the stomatal openings, thereby reducing the loss of water by evapotranspiration. The effect of stomatal closure on canopy temperatures can be calculated using *Equation 3.10* with various values of the canopy resistance r_c. Some results of such calculations are shown in *Figure 3.15* A canopy resistance of zero corresponds to a canopy having water drops on its leaves, evaporating as a free water surface. For this case the canopy-air temperature difference is a function only of external factors. As the canopy resistance increases, the effect of the vapour pressure deficit on the radiative temperature decreases. When the canopy is completely senesced, only the net radiation and the aerodynamic resistance determine the canopy-air temperature difference.

Figure 3.16 shows the effect of increasing stomatal resistance (due to a water deficit) during the course of a day (Idso, Reginato and Farah, 1982). The diagonal line, which was experimentally determined in a previous experiment, represents the relationship between the canopy-air temperature difference and the vapour pressure deficit for well watered cotton at Phoenix, Arizona. The dots represent temperature differences measured at about 1/2 hourly intervals near midday. As water becomes limiting, the stomata close, thus reducing the evaporative cooling of the leaves. The temperature difference is just slightly higher than for well-watered conditions for the lowest set of data on the figure.

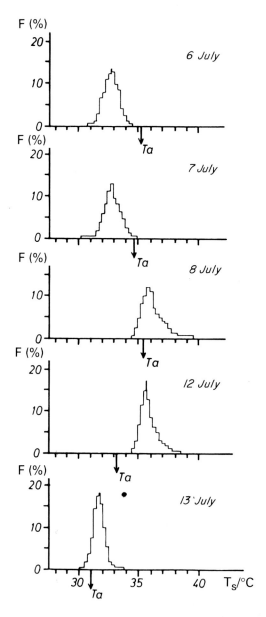

FIGURE 3.12. Temperature histograms of a maize crop as a function of water deficit at Montfavet. Measurements were carried out with an infrared imaging device placed 10 m above the soil over a canopy of maize whose rows were oriented east-west. Irrigation was stopped on the evening of 5 July, and resumed on the evening of 12 July. T_a is the temperature of air above the canopy. Histogram step width, 0.25°C (Montfavet, July 1982)

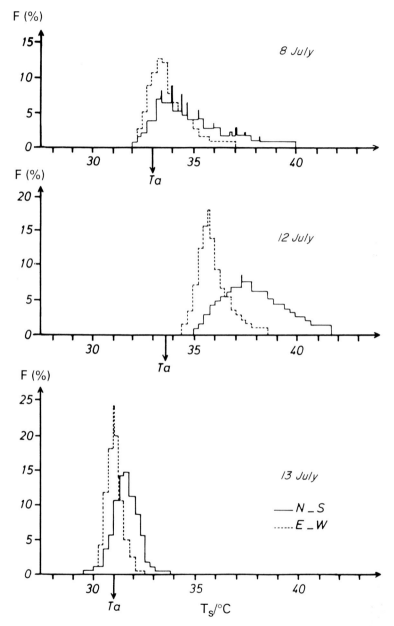

FIGURE 3.13. Histograms for temperature determined on two plots of maize. On one plot the rows were oriented east-west and on the other, north-south. The two plots were subjected the same water treatment as that shown in *Figure 3.12* (Montfavet, July 1982)

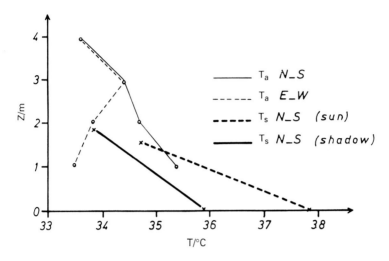

FIGURE 3.14. Vertical profiles of air temperature and of surface temperature determined in the maize plots one week after irrigation (Montfavet, July 1982)

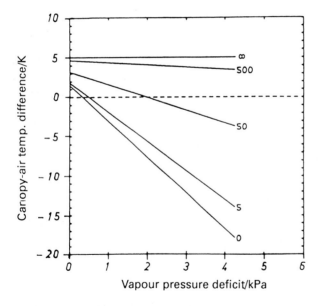

FIGURE 3.15. Theoretical relationship between the canopy-air temperature difference and the vapour pressure deficit for several values of the canopy resistance (values shown at end of lines). All calculations were for an air temperature of 30°C, net radiation of 600 W m^{-2}, and an aerodynamic resistance of 10 s m^{-1} (Based on data of Jackson, 1982b)

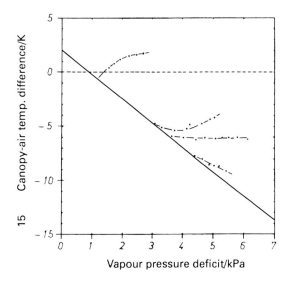

FIGURE 3.16. Departures of canopy-air temperature differences from well-watered conditions during midday showing the effect of physiological factors on radiative temperature (Redrawn from Idso et al., 1982)

The data that appears horizontal at about -6°C remained essentially constant for several hours during the warmest part of the day. The data that appear concave upward indicate that the canopy was warming relative to the air and the rate of transpiration could not meet the evaporative demand. The topmost set of data are for a canopy sufficiently stressed that the evaporative demand could not be met even at the relatively low values of the vapour pressure deficit. The data clearly depict the influence of physiological factors on the radiative temperature.

The radiative temperature measurements were made at an angle of approximately 70° from nadir. At this angle, little if any soil background would be seen and the temperatures can be considered to be those of the vegetation.

Other factors

In addition to external environmental and internal physiological factors, two other factors influence the measurement of radiative temperature. These are the emissivities of the surfaces and absorption of radiation by atmospheric water vapour.

Emissivity of soil and plant surfaces

The calculation of true surface temperature from radiative temperature measurements requires values for the emissivity (ϵ) of the surfaces (*Equation 3.3*) The emissivity of plant canopies is frequently assumed to be 1. The magnitude of the error resulting from this assumption depends upon the canopy emissivity, the amount of soil covered, and the emissivity of the background soil.

The emissivity of soils may range from less than 0.9 to as high as 0.98 (Idso and

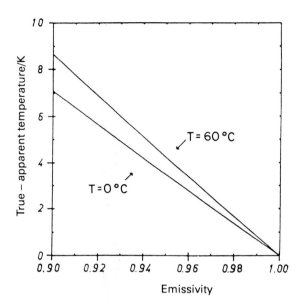

FIGURE 3.17. The maximum increment to be added to the apparent temperature T obtained by assuming a unit emissivity, to obtain the actual surface temperature T_s, as a function of emissivity for two values of the apparent temperature

Jackson, 1969), but for most plant surfaces it may range from 0.95 to near 1. Gates (1964) stated that all plant surfaces have longwave emissivities of 0.95 or more, with most plant leaves in the range 0.97 - 0.98. This was substantiated by Idso, Jackson, Ehrler and Mitchell, (1969), who found only two of 34 plant species that had single-leaf emissivities less than 0.95. The emittance of a canopy should be greater than for individual leaves because of cavities formed by canopy geometry. Direct measurements of canopy emissivity yielded a value of 0.976 (Fuchs and Tanner, 1966; Blad and Rosenberg, 1976). Even with soil exposed between rows of trees, the composite emittance remains high (Southerland and Bartholic, 1977; Becker, 1981; Caselles, Sabrino and Becker, 1988).

Perrier (1971) showed that the maximum difference between true and apparent surface temperatures is given by

$$T_s - T = T(1 - \epsilon^{1/4}), \qquad (3.17)$$

where T_s is the true surface temperature and T is the calculated apparent temperature. Figure 3.17 shows the maximum (T_s - T) versus emissivity over the range from 0.9 to 1.0, calculated using Equation 3.17 for two temperatures. For a canopy emissivity of 0.97, the maximum error involved by assuming $\epsilon = 1$ would be about 2°C. Radiative temperatures measured from nadir into a sparse canopy whose soil background has a low emissivity may differ from the true surface temperatures by several degrees. This error may not always be negligible.

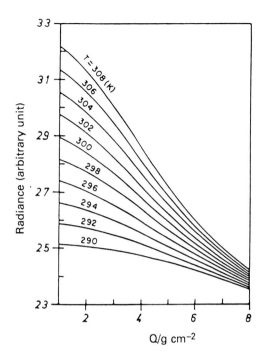

FIGURE 3.18. Radiance received at a satellite sensor versus the amount of atmospheric water vapour **Q** for different ground temperatures (After Kiang, 1982)

Atmospheric effects

For ground-based measurements of radiative temperature such as those described in this report, atmospheric effects play an insignificant role. Within the 8 - 14 μm waveband the influence of water vapour on surface temperature measurements is negligible if the distance to the target is less than 150 m (Lorenz, 1968). However, for data obtained by sensors aboard aircraft or spacecraft, the absorption of thermal infrared radiation by atmospheric water vapour can complicate the interpretation of surface temperatures inferred from such data. An example of this effect is shown in *Figure 3.18*, where the radiance received by a spacecraft sensor is plotted as a function of atmospheric water vapour, for a number of ground temperatures (Kiang, 1982). These results indicate that, as atmospheric water vapour increases, the ability to discriminate surface temperature differences decreases. If atmospheric water vapour content information is available, models can be used to extract reasonable surface temperatures from the satellite data (Lorenz, 1968; Selby, Kneizys, Chetwynd and McClatchey, 1978; Becker, 1987).

Conclusions

The radiative temperature of a plant canopy is complex. The radiation flux that is detected by a radiometer integrates a series of elementary fluxes that arise from various layers of leaves and soil, all of which may be at different temperatures. A number of factors interact to determine the composite radiative temperature of a plant canopy; these can be divided into external, environmental, and internal, physiological factors. Other factors include the emissivity of soil and plant surfaces and, when measurements are made from aircraft and satellite platforms, the absorption of radiation by atmospheric water vapour.

The principal external factors that determine the radiative temperature include components of the surface energy balance, e.g. the net radiation, evapotranspiration, and wind velocity. The vapour pressure deficit of the air plays a role through the evapotranspiration term. Sun elevation and azimuth also affect the energy balance and interact with radiometer view angles to influence the measurement of radiative temperature. The spatial variability of soils and plants, and the amount of soil observed through the canopy also play important roles.

Physiological factors, such as the geometry or architecture of a canopy and stomatal control of water loss from the leaves, influence the radiative temperature in a number of ways. Leaf orientation determines the amount of radiation absorbed and the stomata exert some control over evaporative cooling. Canopy geometry (row orientation, row spacing, plant height) interacts with environmental factors to determine the radiative temperature of the plant canopy.

References

ASTON, A.R. and VAN BAVEL, C.H.M. (1972). Soil surface water depletion and leaf temperature. *Agronomy Journal*, **64**, 368-373

BALICK, L.K. and WILSON, S.K. (1980). Appearance of irregular tree canopies in nighttime high resolution thermal infrared imagery. *Remote Sensing of Environment*, **10**, 299-305

BECKER, F. (1981). Angular reflectivity and emissivity of natural media in the IR thermal bands. In *Proceedings International colloquium on Spectral Signatures of Objects in Remote Sensing*. Avignon, France 8 - 11 September 1981. Les Colloques de l'INRA No.5, INRA, Paris

BECKER, F. (1987). The impact of spectral emissivity on the measurement of land surface temperature from a satellite. *International Journal of Remote Sensing*, **8**, 1509-1522

BLAD, B.L. and ROSENBERG, N.J. (1976). Measurement of crop temperature by leaf thermocouple, infrared thermometry, and remotely sensed thermal imagery. *Agronomy Journal*, **68**, 635-641

BOISSARD, P., BERTOLINI, J.M., VALERY, P. and RENARD, D. (1981). Thermographie de couverts végétaux a courte distance. In *"Signatures spectrales d'objets en télédétection"*. Les Colloques de l'INRA No.5. Avignon, France

BYRNE, G., BEGG, J.E., FLEMING, P.M. and DUNIN, F.X. (1979). Remotely sensed land cover temperature and soil water status - a brief review. *Remote Sensing of Environment*, **8**, 291-305

CASELLES, V., SOBRINO, J.A. and BECKER, F. (1988). Determination of the effective emissivity and temperature under vertical observation of a citrus orchard. Application to frost nowcasting. *International Journal of Remote Sensing*, **9**, 715-727

CHASSERAY, E. (1981). Analyse de la température radiative d'un couvert de mais - traitment informatique. Mémoire de fin d'etudes, E.N.S.A. Rennes, INRA Bioclimatologie Montfavet, France

FUCHS, M. and TANNER, C.B. (1966). Infrared thermometry of vegetation. *Agronomy Journal*, 59, 494-496

FUCHS, M., KANEMASU, E.T., KERR, J.P. and TANNER, C.B. (1967). Effect of viewing angle on canopy temperature measurements with infrared thermometers. *Agronomy Journal*, 59, 494-496

GARDNER, B.R., BLAD, B.L. and WATTS, D.G. (1981). Plant and air temperatures in differentially irrigated corn. *Agricultural Meteorology*, 25, 207-217

GATES, D.M. (1964). Characteristics of soil and vegetated surfaces to reflected and emitted radiation. In *Proceedings 3rd. International Symposium on Remote Sensing of the Environment*. Environmental Research Institute of Michigan, Ann Arbor, Michigan

GOILLOT, Ch. (1976). Apport de la télédétection aéroportee pour l'étude du bocage breton, particulièrement sur le plan du bilan thermique et de l'inventaire des haies. In C.R. table ronde CNRS 5-7 Juillet, 1976. *"Aspects physiques, biologiques et humains des écosystèmes bocagers des régions tempérées humides"*. INRA-ENSA and Université de Rennes, Rennes France

GUYOT, G. (1983). Variabilité angulaire et spatiale des donnees spectrales dans le visible et le proche infra-rouge. In *"Signatures spectrales d'objets en télédétection"*. Les Colloques de l'INRA No.23, Bordeaux, France

GUYOT, G. and CHASSERAY, E. (1981). Analyse de la signification physique et biologique de la température radiative d'un couvert de céréales. In *C.R. Colloque International. "Signatures spectrales d'objets en télédétection."* Les colloques de l'INRA No.5. Avignon, France

HATFIELD, J.L., PERRIER, A. and JACKSON, R.D. (1983). Estimation of evapotranspiration at one time-of-day using remotely sensed surface temperatures. *Agricultural Water Management*, 7, 341-350

HATFIELD, J.L., MILLARD, J.P. and GOETTELMAN, R.C. (1982). Variability of surface temperature in agricultural fields of central California. *Photogrammetric Engineering and Remote Sensing*, 48, 1319-1325

HATFIELD, J.L., VAUCLIN, M., VIEIRA, S.R. and BERNARD, R. (1984). Surface temperature variability patterns with irrigated fields. *Agricultural Water Management*, 8, 429-437

IDSO, S.B. and JACKSON, R.D. (1969). Comparison of two methods for determining infrared emittance of bare soils. *Journal of Applied Meteorology*, 8, 168-169

IDSO, S.B., JACKSON, R.D., EHRLER, W.L. and MITCHELL, S.T. (1969). A method for determination of infrared emittance of leaves. *Ecology*, 50, 899-902

IDSO, S.B., REGINATO, R.J., REICOSKY, D.C. and HATFIELD, J.L. (1981). Determining soil-induced plant water potential depressions in alfalfa by means of infrared thermometry. *Agronomy Journal*, 73, 826-830

IDSO, S.B., REGINATO, R.J. and FARAH, S.M. (1982). Soil- and atmosphere-induced plant water stress in cotton as inferred from foliage temperatures. *Water Resources Research*, 18, 1143-1148

JACKSON, R.D. (1981). Interactions between canopy geometry and thermal infrared measurements. In *"Signatures spectrales d'objets en télédétection*. Les Colloques de l'INRA No.5. Avignon, France

JACKSON, R.D. (1982a). Canopy temperatures and crop water stress. In *Advances in Irrigation, Vol.1*. Ed. by D. Hillel. Academic Press, New York

JACKSON, R.D. (1982b). Soil moisture inferences from thermal infrared measurements of vegetation temperatures. *IEEE Transactions on Geoscience and Remote Sensing*, **GE-20**, 282-286

JACKSON, R.D., REGINATO, R.J. and IDSO, S.B. (1977). Wheat canopy temperature: A practical tool for evaluating water requirements. *Water Resources Research*, **19**, 651-656

JACKSON, R.D., REGINATO, R.J., PINTER, P.J. Jr. and IDSO, S.B. (1979). Plant canopy information extraction from composite scene reflectance of row crops. *Applied Optics*, **18**, 3775-3782

JACKSON, R.D., IDSO, S.B., REGINATO, R.J. and PINTER, P.J. Jr. (1981). Canopy temperature as a crop water stress indicator. *Water Resources Research*, **17**, 1133-1138

KATERJI, N., ITIER, B. and FERREIRA, I. (1988). Etude de quelques critères indicateurs de l'état hydrique d'une culture de tamate en région semi-aride. *Agronomie*, **8**, 425-433

KIANG, R.K. (1982). Atmospheric effects on TM measurements: Characterization and comparison with the effects on MSS. *IEEE Transactions on Geoscience and Remote Sensing*, **GE-20**, 365-370

KIMES, D.S. (1980). Effect of vegetation canopy structure on remotely sensed canopy temperatures. *Remote Sensing of Environment*, **10**, 165-174

KIMES, D.S. (1982). Azimuthal radiometric temperature measurements of wheat canopies. *Applied Optics*, **20**, 1119-1121

KIMES, D.S. and KIRCHNER, J.A. (1983a). Diurnal variation of vegetation canopy structure. *International Journal of Remote Sensing*, **4**, 257-271

KIMES, D.S. and KIRCHNER, J.A. (1983b). Directional radiometric measurements of row crop temperatures. *International Journal of Remote Sensing*, **4**, 299-311

KIMES, D.S., IDSO, S.B., PINTER, P.J. Jr., JACKSON, R.D. and REGINATO, R.J. (1980a). View angle effects in the radiometric measurement of plant canopy temperatures. *Remote Sensing of Environment*, **10**, 273-284

KIMES, D.S., IDSO, S.B., PINTER, P.J. Jr., JACKSON, R.D. and REGINATO, R.J. (1980b). Complexities of nadir looking radiometric temperature measurements of plant canopies. *Applied Optics*, **19**, 2162-2168

LORENZ, D. (1968). Temperature measurements of natural surfaces using infrared radiometers. *Applied Optics*, **7**, 1705-1710

McGUIRE, M.J., SMITH, J.A., BALICK, L.K. and HUTCHINSON, B.H. (1989). Modelling directional thermal radiance from a forest canopy. *Remote Sensing of Environment*, **27**, 169-186

MILLARD, J.P., GOETTELMAN, R.C. and LE ROY, M.J. (1981). Infrared temperature variability in a large agricultural field. *International Journal of Remote Sensing*, **2**, 201-211

MILLARD, J.P., JACKSON, R.D., GOETTELMAN, R.C., REGINATO, R.J. and IDSO, S.B. (1978). Crop water stress assessment using and airborne thermal scanner. *Photogrammetric Engineering and Remote Sensing*, **44**, 77-85

MONTEITH, J.L. (1963). Gas exchange in plant communities. In *Environmental Control of Plant Growth*. Ed. by L.T. Evans. Academic Press, New York

MONTEITH, J.L. (1973). *Principles of Environmental Physics*. Edward Arnold, London

MONTEITH, J.L. and SZEICZ, G. (1962). Radiative temperature in the heat balance of natural surfaces. *Quarterly Journal Royal Meteorological Society*, **88**, 496-507

NIELSEN, D.C., CLAWSON, K.L. and BLAD, B.L. (1984). Effect of solar azimuth and infrared thermometer view direction of soybean canopy temperature. *Agronomy Journal*, **76**, 607-610

PERRIER, A. (1971). Leaf temperature measurement. In *Plant Photosynthetic Production: Manual of Methods*. Ed. by Z. Sestak, J. Catsky and P.G. Jarvis. Junk, The Hague

PERRIER, A., ITIER, B., BOISSARD, P., GOILLOT, C., BELLUOMO, P. and VALERY, P. (1980). *A study of radiometric surface temperatures: their fluctuations, distribution and meaning*. Tellus Newsletter Series No. 13. JRC Ispra, Italy

PRÉVOST, L. (1985). Modélisation des échanges radiatifs au sein des couverts végétaux. Application à la Télédétection. Validation sur un couvert de maïs. Thése Université de Paris 6

SEGUIN, B. (1980a). Détermination de l'évaporation réelle dans les bilans hydrologiques par télédétection en thermographie infra-rouge. *Bulletin des Sciences Hydrologiques*, **25**, 143-153

SEGUIN, B. (1980b). Application de la télédétection dans l'infra-rouge thermique á la détermination de l'évaporation et de l'humidité du sol. In *"Télédétection par satellitè. Application en agroclimatologie et agrométéorologie"*. C.R. d'un cours de formation conjointement organisé par l'ESA, la FAO and l'OMM. October, 1979. ESA SP 1022, 27-39. Rome

SEGUIN, B. and ITIER, B. (1983). Using midday surface temperature to estimate daily evaporation from satellite thermal I.R. data. *International Journal of Remote Sensing*, **4**, 371-383

SEGUIN, B., BAELZ, S., MONGET, J.M. and PETIT, V. (1982). Utilisation de la thermographie I.R. pour l'estimation de l'évaporation régionale. I. Mise au point méthodologique sur le site de la Crau. *Agronomie*, **2**, 7-16

SELBY, J.E.A., KNEIZYS, F.X., CHETWYND, J.H. Jr. and McCLATCHEY, R.A. (1978). Atmospheric transmittance/radiance: Computer code LOWTRAN IV. Air Force Geophysics Laboratory Report No. AFGL-TR-78-0053. Hanscom AFB, Massachusetts

SMITH, J.A., RANSON, K.J., NGUYEN, D., BALICK, L.K., LINK, L.E., FRITSCHEN, L. and HUTCHINSON, B.A. (1981). Thermal vegetation canopy model studies. *Remote Sensing of Environment*, **11**, 311-326

SOBRINO, J.A., CASELLES, V. and BECKER, F. (1988). A theoretical model for interpreting remotely sensed thermal IR measurements obtained over agricultural areas. In *Proceedings Fourth International Colloquium on Spectral Signatures of Objects in Remote Sensing.* Aussois-Modane, France. 18-22 January 1988. ESA. SP-287. European Space Agency, Paris

SOER, G.J.R. (1980). Estimation of the regional evapotranspiration and soil moisture conditions using remotely sensed crop surface temperature. *Remote Sensing of Environment*, **9**, 27-47

SOUTHERLAND, R.A. and BARTHOLIC, J.F. (1977). Significance of vegetation in interpreting thermal radiation from a terrestrial surface. *Journal of Applied Meteorology*, **22**, 153-166

VAN BAVEL, C.H.M. and EHRLER, W.L. (1968). Water loss from a sorghum field and stomatal control. *Agronomy Journal*, **60**, 84-88

II.

CLIMATE AND SOIL

4

DISCRIMINATION AND MONITORING OF SOILS

R. EVANS

*Department of Geography, University of Cambridge,
Downing Street, Cambridge CB2 3EN, U.K.*

Introduction

Soil is the world's most important resource: we depend on it for most of our food. The quality of soils for growing crops varies and we need to identify and protect the most productive soils. Erosion not only degrades soils but its off-site impacts may be severe, indeed more so over the short term than the loss in crop productivity (Clark, Haverkamp and Chapman, 1985). Soil erosion needs to be monitored, therefore, as do the impacts on soils of other degradational processes such as salinisation. Remote sensing techniques are tools to be used to identify and map soils and to monitor their degradation; they are not a substitute for field data collection. In this paper remote sensing techniques are discussed for discriminating and mapping soils, for providing information on the interaction of soils, weather and crop growth, and for monitoring soil erosion and degradation. The techniques discussed range from airborne photography, using different bands and emulsions, multi-spectral imagery, both airborne and spaceborne, thermal scanning, and the use of both active and passive microwave sensors. The literature on these topics is voluminous and is not treated exhaustively; I have concentrated on the last decade. Much of the literature is not easily accessible, nor is it of high quality.

Remote sensing and soil mapping

Much of the World is not covered by detailed topographical maps; only about 5% of the World's surface has been mapped at a scale of 1:25000, 15-20% at 1:100000 and 35% at 1:250000 (Brandenberger, 1984). However, large areas of the World's surface are not worth detailed mapping. Eighty-five per cent of the World is mapped at a scale of 1:1000000. For exploratory purposes Brandenberger (1984) considers topographic maps at scales larger than 1:100000 are needed and there are large parts of the World which still lack this scale of map coverage. Geological maps (Weber, 1985), and soil maps more so, are rare, as a visit to Cambridge University Library, a world copyright library, will show. The Food and Agriculture Organisation of the United Nations has published

(FAO, 1970-78) 1:5000000 scale maps which cover the World: other than this, coverage is sporadic and few countries are completely mapped even at small scales; England and Wales is one area with a map at 1:250000 scale. There is a great need, therefore for soil maps, and remote sensing techniques should play an important role in their production.

Soil map scale and remote sensing

Large scale soil maps

Remote sensing techniques have a bigger role to play in the production of small scale soil maps than of large scale maps. Detailed soil maps, published at scales of 1:50000 or larger, require a large fieldwork input, as at these scales the units delineated on the map are largely based on the inherent characteristics of the soil such as its particle size, drainage characteristics and genesis. To assess these characteristics soils must be examined in the field by inspection of sections in pits, both purpose dug and in gravel pits or quarries, in road side cuttings or drainage ditches, and by augering holes in the ground. At the reconnaissance stage of mapping the range of soils within an area is decided, subsequently, lines are drawn on maps delineating soil units based on augering; the soils are then sampled for their physical and chemical characteristics. This kind of map portrays soil series, or soil phases if one characteristic of the series varies slightly and is mappable, such as stoniness or an erosion phase. Where soil series cannot be separated easily, complexes, where the soil pattern is not easily predicted, or associations, where they are, are mapped. Mapping is time-consuming, even if done by 'free surveying' and not based on some kind of sampling procedure.

At large scales of mapping, unless soil characteristics which are directly relevant to the classification criteria of soils are visible on images or easily and accurately inferrable from the remotely sensed data, remote sensing will be of little use for discriminating between soils directly.

Air photos at scales of 1:10000 - 1:20000 are of most use for detailed mapping of soils in England and Wales (Evans and Carroll, 1976), although in some localities smaller scale maps may be adequate. In England and Wales 1:15000 scale air photos were commissioned where it was thought worthwhile, otherwise air photos commissioned by County Councils were bought, generally at scales of 1:10000 - 1:12000, or the Ordnance Survey (1:20000 - 1:25000). At these scales features as small as 0.3 m or less can be resolved, for example, tile drains within fields; all features of interest can be identified.

Classic air photo-interpretation techniques (Carroll, Evans and Bendelow, 1977; Christian and Stewart, 1968; Goosen, 1967; SCS, 1966; Vink, 1968) relating landforms to soils can work at large scales (Carroll, 1972), but are often more useful for smaller scale maps. Air photo pairs are examined under a stereoscope and the landforms distinguished related to soil types. Generally, this physiographic or feature analysis relies on relating patterns of drainage, relief, land use, tonal features and cultural features to soil characteristics.

For example, in England, a dense network of dry valleys, within a high rolling relief, dominantly under arable cropping, with pale toned slopes and dark toned valley floors and crests, and on the crests dry pits dug for marl, all indicate a locality with soils on chalk. But what is described here is an association of soils, for example, deeper decalcified soils on drift of some kind on the crests, shallow soils over chalk on slopes, and deep soils on drift in valley floors. If the tone changes on the air photos coincide with changes within the soil profile which are used to classify the soil, such as the change in tone from dark to pale coinciding with a change from soils deeper than 30cm to shallower than this (= rendzina, Avery, 1980), then soils can be delineated directly from the air photos. This saves much time when mapping in the field.

In parts of England and Wales air photos taken at the best time (Evans, 1974) portray bare soil or crop patterns which aid in detailed soil mapping (Evans, 1972). The patterns are of both natural (Evans, 1972) and man-made origin (Evans and Jones, 1977; Riley, 1982; Wilson, 1983). They are related to changes in colour of the bare soil surface or to changes in stone content related to changing characteristics with depth of the soil, or to differences in growth patterns of crops, related again to variations in depth of the soil (Jones and Evans, 1975). Often these features are difficult to see even when the surveyor is mapping at the best time, because of his low viewing angle, his direction of view into or away from the sun, and the poorer colour contrasts seen when looking through a layer of air near the ground, which is much hazier than is the equivalent vertical distance between an air photo camera and the ground (Evans, 1979).

The tonal patterns not only enable the surveyor to draw lines on a map, but they also aid his understanding of the distribution of the soil. A soil pattern may be too complex (Evans, 1972) and the variability of soils too rapid to be mapped even at very large scales, but an air photo, whether vertical or oblique, explains the variability. In such instances a decision can be taken by the surveyor to spend less time mapping the patterned area and more time mapping adjacent soils (Seale, personal communication); overall no time is saved but a better quality map results.

Air photo patterns similar to those seen in England and Wales have been recorded in other countries. Striped patterns are not only seen widely on chalk substrates in England (Evans, 1976) but similar periglacial flow patterns are seen in the Arctic regions at present (Frost, 1950; Rudberg, 1988), and a similar pattern, though of differing origin, the gilgai pattern, is seen in the United States and elsewhere (Elberson, 1983). Periglacial polygonal patterns are found also as relicts in Scandinavia (Svensson, 1967a) and France (Agache, 1970) and as active features in the arctic (Black, 1952; Svensson, 1967b). Banded patterns over differing lithologies have been recorded in the USA (Dury, 1987; Lillesand and Kieffer, 1979; Way, 1978), and crop patterns related to archaeology are widely found throughout Europe (Martin, 1971; Scollar, 1964).

Often, however, air photo-interpretation cannot accurately delineate soil map units, for example in Devon, boundaries drawn on air photos coincided with between one-fifth and one-half of mapped boundaries (Harrod and Evans, unpublished). But photo-interpretation familiarises the interpreter with an area, indicates where soils are likely to change, aids in devising the mapping programme and provides an up-to-date base map, a factor of great importance even in a country as well mapped as the United Kingdom. All these factors help make a better map more efficiently. As a portion of the total cost of producing the map, air photos, even when new coverage is commissioned, are a very small part, probably less than 10%.

Medium scale soil maps

Soil maps at scales between 1:50000 and 1:250000 are useful for planning purposes. Often these will be soil association maps. The units portrayed on the maps are defined mostly on their soil characteristics, but geomorphology, geology and land use, that is their physiographic characteristics, are also taken into account. Field work forms a large proportion of the time spent producing the maps.

Small scale air photos are used to help delineate map units in association with any other information that is to hand, for example, topography, geology and existing soil maps at any scales. The 1:250000 soil map of England and Wales was produced with the aid of 1:50000 and 1:60000 scale air photos. The cost of purchasing these was less than 1% of the total map production costs. It took about 90 minutes to carry out the air photo-interpretation of 100 km^2, which took 8 to 10 days to map in the field. Air photo-interpretation helped familiarise the surveyor with the area, and occasionally allowed lines delineating soil associations to be drawn directly, thus saving time. But in many places map unit boundaries were placed only after carrying out field work, and often relating this to 1:50000 (or 1:63360) scale geology maps. In many parts of the World this important supplementary information is not available, nor is there much time for mapping, and hence a different kind of 'soil' map is produced.

Physiographic maps

In areas where there is little or no prior information on soils, or when small scale maps (1:10^6) are required, or when maps of large areas of ground have to be produced rapidly, remote sensing techniques are of major inportance in helping produce 'soil' maps. These are based on physiographic (Vink, 1968), land systems (Christian and Stewart, 1968), land facet (Beckett and Webster, 1965; Webster, 1962) or feature analysis (Carroll, Evans and Bendelow, 1977) of air photos/images to give landscapes where patterns of soils are associated with particular landforms and vegetation types. Field work is carried out rapidly to characterise the soils within the physiographic units delineated on the image. Soils can be sampled along route lines in difficult country or on a grid pattern or a transect where the terrain is more accessible. Air photos, in particular, are helpful in planning the survey.

Physiographic maps can be produced at any scale, even large scales (eg Fagbami and Vega-Catalan, 1985; Rubio, 1988), and are generally the first stage in identifying the suitability of land for development (Cook, 1981; Dwivedi, 1985). At a later stage more detailed maps may be made of smaller areas, identified as being most suited for agriculture. Although air photos will be the primary source of information (Hothmer, 1985), satellite images have an important role to play (Adedeji, 1988; Biswas, 1987; Fagbami, 1986a; Farshad, 1986; King, 1985; Mulders and Epema, 1986; Singh and Dwivedi, 1986; Venkataratnam, 1986; Zonneveld and Surasana, 1988).

Ideally, the interpretation should start using satellite images, eg, Landsat, SPOT or even radar images (Boer, 1983; Evans and Carroll, 1986; King, 1985), to delineate landscape units at a scale of between 1:250000 and 1:10^6. The 5 m resolution Russian satellite imagery, now becoming available (Anon, 1988), could be particularly useful. At this stage it does not matter that the landform and vegetation types which comprise the delineated units cannot be identified (Meijerinck, 1988), it suffices that these areas can be delineated as fairly homogeneous units for sampling. This stage should be followed by stereoscopic interpretation of satellite images (Wheeler, Tarus, Mitchell, King and White, 1988): this is straightforward with SPOT images, but Landsat images are viewed stereoscopically only with difficulty. Even better, larger scale (1:300000) Large Format Camera pictures (Girard, 1986) or small scale air photos (1:50000 - 1:100000) should be examined to produce maps at scales of 1:100000 - 1:250000. The better the resolution of the images/photos the more useful they are for interpretation, as features can be discerned and identified, giving greater confidence in the interpretation. Finally, detailed soil maps can be made.

Maps such as these have been made for many former British colonies. They are also useful for identifying areas at risk of accelerating erosion (Rubio, 1988) if land use is changed (Valenzuela, 1988).

Photo/image type for interpretive studies

It is apposite to discuss the relative merits of multi-band images/photos and the use of different photographic emulsions for interpreting soils and landscapes. A distinction is made here between interpreting images/photos by eye and analysing digital data by machine, the latter is discussed below.

Most air photos used to be, and often still are in humid areas, taken on panchromatic rather than colour film because:

- film speed is faster so image motion is less of a problem;

- filters can be used to cut out the hazy blue end of the spectrum;

- in England and Wales, for example, it can be used on more days in a year (Evans, 1974);

- it is easier and cheaper to process, and wrongly exposed film can be more easily corrected when processing the film.

Colour photos may be easier to interpret because features are more easily recognised in colour than black-and-white, an advantage if the interpreter lacks experience. Often, however, the advantages of ease of interpretation and time saved because of this are small (Dwivedi, 1985; Szilagyi, 1986), and are greatly outweighed by cost.

Colour infrared photos record differences in density of vegetation better than other film because multiple reflections of near-infrared from, and between, overlapping leaves are strongly recorded. The greater density of vegetation may be related to wetter soils, but also to more moisture available to the plant from a deeper soil, to different application of fertiliser, to different drilling rates of seeds or to different vegetation types, among other reasons. There is often little advantage for discriminating soils between colour infrared and panchromatic photography even in largely pastoral areas; Harrod and Evans (unpublished) found that it was quicker, almost twice as fast, to interpret colour infrared photos, but that soil boundary delineation was often (5-10%) less accurate. The extra cost of colour infrared photography was not outweighed by its advantages, therefore.

Multi-band photos, that is, black-and-white photos taken in different parts of the visible spectrum, have little advantage over panchromatic photos, even in the red waveband where tonal contrasts are often higher (Evans, 1975a). The blue and green bands are often too affected by scattering of light back to the camera from aerosols, so that tonal contrasts are reduced, and the infrared band is often not as easy to interpret.

It is at this juncture that techniques of interpreting multi-band photos overlap with digital analytical techniques for discriminating and mapping soils on images. These analytical techniques depend on the relationships of tonal/signal strength response in different wavebands to soil properties, both in the visible and non-visible parts of the electro-magnetic spectrum, what is still called the 'spectral signature' approach (Steven, 1987).

Digital analysis of images for discriminating soils

In the 1960s Purdue University began to promote the use of computer analysis of digitally recorded linescan images taken in different wavebands to identify soil types (Purdue University, 1967); Grody, (1988) presently advocates a similar approach. Groups of soils have been shown to have different spectral responses but this work has mostly been (Condit, 1970; Krinov, 1947), and still is being (Escadafal and Pouget, 1986), done in the laboratory: measuring the reflectance from very small samples of soil with very smooth surfaces. How this data related to field conditions was not explored. The reasoning behind this approach seems to have been:

- that machines could more cheaply take the interpreter's place and produce more consistent results, because interpreters need a high level of expertise, their training is costly and interpretation is a subjective process, a view still held (Greenbaum, 1987);

- that satellite images could not be viewed stereoscopically, therefore image tones and texture became the prime factors for discriminating soils, and machines could be trained to recognise these spectral signatures;

- that the poor resolution (80m) of the original satellite images did not allow identification of many features which could be related to the factor of interest, in this case soil, so image tone and texture became the main classifiers;

- that images could be rapidly scanned and classified;

- that surface features by their different responses to electromagnetic energy, both reflected and emitted, could be characterised.

However, the technique of digital computer analysis of images relies on tones (=signal strength) being related unambiguously to the feature of interest. This is not so.

Evans (1975a) noted the multi-band approach for discriminating soils in eastern England was inadequate, both because different soil types could have similar signatures and the same soil type could image very differently on different occasions because its surface characteristics varied, for example, from a rough ploughed surface to a smooth drilled one, or it could be covered in a growing crop or a harvested stubble. Organic matter, clay content and soil moisture, as well as soil colour, can discriminate between soils (Baumgartner, Silva, Biehl and Stoner, 1985; Evans, Head and Dirkzwager, 1976; Wright and Birnie, 1986), but not with a sufficient accuracy to classify soils. The soils can be as easily separated by eye, and the variation of, for example, organic matter between fields of soils of the same type, is such that dissimilar soils will be identified (Evans, 1975a). The bare soil surface roughness and colour also change over time (Evans, 1979). In other words, soils do not have characteristic reflectances.

Townshend and Hancock (1981) note that even when a reasonable correlation exists in the laboratory between bare soils and their reflectances in different wavebands ($R^2 = 76\%$ for broad groupings), the results will only rarely be matched in the field. Kijowski

(1988) states that bare soil factors form only a 'tentative' basis for discriminating soils by means of sliced densitometrically analysed images. Recent work in Northern Ireland (Cruickshank and Tomlinson, 1988) also shows that it is not easy to discriminate and identify consistently soils from Thematic Mapper satellite images. Doubt has also been cast on the technique by Fagbami (1986b), Huete and Jackson (1988), and Wayumba and Philipson (1985). In many localities this is because the soils are covered by vegetation and the tonal/signal characteristics relate to those features not the soil.

In a locality in Texas where land use/vegetation patterns do relate to soils, Thompson and Henderson (1984) found only 56% of their soil boundaries obtained by digitally analysing Thematic Mapper data coincided with mapped boundaries. In Wales, Greenbaum (1987) found that on linescan images vegetation types related better to soils on drift than rock, and could be used to discriminate solid and drift lithologies. But the technique is complementary to the use of air photos, and a better understanding is needed of the complex relationships between geology and vegetation communities, and the images they provide.

Soils do not have consistent characteristic emittances in the longer wavebands in the thermal infrared. Although there is a strong relationship between ground surface temperature and thermal emittance recorded by infrared linescan imagery (Evans, 1975b), this varies throughout the day (Axelsson and Lunden, 1986; Evans, 1975b; Lamers, 1985; Naert, 1986) and season (Lynn, 1986), and is often related more to surface roughness, ground cover, and soil colour (albedo) than inherent properties used to classify soils. This presumably holds true also for emittances at microwave frequencies, as surface roughness and vegetation cover have been found to affect the radiometric characteristics of soils (Promes, Jackson and O'Neill, 1988; Schmugge and O'Neill, 1986; Schmugge, Wang and Asrar, 1988; Theis and Blanchard, 1988).

To incorporate microwave backscatter from soils into a 'spectral signature' approach also does not seem sensible, as again the strength of the returned signal to air- or spacecraft will vary not just with soil surface roughness but also with surface roughness due to cover by crops and other vegetation types (eg Evans, 1985).

The number of variables to contend with in the 'spectral signature' approach is, therefore, so great that even now, twenty years or so after the initial proposals, it does not seem a feasible approach to discriminate and identify soils over large parts of the Earth's surface. Only in arid and semi-arid areas little disturbed by man is it likely to produce satisfactory results, and these are the areas where the economic returns are least. For this technique to work at all, images will have to be taken in the visible and near-infrared wavelengths at the best time of year, when the maximum area of bare dry soil is exposed. But even then there is still an almost insuperable problem. This is that the amount of haze or aerosol in the atmosphere varies greatly both over the short and long term and affects the spectral response recorded both close to the ground (Deering and Eck, 1987; Evans, 1979) and from air- and space-craft (Djavadi and Anderson, 1987; Evans, 1979; Huete and Jackson, 1988; Steven, 1987; Steven and Rollin, 1986). Correcting the spectral response by using mean values of visibility will not be satisfactory because of the great variability over time and space of aerosol content (Evans, 1979). The American response to the lack of progress on discriminating soils by digital analysis has been to propose even more bands for future satellites (Baumgardner, lecture at ITC, Enschede, 1985). This does not seem a fruitful approach: it will result in more data to compress and handle and the need for more sophisticated computer hardware and software. Steven

(1987) is only slightly less pessimistic about the use of spectral signatures.

Platform stability is not a problem for satellites obtaining multi-spectral data, but it is for aircraft mounted scanners (Evans, 1975b; Greenbaum, 1987). Distortions of the linescan caused by aircraft movement have to be corrected before data from different flights can be compared with each other or with maps, for instance, and this again is a problem which is taking a long time to overcome.

As Hothmer (1985) has noted, the less developed world is not using satellite imagery to its maximum extent, if digital analysis is considered the best way of analysing images. This is because of:

- the cost of obtaining the computer tapes and the hardware and software to go with them is too high;

- the lack of trained personnel;

- lack of foreign currency;

- higher priority allocation of scarce monetary resources to other more pressing needs (Ihemadu, 1987; Specter, 1988; Wayumba and Philipson, 1985; Wheeler et al., 1988).

These pressures have been exacerbated as aid from the developed world for resource evaluation has declined (Hothmer, 1985).

I consider that digital analysis of images for discriminating and mapping soils is not a useful way forward. Where the technique has been shown to work well (eg Evans, Head and Dirkzwager, 1976) the tonal patterns recorded on the image could easily be discerned and mapped by the interpreter, or, if the pattern was too complex, the image itself used to portray this.

Remote sensing for recording the interactions of crop growth with soils and weather

If air photos are to be used for mapping soils with maximum effect and economy, they must be taken at the best time. For interpreting pastoral areas of England, for instance, October is best (Harrod and Evans, unpublished). For some soil patterns the best time is when the soils are bare (Evans, 1972; 1974). Other patterns show only in crops (Evans and Jones, 1977; Jones and Evans, 1975), and in England one group of patterns on chalky soils is visible throughout the year (Evans and Catt, 1987). The scale of the patterns is such that most cannot be seen on satellite images and are best recorded on large and medium scale as well as oblique air photos. These patterns, both natural (Evans, 1972) and archaeological (Agache, 1970; Evans and Jones, 1977; Riley, 1982; Wilson, 1983), also help our understanding of the interactions of crop growth and yield with soils and the weather, so providing us with better guidelines for assessing the quality of soils.

Those patterns recorded throughout the year have pale coloured subsoils exposed at the surface. Temperature varies between the pale and dark coloured soils due to the effect of colour on the absorption of solar radiation, the dark soils attaining higher daily maximum temperatures so that seed germination, crop growth and yield are all better on these (Evans and Catt, 1987). As soils dry out it is often the palest coloured soils which change colour most (Evans, 1979) and temperature differences rapidly become more marked between soils of different colour (Evans and Catt, 1987; Lamers, 1985).

Other patterns only become visible part way through the growing season, and the visibility of the pattern can vary from year to year. These patterns relate to:

- differences in availability of soil moisture to the plant, which in turn depends on the variation in the depth of soil to hard rock, sand or gravel, or to some layer which impedes rooting, for instance a very acid layer or one with a high subsoil density;
- to variations in weather from year to year.

If the distribution of soil types in a field or locality is known, the amount of water held in the soil profile which is available to the plant can be estimated (Hodgson, 1976). This value in turn can be related to the amount of rain which has fallen during the growing season and to how much water the crop has transpired (Jones and Evans, 1975; Evans and Catt, 1987). This difference, the potential soil moisture deficit (PSMD), has to be extracted from the soil. If moisture is available there is no check on growth. Most crop patterns appear in England and Wales when PSMDs are greater than 50 mm (Jones and Evans, 1975); that is about the amount held in the plough layer. If the soil is directly on hard rock that is all the water available to the plant. To record crop patterns, therefore, air photos should not be taken before PSMDs reach about 50 mm. In the UK the Meteorological Office can provide data on PSMD. PSMD can also be assessed by monitoring rainfall at a site and estimating evaporation using the system described in MAFF Bulletin No. 16 (MAFF, 1967). Archaeologists now use data from the Met. Office to assess if it is worthwhile flying to record archaeological crop marks (Douglas, pers. comm.). Rainfall amounts vary from year to year and in wetter years the appearance of crop patterns may be delayed or they may not appear at all. When soils are less than 40 cm deep crop patterns can occur every year in eastern and southern England, but when soils are deeper than 80 cm they occur infrequently, say one year in ten (Jones and Evans, 1975). These depth limits are used to classify soil series (Avery, 1980); the same soil series can therefore suffer from drought every year or rarely.

Patterns show for other reasons (Evans and Catt, 1987; Evans and Jones, 1977), but most can be attributed to the above outlined causes. Where crop patterns occur the soils have inherent limitations on plant growth and yield. Even if these soils are fertilised or crop varieties improved they will not yield as well as the adjacent deeper or darker soils. Crop patterns commonly occur on soils on sands and gravels, limestones, sandstones and slates (SSEW, 1983) which cover large parts of England and Wales (13%). In these localities the variability of growth and yield in crop trials is more likely to be related to variability of soil type than to the agricultural practices being tested. The extrapolation of data from such a site will have to be done with caution, therefore.

If the distribution of soils is known, estimates of the amounts of water available to the crop can be made, and relating these to weather and rates of evaporation may give a guide to potential crop yields. However, a scheme relating cereal height, a surrogate for yield, to weather and soils over a series of years shows that the likely accuracy of yield prediction is not sufficient to displace the present yield prediction scheme, because yield is also affected by factors such as disease, date of drilling and crop variety, for instance (Evans, unpublished). Russell (1988) also notes that biomass is not a good indicator of yield.

Remote sensing of soil moisture

Much has been written about estimating soil moisture content by remote sensing techniques: by assessing variations and changes in soil colour; by estimating thermal inertia using infrared linescan imagery or microwave radiometry; or by measuring radar backscatter.

As soils dry out they do change colour, but this change usually takes place over a very restricted range of soil moisture content. Soil surface roughness also affects colour. For example, the rougher the surface the darker the tone because of shading. The relationship of soil colour to soil moisture content is a complex one (Evans, 1979; Evans, Head and Dirkzwager, 1976; Kijowski, 1988; Musick and Pelletier, 1986), therefore, and it is unlikely to yield accurate estimates of soil moisture content (Evans, 1979).

The temperature of soils of similar texture and moisture content but varying colour, is related to colour not to moisture content (Evans, 1975b; Evans and Catt, 1987). In the English Fenland dark coloured soils with a high organic matter content hold more water than adjacent soils with lower organic matter contents (Evans, Head and Dirkzwager, 1976), but they are also warmer during the day (Evans, 1975b). Surface temperatures also fluctuate more under cloudy conditions in these dark organic rich soils. The relationships between soil temperature, thermal inertia and soil moisture are complex, and may not be strong, therefore, at the field or within-field scale (Axelsson and Lunden, 1986; Berge and Stroosnijder, 1987; Evans, 1975b; Lamers, 1985).

At microwave wavelengths, ground, air- and space-borne scatterometer data show the strength of the return backscatter is proportional to the amount of free water in the top 5 to 10 cm of the soil (Schmugge, 1983). The dielectric constant is directly proportional to the amount of unbound water in the soil, and hence the higher the soil moisture content the greater the radar return. There is discussion whether the radar backscatter relates best to soil moisture measured by weight, as percentage of field capacity or volumetrically, taking into account the soil's bulk density (Dobson, Kouyate and Ulaby, 1984; Hallikainen, Ulaby, Dobson, El-Rayes and Wu, 1985); this is because the statistical relationships vary between soils of different texture. Although statistical relationships are often good between backscatter and soil moisture content, they do vary and are not always consistent (Mawser and Greichgauer, 1988; Soares, Bernard and Vidal-Madjar, 1987; Zotova and Geller, 1985).

This lack of consistency is even more marked when imaged radar backscatter is compared with data on soil moisture collected at the time of overflight (Evans, 1985). Synthetic aperture radar images record ground surface roughness, be it of the bare soil surface or vegetation and crops, rather than soil moisture content. The technique may work best when surface conditions are uniform (Blyth and Evans, 1985), preferably short grass turf.

Microwave emissivity can be strongly related to the amount of unbound water in the soil (Schmugge, 1983; Schmugge and O'Neill, 1986). However, relationships vary with soil particle size, organic matter content and bulk density (Dobson, Kouyate and Ulaby, 1984; Jackson and O'Neill, 1988; Wang, O'Neill, Jackson and Engman, 1983). Emissivity is a complex function also of surface roughness (Mo, Schmugge and Wang, 1987) and vegetation density (Promes, Jackson and O'Neill, 1988; Schmugge and O'Neill, 1986; Theis and Blanchard, 1988; Ulaby, Razani and Dobson, 1983), as radiation is scattered by these before reaching the radiometer.

There are relationships between soil colour, thermal inertia and microwave backscatter and emittance and soil moisture content, therefore, but they are often of an indirect and complex nature. It seems unlikely that unless much is known of the ground's surface at the time imagery is obtained, that it will be possible to estimate accurately the amount of soil moisture held in the top 5 to 10 cm of the soil.

Even if there is a relationship, this begs the question 'Who wants this information'? In the agricultural community the answer may be 'very few people'. The amount of moisture held in the top 5 cm or so of soil governs whether the land can be cultivated or drilled. This is easily assessed on the ground by the farmer. What is of greater interest to the farmer is the amount of moisture held in the soil profile, as it is this which will determine, for example, the yield of his crop, or if he needs to irrigate his crop. There may be an indirect relationship between topsoil moisture content and that at depth, but not invariably so. In localities where the soil's characteristics are known or can be predicted from soil maps, there are other ways of estimating the moisture available in the soil to the plant, and relating this to rainfall and evaporation (see above).

Remote sensing and soil degradation

Degradation of soils by salinisation resulting from irrigation is a growing problem in many semi-arid parts of the world. Satellite images have been used successfully to assess its extent in parts of India (Sharma and Bhargava, 1988) and the USA (Chaturvedi, Carver, Harlan, Hancock, Small and Dalstead, 1983), for example. Unfortunately, incipient salinisation cannot be identified on air photos or images (Chaturvedi et al., 1983).

Man, by his unsuitable use of the land, has initiated erosion everywhere. Even in climates such as that in Great Britain, with its mostly gentle rainfall, erosion of arable land takes place frequently (Evans, Bullock and Davies, 1988). In many semi-arid parts of the world rangeland has been converted to arable, with devastating effects, or has been severely overgrazed. Remote sensing techniques can aid in assessing the extent, frequency and rates of erosion, although presently their usefulness is severely underestimated.

Generally, air photos will be most useful for monitoring erosion. In semi-arid Kenya 1:20000 scale air photos were much better for identifying erosion and overgrazing than 1:50000 and 1:60000 photos (Wayumba and Philipson, 1985). However, Landsat images are being used to assess and forecast erosion, often initiated by overgrazing, of arid land in central Australia (Pickup and Chewings, 1988a; 1988b) and Burkino Faso (Folving, 1988). If gullies can be seen on satellite images they are enormous, and often it will be most useful to identify, monitor and combat erosion before it gets to this stage.

Although rill and gully erosion are often seen on air photos (eg Lillesand and Kieffer, 1979; SCS, 1966) their presence is not always considered noteworthy, and there are few studies where air photos have been used specifically to map or monitor erosion. Garland (1982) considers 1:20000 scale panchromatic air photos are better for mapping erosion in South Africa than are black-and-white infrared ones, and Frazier and Hooper (1983) consider 35 mm black-and-white chromogenic air photos useful for recording erosion in Washington State, USA. Large (230 × 230 mm, 1:10000 scale) and small format (70 × 70 mm, 1:5000) photos of various emulsion types allowed rapid assessment of erosion in remote hill country in North Island, New Zealand (Stephens, Hicks and Trustrum, 1981); panchromatic large format photos were the most useful and economic overall.

Keech (1968) mapped gullies in Zimbabwe and related these to land use and soils.

Gully and rill networks on eroding burnt moorland in North Yorkshire, England were mapped from air photos (Alam and Harris, 1987). Evans (1980) and Evans and Cook, (1987) identified eroded fields in England on oblique and vertical air photos and related these to soils, relief and land use. Stephens, MacMillan, Daigle and Cihlar (1985) interpreted 1:27000 colour infrared air photos to delineate erosion mapping units in New Brunswick, Canada, and used these to estimate soil losses. Large scale photos have been used to measure erosion rates in the USA photogrametrically (Thomas, Welch and Jordan, 1986). The extent of wind erosion has been assessed in a small part of Texas (Lyon, McCarthy and Heinen, 1986), and the encroachment of dunes measured in a semi-arid locality in Russia (Vinogradov, 1988). Air photos have also been used to identify eroded land as source areas of sediment in Sumatra (Meijerinck, Wijngaarden, Asrun and Maathuis, 1988), and to identify areas at risk of erosion (Bergsma, 1983). Bergsma stresses the usefulness of sequential photography for monitoring erosion, and Wayumba and Philipson (1985) monitored overgrazing and erosion in semi-arid Kenya between 1960 and 1980 using air photos.

Possibly the only scheme specifically designed to monitor the extent, frequency and rates of water erosion was that operating in England and Wales between 1982 and 1987 (Evans and Cook, 1986; Evans, Bullock and Davies, 1988). It was funded by the Ministry of Agriculture and ran for four of the six years under the full control of the Soil Survey of England and Wales (now Soil Survey and Land Research Centre). Panchromatic air photos, mostly at 1:10000 scale, were taken by the Ministry's Air Photo Unit; up to 17 localities were photographed in a year, covering more than 700 km^2. The photos were interpreted and eroded fields identified. Field work was then carried out to check the interpretation and to measure rates of erosion. Rills over 15 cm wide, and occasionally as narrow as 10 cm, were identifiable, but the most easily interpreted features were depositional fans down to as small as about 0.3 m by 1 or 2 m.

The full results for the first three years of the project are given by Evans (1988), with interim results in Evans and Cook (1986), Evans and Skinner (1987) and Evans, Bullock and Davies (1988). Briefly, coarse silty soils erode most rapidly, but erosion is most widespread and frequent on sandlands growing a wide range of crops. Most erosion occurs in fields of autumn-sown cereals, although the proportions vary from year to year and between localities. Fields down to sugar-beet, market garden crops and soft fruits, erode on average three times faster or more than do those under winter cereals. However, most rates of erosion are low, $< 1 m^3$ ha^{-1} in a year, although as erosion often affects only a small part of the field, rates there are much higher, often > 40 m^3 ha^{-1}; a surface lowering of > 4 mm. Much soil is carried out of fields to be deposited on roads or in streams.

The scheme allowed rapid surveillance of large areas of land, and most erosion was identified. Other than very small events, only those happening between the time of air photography and field work were likely to be missed. With regard to manpower requirements, these were small, one man interpreted the photos, and was aided by up to five others in carrying out the field work. Overall about 50 days were spent each year interpreting the photos and gathering information. And real data has been collected, not synthesised from a model of possibly dubious validity, as can happen in other erosion assessment schemes, for example, in the USA (SCS, 1977).

Conclusions

Remote sensing techniques have an important role to play in helping produce maps of soil and physiography for which there is still a large need; satellite images for small scale maps ($1:10^6$), air photos for larger scale maps. Remote sensing techniques can help:

- speed up soil mapping;
- produce better quality soil maps;
- improve our understanding of the interaction of soil, weather and crop growth.

The next generation of satellite images, providing they can be viewed stereoscopically, will be used for mapping at larger scales. Only when image resolution improves even further will the full potential of satellites be achieved. Air- and space-borne images can also help locate and monitor erosion, which is a world-wide problem. But, again, until satellite resolution improves so that erosional features, which are often small, can be discerned and identified, air photos will be mainly used for this task. It seems unlikely that digital analysis of images will have a large role to play in discriminating and identifying soils. Nor will thermal and microwave images be able to monitor soil moisture content with sufficient accuracy. Radar images will be of little use to the soil surveyor.

Unless remote sensing techniques are cheap to use their uptake will be limited, and there is a serious need to train interpreters (Ihemadu, 1987; Specter, 1988). Presently the pace of technology is outstripping the demand for its uses, and it must be asked 'Do we need all this technology' (Voûte, 1987)? Remote sensing techniques will only be used widely when governments decide, for example, they need soil maps or need to monitor erosion and degradation. The remote sensing community will better spend its time demonstrating that such information is needed, than producing techniques of dubious use.

References

ADEDEJI, A. (1988). Remote sensing and African development. *Photogrammetria*, **43**, 17-24

AGACHE, R. (1970). Detection Aérienne. De vestiges protohistoriques Gallo-Romains et Médiévaux dans la basin dè la Somme et ses abords. *Numéro Spécial du Bulletin de la Société de Prehistoire du Nord*. No.7. Musee d'Amiens, France

ALAM, M.S. and HARRIS, R. (1987). Moorland soil erosion and spectral reflectance. *International Journal of Remote Sensing*, **8**, 593-608

ANON (1988). Russian satellite images. *International Journal of Remote Sensing*, **9**, 1149

AVERY, B.W. (1980). *Soil classification for England and Wales. (Higher categories)*. Technical Monograph No. 14. Soil Survey, Harpenden

AXELSSON, S. and LUNDEN, B. (1986). Experimental results in soil moisture mapping using IR thermography. *ITC Journal*, **1986-1**, 43-50

BAUMGARTNER, M.F., SILVA, F.L., BIEHL, L.L. and STONER, E.R. (1985). Reflectance properties of soils. *Advances in Agronomy*, **38**, 1-44

BECKETT, P.T.H. and WEBSTER, R. (1965). *A classification system for terrain*. M.E.X.E. No. 872. Military Engineering Experimental Establishment, Christchurch

BERGE, H. TEN, and STROOSNIJDER, L. (1987). Sensitivity of surface variables to changes in physical soil properties: limitations to thermal remote sensing of bare soils. *IEEE Transactions on Geosciences and Remote Sensing*, **GE-25**, 702-708

BERGSMA, E. (1983). Rainfall erosion surveys for conservation planning. *ITC Journal*, **1983-2**, 166-174

BISWAS, R.R. (1987). A soil map through Landsat satellite imagery in a part of the Auranga catchment in the Ranchi and Palamou districts of Bihar, India. *International Journal of Remote Sensing*, **8**, 541-543

BLACK, R.F. (1952). Polygonal patterns and ground conditions from aerial photographs. *Photogrammetric Engineering*, **18**, 123-124

BLYTH, K. and EVANS, R. (1985). Results in hydrology and soils. In *Investigators Final Report*. **1**, 143-147. Ed. by Trevett, J.W. Proceedings SAR-580 Investigators Workshop. Joint Research Centre, Ispra, Italy, May 1984. Joint Research Centre, Commission of the European Communities/ European Space Agency, Ispra, Italy

BOER, Th. A. de. (1983). Visual interpretation of SAR images of two areas in the Netherlands. In *Satellite microwave remote sensing*, Ed. by Allen, T.D. Ellis Horwood. Chichester

BRANDENBERGER, A.J. (1984). Economic impact of world-wide mapping. *Photogrammetric Engineering and Remote Sensing*, **50**, 1185-1189

CARROLL, D.M. (1972). Air photo interpretation for soil mapping in the Yorkshire Pennines. *East Midland Geographer*, **5**, 296-303

CARROLL, D.M., EVANS, R. and BENDELOW, V.C. (1977). *Air photo interpretation for soil mapping*. Technical Monograph No. 8. Soil Survey, Harpenden

CHATURVEDI, L., CARVER, K.R., HARLAN, J.C., HANCOCK, G.D., SMALL, F.V. and DALSTEAD, K.J. (1983). Multispectral remote sensing of saline seeps. *IEEE Transactions on Geoscience and Remote Sensing*, **GE-21**, 239-251

CHRISTIAN, C.S. and STEWART, G.A. (1968). Methodology of integrated surveys. In *Areal surveys and integrated studies*. Proceedings Toulouse Conference, 1964. Natural Resources Service. UNESCO 6

CLARK, E.H., HAVERKAMP, J.A. and CHAPMAN, W. (1985). *Eroding soils; the off-farm impact*. Conservation Foundation, Washington D.C.

CONDIT, H.R. (1970). The spectral reflectance of American soils. *Photogrammetric Engineering*, **36**, 955-966

COOK, A. (1981). Integrated resource survey as an aid to soil survey in a tropical flood plain environment. In *Terrain Analysis and Remote Sensing*, Ed. by Townshend, J.R.G. George Allen and Unwin, London

CRUICKSHANK, M.M. and TOMLINSON, R.W. (1988). An assessment of the potential of SPOT reflectance data for soil survey in Northern Ireland. *Proceedings Royal Irish Academy*, **88B**, 45-60

DEERING, D.W. and ECK, T.F. (1987). Atmospheric optical depth effects on angular anisotropy of plant canopy reflectance. *International Journal of Remote Sensing*, **8**, 893-916

DJAVADI, D. and ANDERSON, J.M. (1987). Atmospheric correction of thermal infrared data using multi-height data acquisition. *International Journal of Remote Sensing*, **8**, 1879-1884

DOBSON, M.C., KOUYATE, F. and ULABY, F. (1984). A re-examination of soil textural effects on microwave emission and backscattering. *IEEE Transactions on Geoscience and Remote Sensing*, **GE-22**, 530-536

DURY, S.A. (1987). *Image interpretation in geology.* Allen and Unwin, Hemel Hempstead

DWIVEDI, R.S. (1985). The utility of data from various airborne sensors for soil mapping. *International Journal of Remote Sensing*, **6**, 89-100

ELBERSON, W. (1983). Non-glacial types of patterned ground that develop from erosion. *ITC Journal*, **1983-4**, 322-333

ESCADAFAL, R. and POUGET, M. (1986). Luminance spectrale et caractéres de la surface des sols en region aride mediteraneene (sud Tunisia). *ITC Journal*, **1986-1**, 19-23

EVANS, R. (1972). Air photographs for soil survey in lowland England; soil patterns. *Photogrammetric Record*, **7**, 302-320

EVANS, R. (1974). The time factor in aerial photography for soil surveys in lowland England. In *Environmental Remote Sensing: Applications and Achievements*, Ed. by Barrett, E.C. and Curtis, L.F. Edward Arnold, London

EVANS, R. (1975a). Multiband photography for soil survey in Breckland, East Anglia. *Photogrammetric Record*, **8**, 297-308

EVANS, R. (1975b). Infra-red linescan imagery and ground temperature. In *Science and environmental management.* Ed. by Hey, R.D. and Davies, T.D. Lexington Books, Massachusetts

EVANS, R. (1976). Observations on a stripe pattern. *Biuletyn Peryglacjalny*, **25**, 9-22

EVANS, R. (1979). Air photos for soil survey in lowland England: factors affecting the photographic images of bare soils and their relevance for assessing soil moisture content and discrimination of soils by remote sensing. *Remote Sensing of Environment*, **8**, 39-63

EVANS, R. (1980). Characteristics of water-eroded fields in lowland England. In *Assessment of erosion.* Ed by de Boodt, M. and Gabriels, D. John Wiley and Son, Chichester

EVANS, R. (1985). Synthetic aperture radar imagery for soil survey in lowland England. In *Investigators Final Report, Volume 2.* Proceedings of SAR-580 Investigators workshop. Joint Research Centre, Ispra, Italy. May, 1984, 399-413. Ed. by Trevett, J.W. Joint Research Centre, Ispra, Commission of the European Communities/ European Space Agency, Ispra

EVANS, R. (1988). Water erosion in England and Wales. Report to Soil Survey, Silsoe

EVANS, R., BULLOCK, P. and DAVIES, D.B. (1988). Monitoring erosion in England and Wales. In *Agricultural Erosion assessment and modelling.* Ed. by Morgan, R.P.C. and Rickson, R.J. Directorate-General for Agriculture, Co-ordination of Agricultural Research, Commission of the European Communities, EUR 10860 EN, Luxembourg

EVANS, R. and CARROLL, D.M. (1976). Remote sensing for small scale soil mapping in England and Wales. In *Land use studies by remote sensing*. Ed. by Collins, W.G. and van Genderen, J.L. Remote Sensing Society, Reading

EVANS, R. and CARROLL, D.M. (1986). Radar images for soil survey in England and Wales. *ITC Journal*, **1986-1**, 88-93

EVANS, R. and CATT, J.A. (1987). Causes of crop patterns in eastern England. *Journal of Soil Science*, **38**, 309-324

EVANS, R. and COOK, S. (1986). Soil erosion in Britain. *SEESOIL*, **3**, 28-59

EVANS, R., HEAD, J. and DIRKZWAGER, M. (1976). Air photo-tones and soil properties. Implications for interpreting satellite imagery. *Remote Sensing of Environment*, **4**, 265-280

EVANS, R. and JONES, R.J.A. (1977). Crop marks and soils at two archaelogical sites in Britain. *Journal of Archaeological Science*, **4**, 63-76

EVANS, R. and SKINNER, R.J. (1987). A survey of water erosion. *Soil and Water*, **13**, 28-31

FAGBAMI, A. (1986a). Remote sensing options for soil survey in developing countries. *ITC Journal*, **1986-1**, 3-8

FAGBAMI, A. (1986b). Reflectance characteristics in the classification of Nigerian Savanna soils. *ITC Journal*, **1986-1**, 94

FAGBAMI, A. and VEGA-CATALAN, F. (1985). An evaluation of the physiographic soil map of the Benue valley at Makurdi. *ITC Journal*, **1985-4**, 268-274

FAO (1970-78). Soil map of the world. Rome, FAO/UNESCO

FARSHAD, A. (1986). A general soil inventory of Malayer-Shazard region (northwestern Iran), using multi-temporal Landsat images and intermediate technology. *ITC Journal*, **1986-1**, 96

FOLVING, S. (1988). Assessment of land/soil degradation in northern Burkino Faso. In *Proceedings International Geoscience and Remote Sensing Symposium IGARSS 88*, Edinburgh, Scotland. Vol.2, ESA SP-284 (IEEE 88 CH 2497-6). European Space Agency, Paris

FRAZIER, B.E. and HOOPER, G.K. (1983). Use of a chromogenic film for aerial photography of erosion features. *Photogrammetric Engineering and Remote Sensing*, **49**, 1211-1217

FROST, R.E. (1950). *Evaluation of soils and permafrost conditions in the territory of Alaska by means of aerial photographs*. Vols 1. and 2. Engineering Experimental Station, Purdue University. U.S. Army, Corps of Engineers, St.Paul

GARLAND, G.G. (1982). Mapping erosion with air photos: panchromatic or black and white infrared. *ITC Journal*, **1983-3**, 309-312

GIRARD, M.C. (1986). Interpretation pédologique des photographies prises par Spacelab 1. *ITC Journal*, **1986-1**, 14-23

GOOSEN, D. (1967). *Aerial photo interpretation in soil survey*. Soils Bulletin No. 6. Rome, FAO

GREENBAUM, D. (1987). Lithological discrimination in central Snowdonia using airborne multispectral scanner imagery. *International Journal of Remote Sensing*, **8**, 799-816

GRODY, N. (1988). Surface identification using satellite microwave radiometers. *IEEE Transactions on Geoscience and Remote Sensing*, **GE-26**, 850-859

HALLIKAINEN, M.T., ULABY, F.T., DOBSON, M.C., EL-RAYES. M.A. and WU, LIN-KUN. (1985). Microwave dielectric behaviour of wet soil - Part 1: Empirical models and experimental observation. *IEEE Transactions on Geoscience and Remote Sensing*, **GE-23**, 25-34

HODGSON, J.M.H. (ed.) (1976). *Soil Survey field handbook*. Technical Monograph No. 5. Soil Survey, Harpenden

HOTHMER, J. (1985). Photogrammetry and remote sensing within the United Nations system. *Photogrammetria*, **40**, 53-63

HUETE, A.R. and JACKSON, R.D. (1988). Soil and atmospheric influences on the spectra of partial canopies. *Remote Sensing of Environment*, **25**, 89-105

IHEMADU, S.O. (1987). The problems and prospects of remote sensing applications in developing countries. *ITC Journal*, **1987-4**, 284-291

JACKSON, T.J. and O'NEILL, P.E. (1988). Observed effects of soil organic matter content on the microwave emissivity of soils. In *Proceedings International Geoscience and Remote Sensing Symposium IGARSS 88*. Edinburgh, Scotland. ESA SP-284 (IEEE 88 CH 2497-6), Vol. 2. 673-676. European Space Agency, Paris

JONES, R.J.A. and EVANS, R. (1975). Soil and crop marks in the recognition of archaeological sites by air photography. In *Aerial reconnaissance for archaeology*. Ed. by Wilson, D.A. Council for British Archaeology. Research Report No. 12. London

KEECH, M.A. (1968). Soil erosion survey techniques. *Proceedings and Transactions Rhodesia Scientific Association*, **53**, 13-16

KIJOWSKI, A. (1988). Interpretation of aerial photographs as a method of studying variations in lithology and soil moisture of the surface ground layers: detailed study of the Mosina International Study Area. *Quaestiones Geographicae*, **10**, 47-70

KING, R.B. (1985). Comparison of SLAR, SIR and Landsat imagery for mapping land systems in Kalimantan, Indonesia. In *Advanced technology for monitoring and processing global environmental data*. 381-390. Remote Sensing Society/CERMA. Reading

KRINOV, E.L. (1947). *Spectral reflectance of natural formations*. Laboratoria Aerometodov, Akad Nauk SSSR, Moscow. (Natural Research Council of Canada, Technical Translation TT 439, G. Belkov)

LAMERS, J.G. (1985). Soil slaking and the possibilities to record with infrared line scanning. *International Journal of Remote Sensing*, **6**, 153-165

LILLESAND, T.M. and KIEFFER, R.W. (1979). *Remote Sensing and image interpretation*. Wiley, New York

LYNN, D.W. (1986). Timely thermal infrared data acquisition for soil survey in humid temperate environments. *ITC Journal*, **1986-1**, 68-76

LYON, J.G., McCARTHY, J.F. and HEINEN, J.T. (1986). Videodigitisation of aerial photographs for measurement of wind erosion damage on converted rangelands. *Photogrammetric Engineering and Remote Sensing*, **52**, 373-377

MAFF (1967). *Potential transpiration.* MAFF Technical Bulletin No. 16, HMSO, London

MARTIN, A-M. (1971). Archaeological sites - soils and climates. *Photogrammetric Engineering,* **37**, 353-357

MAWSER, W. and GREICHGAUER, T. (1988). Correlations between agricultural plant parameters and multitemporal radar scatterometer data. First results from the European Agriscatt 87 Campaign. In *Proceedings International Geoscience and Remote Sensing Symposium IGARSS 88.* Edinburgh, Scotland. ESA SP-284 (IEEE 88-ch-2497-6) Vol.2. European Space Agency, Paris

MEIJERINCK, A.M.J. (1988). Data acquisition and data capture through terrain mapping units. *ITC Journal,* **1988-1**, 23-44

MEIJERINCK, A.M.J., WIJNGAARDEN, W. VAN, ASRUN, S.A. and MAATHUIS, B.H. (1988). Downstream damage caused by upstream land degradation in the Komering river basin. *ITC Journal,* **1988-1**, 96-108

MO, T., SCHMUGGE, J, and WANG, J. (1987). Calculations of the microwave brightness temperature of rough soil surfaces in bare fields. *IEEE Transactions on Geoscience and Remote Sensing,* **GE-25**, 47-54

MULDERS, M.A. and EPEMA, G.F. (1986). The Thematic mapper: a new tool for soil mapping in arid areas. *ITC Journal,* **1986-1**, 25-29

MUSICK, H.B. and PELLETIER, R.E. (1986). Response of some thematic band ratios to variation in soil water content. *Photogrammetric Engineering and Remote Sensing,* **52**, 1661-1668

NAERT, B. (1986). Teledetection de modes de fonctionnemont hydrique dans les sols sableaux des formations littorale dunaires. *ITC Journal,* **1986-1**, 59-67

PICKUP, G. and CHEWINGS, V.H. (1988a). Forecasting patterns of soil erosion in arid lands from Landsat MSS data. *International Journal of Remote Sensing.* **9**, 69-84

PICKUP, G. and CHEWINGS, V.H. (1988b). Estimating the distribution of grazing patterns of cattle movement in a large arid zone paddock. *International Journal of Remote Sensing,* **9**, 1469-1490

PROMES, P.M., JACKSON, T.J. and O'NEILL, P.E. (1988). Significance of agricultur l row structure on the microwave emissivity of soils. *IEEE Transactions on Geoscience and Remote Sensing,* **GE-26**, 580-589

PURDUE UNIVERSITY (1967). *Remote multispectral sensing in agriculture.* Laboratory for Agricultural Remote Sensing, Vol. 2. (Annual Report) Research Bulletin No. 832. Purdue University Agricultural Experimental Station, Lafayette, Indiana

RILEY, D.N. (1982). *Aerial archaeology in Britain.* Shire Publications, Princes Risborough

RUBIO, J.L. (1988). Erosion risk mapping in areas of the Valencia Province (Spain). In *Agriculture. Erosion assessment and modelling.* Ed. by Morgan, R.P.C. and Rickson, R.J. Directorate-General for Agriculture, Co-ordination of Agricultural Research, Commission of the European Communities, EUR 10860 EN, Luxembourg

RUDBERG, S. (1988). High arctic landscapes: comparisons and reflexions. *Norsk Geografisk Tijdskrift,* **42**, 255-264

RUSSELL, G. (1988). Problems in the estimation of barley yields by remote sensing. In *Proceedings International Geoscience and Remote Sensing Symposium IGARSS 88*. Edinburgh, Scotland. ESA SP-284 (IEEE 88 CH 2497-6) Vol.2. European Space Agency, Paris

SCHMUGGE, T.J. (1983). Remote sensing of soil moisture: recent advances. *IEEE Transactions on Geoscience and Remote Sensing*, **GE-21**, 336-344

SCHMUGGE, T.J. and O'NEILL, P.E. (1986). Passive microwave soil moisture research. *IEEE Transactions Geoscience and Remote Sensing*, **GE-24**, 12-22

SCHMUGGE, T.J., WANG, J.R. and ASRAR, G. (1988). Results from the push broom microwave radiometer flights over the Konza Prairie in 1985. *IEEE Transactions on Geoscience and Remote Sensing*, **GE-26**, 590-596

SCOLLAR, I. (1964). Physical conditions tending to produce crop sites in the Rhineland. In *Proceedings Colloque International D'Archaeologie Aerienne*. 31 Aug-3 Sept, 1963. 39-47. S.E.V.P.E.N., Paris

SCS. (1966). *Air photo-interpretation in classifying and mapping soils*. Handbook No. 294. Soil Conservation Service, United States Department of Agriculture, Washington D.C.

SCS. (1977). *Erosion inventory. Primary sample unit and point data worksheet*. Iowa State University State Laboratory Soil Conservation Service, United States Department of Agriculture, Washington D.C.

SHARMA, R.C. and BHARGAVA, G.P. (1988). Landsat imagery for mapping saline soils and wet lands in north-west India. *International Journal of Remote Sensing*, **9**, 39-49

SINGH, A.N. and DWIVEDI, R.S. (1986). The utility of LANDSAT imagery as an integral part of the data base for small-scale soil mapping. *International Journal of Remote Sensing*, **7**, 1099-1108

SOARES, J.V., BERNARD, R. and VIDAL-MADJAR, D. (1987). Spatial and temporal behaviour of a large agricultural area as observed from airborne c-band scatterometer and thermal infrared radiometer. *International Journal of Remote Sensing*, **8**, 981-996

SPECTER, C. (1988). Managing remote sensing technology transfer to developing countries: a survey of experts in the field. *Photogrammetria*, **43**, 25-36

SSEW. (1983). *Soil map of England and Wales*. Soil Survey of England and Wales, Harpenden

STEPHENS, P.R., HICKS, D.L. and TRUSTRUM, N.A. (1981). Aerial photographic techniques for soil conservation research. *Photogrammetric Engineering and Remote Sensing*, **47**, 79-97

STEPHENS, P.R., MacMILLAN, J.K., DAIGLE, J.L. and CIHLAR, J. (1985). Estimating universal soil loss equation factor values with aerial photography. *Journal of Soil and Water Conservation*, **40**, 293-296

STEVEN, M.D. (1987). Ground truth: An underview. *International Journal Remote Sensing*, **8**, 1033-1038

STEVEN, M.D. and ROLLIN, E.M. (1986). Estimation of atmospheric corrections for multiple aircraft imagery. *International Journal Remote Sensing*, **7**, 481-497

SVENSSON, H. (1967a). Studies of a ground pattern. *Geografiska Annaler*, **49A**, 344-350

SVENSSON, H. (1967b). A tetragon patterned block field, in polygonal ground and solifluction features. Photographic interpretation and field studies in northernmost Scandinavia. *Lund Studies in Geography, A*, **40**, 8-23

SZILAGYI, A. (1986). Results of photo interpretation for soil mapping in a hilly area in Hungary. *ITC Journal*, **1986-1**, 94

THEIS, S.W. and BLANCHARD, A.J. (1988). The effect of measurement error and confusion from vegetation on passive microwave estimates of soil moisture. *International Journal of Remote Sensing*, **9**, 333-340

THOMAS, A.W., WELCH, R. and JORDAN, T.R. (1986). Quantifying concentrated-flow erosion on cropland with aerial photogrammetry. *Journal of Soil and Water Conservation*, **41**, 249-252

THOMPSON, D.R. and HENDERSON, K.E. (1984). Evaluation of Thematic Mapper for detecting soil properties under grassland vegetation. *IEEE Transactions on Geoscience and Remote Sensing*, **GE-22**, 319-323

TOWNSHEND, J.R.G. and HANCOCK, P.J. (1981). The role of remote sensing in mapping surficial deposits. In *Terrain analysis and remote sensing*. Ed. by Townshend, J.R.G. George Allen and Unwin, London

ULABY, F.T., RAZANI, M. and DOBSON, M.C. (1983). Effects of vegetation cover on the microwave radiometric sensitivity to soil moisture. *IEEE Transactions on Geoscience and Remote Sensing*, **GE-21**, 51-61

VALENZUELA, C.R. (1988). ILWIS overview. *ITC Journal*, **1986-1**, 4-14

VENKATARATNAM, L. (1986). Utility of Landsat MSS data for soil mapping in arid and semi-arid tropics. *ITC Journal*, **1986-1**, 98

VINK, A.P.A. (1968). Aerial photographs and the soil sciences, in aerial surveys and integrated studies. *Natural Resources Research*, **6**, 81-141

VINOGRADOV, B.V. (1988). Aerospace monitoring of ecosystem dynamics and ecological prognoses. *Photogrammetria*, **43**, 1-16

VOÛTE, C. (1987). Using outer space for managing matters on earth: a dream come true, or a nightmare in the making. *ITC Journal*, **1987-4**, 292-299

WANG, J.R., O'NEILL, P.E., JACKSON, T.J. and ENGMAN, E.T. (1983). Multifrequency measurements of the effects of soil moisture, soil texture and surface roughness. *IEEE Transactions on Geoscience and Remote Sensing*, **GE-21**, 44-51

WAY, D.S. (1978). *Terrain analysis*. Dowden Hutchinson and Ross/McGraw-Hill. Stroudsberg/New York

WAYUMBA, G.O. and PHILIPSON, W. (1985). Remote sensing of shifting cultivation and grazing patterns in Kenya's semi-arid region. *ITC Journal*, **1985-4**, 261-267

WEBER, C. (1985). Geological remote sensing: *quo vadis. ITC Journal*, **1985-4**, 227-241

WEBSTER, R. (1962). The use of basic physiographic units in air photo-interpretation. *International Archives Photogrammetry*, **14**, 143-148

WHEELER, J.R., TARUS, C., MITCHELL, A.J.B., KING, R.B. and WHITE, R.J. (1988). Evaluation of Landsat TM and SPOT imagery for agricultural land use planning in less developed countries. *Proceedings International Geoscience and Remote Sensing Symposium IGARSS 88*. Edinburgh, Scotland. ESA SP-284 (IEEE 88 CH 2497-6), Vol.1. European Space Agency, Paris

WILSON, D.R. (1983). *Aerial archaeology*. Batsford, London

WRIGHT, G.G. and BIRNIE, R.V. (1986). Detection of surface soil variation using high-resolution satellite data: results from the U.K. SPOT-simulation exercise. *International Journal of Remote Sensing*, **7**, 757-766

ZONNEVELD, I.S. and SURASANA, E. (1988). Ecosystem inventory/vegetation survey. *ITC Journal*, **1988-1**, 67-75

ZOTOVA, E.N. and GELLER, A.G. (1985). Soil moisture content estimation by radar survey data during the sowing campaign. *International Journal of Remote Sensing*, **6**, 353-364

5

ESTIMATION OF RAINFALL USING GEOSTATIONARY SATELLITE DATA

J.R. MILFORD AND G. DUGDALE

Department of Meteorology, University of Reading, U.K.

Introduction

There have been many reviews of the wide range of methods which may be used for estimating rainfall using observations made from satellites, including those by Barrett and Martin (1981) and more recently by Barrett (1989). Such reviews set out the principles of the various methods, and experiments which have shown their feasibility. The purpose of this paper is different: it discusses one particular method in more detail in order to show how far it may satisfy the needs of operational users of information on rainfall and water budgets. What governs the choice of method is the practicality of providing routine outputs in real time and with adequate accuracy, rather than the elegance of the science and the exactitude of the result under selected, experimental conditions.

For a new source of information to be adopted as routine it must show a net advantage over existing sources in quality, quantity, timeliness and cost. To establish the operational utility of a remote sensing methodology we must therefore establish the accuracy which is attainable in all circumstances, and see whether this will meet the needs of those who have to make decisions. We have to consider quite specific activities in agriculture, hydrology or elsewhere, and identify the ways in which economically significant decisions are taken at present: only then can we state whether our new information is competitive. Examples of this approach within food early warning systems and water management are given below.

Any routine monitoring of precipitation from space requires frequent views of the weather, and to date these can only be provided by geostationary satellites. Although these provide excellent temporal coverage, up to 48 images per day, they are restricted in their spatial resolution, their poor view of high latitudes and their limited spectral discrimination. In spite of these restrictions data from Meteosat provide operational estimates of rainfall over much of Africa. We shall discuss their potential and limitations below.

Most of the work presented here is from the TAMSAT[1] group in Reading. Since 1981 we have sought to improve the information on water balances for agricultural and hydrological purposes in Africa which may be obtained from Meteosat. The aim has been

---

[1] Tropical Applications of Meteorology using SATellite and other data

to develop methods which are objective, are fully pre-calibrated and may be automated to give an operational, real time product.

Life history studies do not yet meet these criteria, and we have concentrated on the use of cold cloud statistics. Most success has been gained from analysis of thermal infra-red (TIR) data at full time and space resolution to give maps of the duration of cold cloud using various temperature thresholds to define "cold". These durations are then regressed against rainfall measured in surface gauges to calibrate their conversion to rainfall estimates.

In the following sections we discuss the ways in which the regressions which provide the calibrations for this method vary in space and time. When the variations in the regressions become significantly larger than the uncertainty in the rainfall estimates it may be appropriate to localise the calibrations further. In addition, to assess the competitiveness of the information which the satellite can supply we must establish the performance of the alternative, conventional raingauge measurements. We have done this up to a scale of some 25 km (Flitcroft, Milford and Dugdale, 1989), showing that for averages over Meteosat pixels the gauges usually provide about half of the total uncertainty.

In the last section of the paper we discuss some cases in which the satellite estimates of rainfall may be competitive with information from other sources. Of course, the precise requirements of the final user must be known if we are to choose the optimum procedure and parameter values, but examples of specific information which can be provided from Meteosat TIR data with reasonable accuracy include the length of dry spells affecting rainfed agricultural areas, the location of isolated showers which may encourage locust breeding, the prediction of growth in extensive rangelands and forecasts of runoff from large catchments.

Rainfall estimates from cold cloud statistics

The basic methodology of the cold cloud statistics procedures is simple. A regular series of TIR images of an area is received, pixels with apparent temperatures lower than some predetermined threshold are classified as "cold cloud", and their characteristics accumulated over some period. The resultant map is converted to a rainfall estimate, possibly with the help of information from contemporary sources (which may be other satellite sensors or observations from the Earth's surface). The procedures adopted and the form of the algorithms are regarded as a statistical model, which is calibrated through comparisons between the cold cloud characteristics and sets of conventional raingauge data. To establish the utility of the method, it must subsequently be validated by comparing estimates from some area or period distinct from that used for the calibration.

Within this sequence there are many decisions to be taken and numerical values to be assigned. We shall assume here that the radiometer is accurately calibrated (or corrected) and that navigation is also accurate. When looking for the limit of spatial resolution, however, we must remember that the field of view of a radiometer is not sharp-edged, and that pixels will always overlap those of other images: even the best co-location leaves a residual uncertainty of the size of a pixel. A list of the factors to be considered includes the following:

- the type of regression model employed (linear, non-linear or multivariate);

- the interval between images (slots);
- time averaging period;
- the space averaging scale;
- the threshold temperature adopted;
- data treatment (e.g. linear or temperature weighted accumulation);
- additional data incorporated (e.g. water vapour channel, visible channel or contemporary surface raingauge measurements);
- localisation of calibration (variation with geographic location, time of year, character of season, topography and local storm climatology).

Since Lethbridge (1967) published the earliest statistics on the relation between rainfall and bi-spectral characteristics of clouds, a multitude of results have appeared, especially for the GATE area of the N. Atlantic, or the USA. However, the relation between rainfall and cold cloud duration depends on storm characteristics and is therefore climate-specific. This means that we cannot assume that results are transferable but must prove them to be so in every case before we use them outside the original study area. As an example, threshold temperatures appropriate for the semi-arid regions of Africa are substantially lower than those used over the oceans, or over N. America.

Optimisation of the factors listed is complex because they are interdependent and also because the optimum result depends on the user's viewpoint. In addition, to establish even a simple relationship we need a considerable size of sample, typically a minimum of 100 data pairs. It is therefore not surprising that not all these factors have been evaluated, even for a small part of Africa.

In the next section we describe the initial calibration and validation procedure which we have used. Following that we quote the results of some experiments which justify such a simplistic procedure, show the success and limitations of the cold cloud statistics methods to date and indicate how far they can sensibly be refined.

Calibration and validation

Choice of threshold temperature (T_t)

A number of analyses may be carried out once we have a sufficiently large sample of data pairs, i.e. Cold Cloud Duration (CCD) and rainfall values referring to similar periods and locations. In the TAMSAT procedure we construct contingency tables of CCD against rain in arbitrarily chosen classes for a number of threshold temperatures. (For the rainfall class limits, see *Table 5.3* below.) The contingency tables show which threshold discriminates best between days with and without significant rain: note that the minimum amount of rain which is significant may vary according to the use which will be made of the information. *Table 5.1* shows an example used to choose T_t to discriminate between rain over ten-day periods and no rain: a lower value would be used to distinguish significant rain, e.g. above 10 mm (*Table 5.2*).

TABLE 5.1. Dekad contingency tables for zone 2 in *Figure 5.4*, using three temperature thresholds. For -50°C occasions with rain and no cloud are almost equal in number to those with cloud and no rain. The sum is close to the minimum. -50°C is therefore used as threshold.

	No Cloud	Cloud		No Cloud	Cloud		No Cloud	Cloud
No rain	4	18	No rain	7	15	No rain	11	11
Rain	7	295	Rain	16	286	Rain	30	272
T_t	-40°C			-50°C			-60°C	

TABLE 5.2. As *Table 5.1* but for discrimination between rainfall amounts above and below 10mm. A threshold of -60°C is indicated

	cloud < 5h	> 5h		cloud < 5h	> 5h		cloud < 5h	> 5h
Rain			Rain			Rain		
< 10 mm	27	68	< 10 mm	47	48	< 10 mm	68	27
> 10 mm	22	207	> 10 mm	9	220	> 10 mm	26	203
T_t	-40°C			-50°C			-60°C	

Calibration

When the optimum threshold has been chosen for a given area, ten-day rainfalls are regressed against CCD, omitting periods with zero cold cloud. The best discriminating threshold also tends to give the smallest intercept on these regressions. In practice, because of the skewness of the data we prefer to regress the median values of the data grouped into classes (as in *Table 5.3*).

To assess the reliance which may be placed on individual values of the estimates, the quartile range of observed rainfalls corresponding to each CCD class is found. Alternative calibrations may also be compared through the correlation coefficients between individual CCD and rainfall values, usually excluding zero values of CCD. A characteristic data set is shown in *Table 5.3*.

Validation

When a relationship has been proposed for operational use it is essential that it be tested with independent data sets, using contingency tables as above, or regressing estimated against actual rainfall. The stability of a calibration may be tested by using the calibration derived from one year in subsequent years. Similar tests, comparing data from different sites, show how far a given set of parameters are applicable beyond the limits of the area containing the stations used for calibration.

TABLE 5.3. (a) Contingency table; cold cloud duration against rainfall class, for classes of 10 or 20 mm, $T_t = -50°C$. Data for July, 1985, 1986 and 1987, Zone 2 (in *Figure 5.4*)

Actual Rainfall (mm)	Cold cloud duration (hrs)											
	0.0 - 0.0	0.5 - 2.0	2.5 - 4.5	5.0 - 7.0	7.5 - 9.5	10.0 - 12.0	12.5 - 14.5	15.0 - 17.0	17.5 - 19.5	20.0 - 22.0	22.5 - 24.5	25.0 - 27.0	27.5
0-0	10	12	3	2	3	3	4	1	0	0	0	0	0
1-10	15	16	31	36	31	24	17	11	3	2	0	0	0
11-20	1	2	9	15	18	21	30	15	16	4	4	2	1
21-30	1	1	11	9	15	21	18	22	15	4	4	3	1
31-40	0	1	2	9	7	17	6	12	14	7	4	1	1
41-60	0	1	1	11	14	18	20	16	23	16	5	5	2
61-80	0	0	1	4	1	7	6	8	11	6	4	5	0
81-100	0	0	0	1	1	2	3	5	5	2	4	4	5
101-120	0	0	0	0	0	0	0	3	2	4	1	2	1
121-140	0	0	0	1	0	0	0	1	0	1	0	2	0
141-160	0	0	0	0	0	0	0	0	0	1	0	0	1
161-180	0	0	0	0	0	0	0	0	0	0	0	0	0
181-200	0	0	0	0	0	0	0	0	0	0	0	0	0
200+	0	0	0	0	0	0	0	0	0	0	0	0	0

TABLE 5.3. (b) Interpolated quartile values for July data, as above.

		Quartile Table at -50°C		
CCD Range	N	Rainfall in mm		
		25%	Median	75%
0.0 - 0.0	27	0.0	0.8	2.0
0.5 - 2.0	33	0.0	1.5	6.2
2.5 - 4.5	58	3.3	7.9	21.3
5.0 - 7.0	90	4.1	13.3	35.4
7.5 - 9.5	90	6.4	16.2	31.6
10.0 - 12.0	113	11.1	24.3	39.5
12.5 - 14.5	104	12.6	20.8	43.3
15.0 - 17.0	94	19.1	29.3	50.7
17.5 - 19.5	89	24.9	36.0	55.3
20.0 - 22.0	47	33.2	46.3	67.4
22.5 - 24.5	26	26.6	42.3	72.8
25.0 - 27.0	24	38.2	62.6	88.1
27.5 +	12	36.6	82.8	92.0

Selected results

Regression models

The simplest possible model is to use a single threshold and a linear relationship between CCD and rainfall. Limited sets of data have been processed in more elaborate ways, including the use of a quadratic regression (Jacobs, 1987), a regression including the minimum temperature of the cloud as well as CCD (Snijders, 1990) and the radiance in the water vapour channel (Turpeinen, 1987). The 'cloud volume' at temperatures below the threshold has also been used (Cadet, 1988, personal communication). None of these has added greatly to the accuracy of the rainfall estimates within the calibration data set used, and none has been shown to have an advantage when used in the operational, or pre-calibrated mode. Applying a linear regression model to a set of data pairs of CCD and rainfall directly is not recommended, because of the skewness of the distribution of rainfall amounts and the inclusion of many zeroes in some data sets. We have used regressions of all points other than (0,0), but more lately have used regressions on the medians of data pairs grouped according to the CCD. This gives more weight to the important larger falls and somewhat improves the estimates, whether judged by contingency tables or correlation coefficients.

Interval between images

Identification of rain days, and use of mean rain per rain day has been used as an extreme of the range of intervals between images (Barrett, 1989). A number of daytime images, including visible as well as TIR, may be used to classify raindays. Alternatively, the NOAA polar orbiters have been used to give four images per day and better spatial resolution than Meteosat. However, the fact that many of the smaller storms in drier regions contain cold clouds for less than 3 hours during their lifetimes (see e.g. Rowell, 1988) indicates that a higher scene frequency is required, and that we should rely on geostationary satellites, at least at latitudes below 45°. For the rest of this paper only results from Meteosat will be discussed.

Figure 5.1 shows how the information from Meteosat is degraded as the interval between images increases. The data are from a ten-day period in July 1986 over West Africa. It is seen that relatively little information is lost with a 2 hour interval, but beyond that the loss becomes significant. Using hourly data allows for occasional interruption from faults or the necessity for system maintenance, but with an automatic pixel classification routine there is no particular disadvantage in using all the available slots.

Averaging time and area

Because we depend for calibration on raingauges which are read daily, the maximum time resolution available for rainfall estimates based on Meteosat data is one day. The scatter on the calibration relationships on a daily basis and for the individual pixels is large, partly because of the nature of the rainfall, but also because falls may happen to be recorded on an adjacent day rather than that on which the cold cloud was observed. Averaging over a longer period or a larger area is expected to improve the accuracy of the estimate by smoothing small scale variability, but the limit which is acceptable is likely to be set by the user. Large scale agricultural climatological data are already based on dekads, so this is commonly adopted.

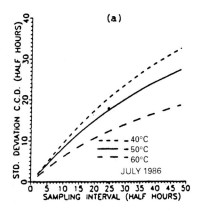 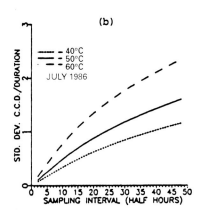

FIGURE 5.1. Absolute (a) and relative (b) increase in scatter of cold cloud durations as sampling interval increases. Meteosat TIR data for 21-30 July 1986 at 14°N in W. Africa, and for three temperature thresholds, as indicated.

Table 5.4 shows how the coefficient of variation for rainfall estimates reduces with increasing averaging time up to 30 days for a rather small sample of data. It is not known how far this result may be generalised. It is known that about half of the r.m.s. difference between actual and estimated rainfall is due to the inability of a single raingauge to represent the average rainfall over a pixel of 5 km × 5 km (Flitcroft et al., 1989).

The corresponding increase in accuracy as the area over which estimates are averaged increases is not so easy to calculate: this is because the data are not independent when the whole area is affected by a travelling weather system. If stations are so far apart that they provide independent data, the result will be equivalent to time averaging. Experiments to relate the estimates to the true area average rainfall will best be done through catchment studies. Runoff will be the best measure of the total volume of rain which has fallen on the catchment, provided that the hydrological characteristics which control it are sufficiently predictable.

Localisation of calibrations

The validation procedure also shows how widely a given calibration may be used. As more data become available we can reduce the area for which each calibration is calculated while maintaining our criterion of a minimum of 100 points to establish any one regression. The same type of comparison between estimated and actual rainfalls will also show sub-areas within the original region where estimates are consistently high or low: if these deviations are statistically significant they may show topographic influences. There appears to be useful stability from one year to another, as shown in *Table 5.5*.

Figures 5.2 and *5.3* illustrate the way in which the effect of the choice of T_t varies in a relatively homogeneous area, Niger in the West African Sahel. The dependence on latitude and month is marked. In this region isohyets and latitude are closely correlated, and it may be that calibrations should be related to the rainfall amount for a particular year rather than latitude. The improvement such a refinement might bring has yet to be investigated. *Figure 5.4* shows the zones used at present for our rainfall estimates

TABLE 5.4. Accuracy of rainfall estimates on different time scales. Data for July, 1986, Niger, 13-15°N. Estimates, for 153 stations, were derived from regressions for these data so that the mean error is zero.

Period	Observed rainfall		RMS error	RMS/Mean
	Mean	Standard Deviation	(Observed -Estimated)	
	mm	mm	mm	(CV)
Daily	6.8	12.3	11.9	1.75
5 days	16.0	18.7	17.0	1.06
10 days	28.6	26.2	22.1	0.77
15 days	43.9	33.0	27.6	0.63
20 days	54.7	39.4	27.8	0.51
30 days	86.2	48.8	38.0	0.44

TABLE 5.5. Comparison of actual and estimated dekad rainfall for Niger, July 1985, 1986 and 1987 using a single calibration for all latitudes. $R = 0$ when $D = 0$, otherwise $R = 4.52D + 5.1$ mm, where D is cold cloud duration in hours below -60°C.

Year	Number of observations	Exact category	1 cat. out	2 or 3 cat. out	>3 cat. out	% ≤ 1 cat. out	r
1985	327	69	106	132	20	54	0.53
1986	309	68	106	103	26	56	0.48
1987	313	60	121	109	28	57	0.60

Year	Total Actual Rain	Total Estimated Rain	Mean Actual Rain	Mean Estimated Rain
1985	8931	11119	27.3	34.0
1986	11155	11201	36.1	36.3
1987	6899	10289	21.7	32.3

Categories used:
Rainfall (mm) 0,1-10, 11-20, 21-30, 31-40, 41-60, 61-80, etc.
Cold cloud duration (h) 0, 0.5-2.0, 2.5-4.5, 5.0-7.0, 7.5-9.5, etc.

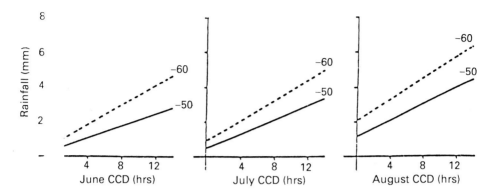

FIGURE 5.2. Regressions of rainfall on cold cloud duration (CCD) for dekads, excluding periods with no cold cloud, Niger 1985. These illustrate the difference between months.

from Meteosat. These were chosen to have reasonably uniform calibrations, assessed subjectively; in due course an objective clustering analysis should be used.

Improvements in the methodology

The emphasis to date has been to produce operationally implemented systems for rainfall estimation, and to establish their accuracy. The actual reliability of the estimates, and particularly our knowledge of this reliability, will improve further as we analyse more data from more stations, and particularly from further years with different rainfall anomalies.

Improvements in the methodology will come first from more objective methods of establishing the homogeneity of areas to which we can apply a single calibration. For example, a preliminary view of data from Ethiopia suggests that we shall have to use rather small areas and our knowledge of orographic influences on rainfall in order to get estimates for local use in such a mountainous area. Investigations in regions where various types of weather system produce rain are needed to establish whether cloud top temperature, perhaps combined with knowledge of the weather systems themselves, can lead to results comparable to those found where convective rainfall predominates. It would also be a great disappointment to meteorologists if knowledge of the convective systems themselves could not lead to better information on the rainfall which they produce!

Utility of estimates

In order to judge whether the accuracy of the estimates is adequate, a specific purpose must first be declared. For example, rainfall estimates are prepared from cold cloud statistics in FAO, Rome, using the methods described above. This is within the ARTEMIS system (Africa Real Time Environmental Monitoring using Imaging Satellites) for which the original motivation was the identification of areas of rainfall in dry areas which might provide breeding conditions for locusts. Interest in rainfall estimates also comes from Food Early Warning Systems: rain-fed crops will be put at risk if a dry period of two or more dekads occurs after planting rains have fallen. In either case the need is to

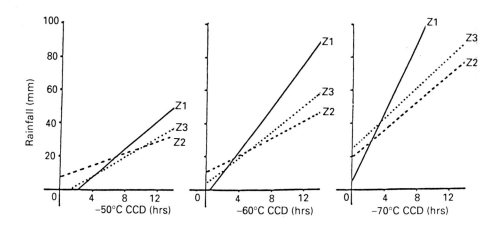

FIGURE 5.3. Regression of rainfall on cold cloud duration (CCD) for dekads, excluding periods with no cold cloud, Niger 1985. These illustrate the differences between latitudes. Z1 is N of 15°N, Z2 is from 13 to 15°N, and Z3 is S of 13°N

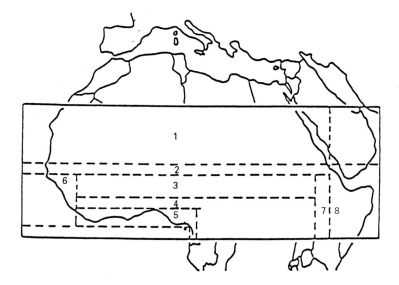

FIGURE 5.4. Calibration zones recommended, based on 1988 TAMSAT rainfall estimates. (Map on Meteosat projection)

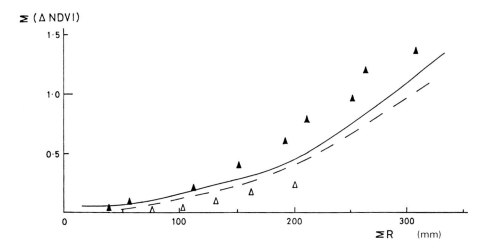

FIGURE 5.5. Comparison of accumulated **NDVI** (above a dry season minimum) and accumulated rainfall estimated from Meteosat TIR (after Justice *et al.*, 1990). Data are means for areas approximately 50 km across centred on 15°N: △ 1985, centred on 6°E; ▲ 1986, centred on 6°E; ---- 1985, centred on 0, 3, 6°W (averaged); ──── 1986, centred on 0, 3, 6°W (averaged)

identify falls of more than 15 mm (say), since less than this will be evaporated before it can have an observable agricultural effect. *Table 5.2* above suggests that this can be done with a rather low value of T_t. Extensive biomass models need information on the rainfall amount. The proper averaging procedure for soil and vegetation characteristics over tens of kilometres has not been established, but rainfall on this scale may be useful as a predictor of **NDVI** (Normalized Difference Vegetation Index), mapped at the 'GAC' scale (Global Area Coverage, some 25 km). *Figure 5.5*, after Justice, Narracott, Dugdale, Townshend and Kumar (1990), suggests that in certain areas rain estimates may help forecast seasonal biomass production (which is related to 'accumulated' **NDVI**). Whether this can then be used for operational decisions remains an open question, especially since **NDVI** only shows up vegetation changes at all sensitively over a limited range of the density of active biomass.

In areas where rain-fed agriculture predominates, knowledge of dry spells during a growing season may provide a particularly important warning for those with national responsibilities. Even where the estimates have such wide uncertainties that they are quantitatively of very limited use they are good indicators of dry dekads, if the threshold is appropriately chosen. *Plate 5.1* illustrates a typical image of the Sudan: it clearly shows dry areas south of Khartoum and around El Fasher in the West. Two or three dekads with coincident dry areas should alert the authorities to potential crop failure, even in the absence of raingauge data, and will show the extent of the region at hazard better than any feasible gauge network.

For evaluating water resources, to support either planning or management decisions, it is the available water from a catchment which is most often required. This will be derived from an area average of rainfall minus evaporation, and uncertainties in actual

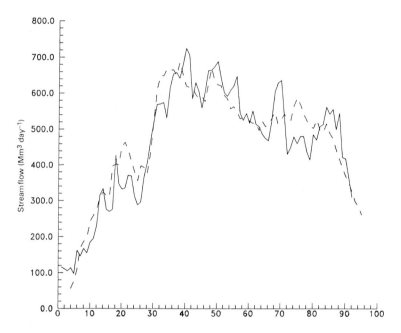

FIGURE 5.6. Streamflow forecasts at Roseires, Sudan: day 1 is July 1, 1988. Solid line is recorded streamflow, dashed is modelled flow using average cold cloud duration over the Blue Nile catchment as input. Lag between input and the catchment output at Roseires is up to four days

evaporation must now be included. For day-to-day management operations the daily runoff may be needed to the equivalent of a few mm of precipitation, and preliminary results from Senegal (Hardy, Dugdale, Milford and Sutcliffe, 1989) suggest that this may be attainable, or at least that the satellite-derived estimates have as good a chance as any practicable network of raingauges.

Finally, we may note that flood forecasting will have to work on daily or shorter periods. Experience with the Blue Nile catchment in 1988 suggests that this may again be approaching practicality for seasonal floods, but quantification of extreme storms which give rise to the most damaging, local flooding will require a much more substantial meteorological input. *Figure 5.6* shows the potential of the rainfall estimates as inputs to hydrological models for forecasting the available water, or floods a few days in advance. Daily cold cloud durations over the whole Blue Nile catchment in Ethiopia are totalled and used as inputs to the conceptual hydrological model proposed by Sutcliffe and Green (1986).

Conclusions

The fundamental limits of the methods using cold cloud statistics to estimate rainfall should be recognised before effort is spent on fruitless attempts at refinement. The strengths of the method include its ability to give real-time estimates over wide areas, particularly for discriminating between days with rain above and below fairly small rainfall thresholds and to provide some quantification of rainfall in areas of size 10 km upwards. When integrated over a larger area, such as a catchment or the 25 km pixel of the NOAA GAC vegetation index, the rainfall estimates on a daily or ten-daily basis become more accurate, and may be used directly in hydrological or agrometeorological models.

To date we have concentrated on a single, thermal infra-red channel because this has been the only one available on an hourly basis with significant information. Future satellites will carry radiometers with better spatial resolution, and others working at a variety of microwave wavelengths, and these should give useful information if they can be developed for use on geostationary platforms. All these may eventually give better estimates of the rainfall, but they are many years from operational status.

We would hope that the purely statistical approach is only the first step towards the measurement of rainfall from satellites, and that the application of our knowledge of atmospheric dynamics and thermodynamics will help to improve estimates in future. Although we have a reasonably clear idea of the structure of the storms which produce rain in W. Africa this has so far helped little in the quantification of the rainfall. As an intermediate step the TAMSAT group has examined factors such as the size, rate of growth, speed of travel, minimum cloud temperature and sharpness of the front. These were incorporated into regressions to try to improve daily rainfall estimates from Meteosat. Positive correlations were found, but the amount of the variance which they explained was small, whether the effects were taken singly or together.

The ultimate limit to the accuracy of rainfall estimates for local use is set by the nature of the rain-producing systems. For areas up to 500 km^2 or so the relationship between the area-average rainfall and that recorded by a single gauge has been calculated (Flitcroft *et al.*, 1989). This analysis is confined to cases where the whole area has been affected by a single storm and the statistics required to produce the similar results for larger areas, such as the grid square of a general circulation model, look very different. Work on these is being undertaken both at ORSTOM, Paris, and at Reading.

Acknowledgements

The work summarized here has been supported by the UK Overseas Development Administration, FAO, and the EEC. A large number of individuals have contributed, notably Mrs V. McDougall, and Messrs I.D. Flitcroft, S. Hardy and M. Saunby in the recent past.

References

BARRETT, E.C. (1989). Satellite remote sensing of rainfall. In *Applications of remote sensing to agro-meteorology*. Ed. by F. Toselli. Kluwer, Dordrecht

BARRETT, E.C. and MARTIN, D.W. (1981). *The use of satellite data in rainfall monitoring.* Academic Press, London

FLITCROFT, I.D., MILFORD, J.R. and DUGDALE, G. (1989). Relating point to area average rainfall in semi-arid West Africa and the implications for rainfall estimates derived from satellite data. *Journal of Applied Meteorology*, **28**, 252-266

HARDY, S., DUGDALE, G., MILFORD, J.R. and SUTCLIFFE, J.V. (1989). The use of satellite derived rainfall estimates as inputs to flow prediction in the river Senegal. In *Proceedings IAHS Third Scientific Assembly*. Baltimore, May 1989. In press

JACOBS, C.M.J. (1987). Preliminary report on the applicability of the Meteosat system in rainfall mapping over Zambia. MARS project report, Staring Centrum, Wageningen Agricultural University, Wageningen, Netherlands

JUSTICE, C.O., NARRACOTT, A.S., DUGDALE, G., TOWNSHEND, J.R.G. and KUMAR, M. (1990). The synergism of AVHRR and Meteosat data for studying vegetation development in semi-arid Africa. *International Journal of Remote Sensing*, In press

LETHBRIDGE, M. (1967). Precipitation probability and satellite radiation data. *Monthly Weather Review*, **95**, 487-490

ROWELL, D.P. (1988). Short range rainfall forecasting in the West African Sahel. Ph.D. Thesis, University of Reading

SNIJDERS, F. (1990). Rainfall monitoring based on Meteosat data. *International Journal of Remote Sensing*. In press

SUTCLIFFE, J.V. and GREEN, C.S. (1986). Water balance investigation of recharge in Madhya Pradesh, India. *Hydrological Science Journal*, **31**, 383-394

TURPEINEN, O.M. (1987). Validation of the ESOC precipitation index. In *Proceedings of 6th Meteosat Scientific Users Meeting*. Amsterdam, November 1986. (EUM P 01) EUMETSAT, Darmstadt, FRG

6

APPLICATION OF REMOTE SENSING AND GEOGRAPHICAL INFORMATION SYSTEMS IN WATER MANAGEMENT

G.J.A. NIEUWENHUIS, J.W. MILTENBURG and H.A.M. THUNNISSEN

The Winand Staring Centre for Integrated Land, Soil and Water Research,
P.O. Box 125, 6700 AC Wageningen, The Netherlands.

Introduction

Information on regional crop transpiration is important for optimal water management in agriculture. In principle, there are two methods to obtain information on actual crop transpiration. First, application of agro-hydrological simulation models and, second, application of remote sensing techniques.

Simulation models have the advantage that evolution in time can be described accurately but they have the disadvantage that actual field conditions must be schematized. Two-dimensional models, especially, suffer from schematization, because simulations are performed for large, regular grid-cells (e.g. de Laat and Awater, 1978). This implies that for each grid-cell (with a minimum size of 100 m × 100 m) only mean values are obtained. With Geographical Information Systems (GIS) only soil profiles have to be schematized. Regional information is obtained by combining one-dimensional model calculations with digitized thematic maps, like soil, land use and drainage maps.

Remote sensing has proven to be a valuable tool to estimate regional transpiration. Crop temperatures derived from thermal images can be transformed into daily transpiration values with surface energy balance models (Jackson, Reginato and Idso, 1977; Soer, 1980; Hatfield, Reginato and Idso, 1984). Jackson et al. (1977) proposed an empirical relation between midday surface-air temperature difference and actual daily evapotranspiration. Nieuwenhuis, Smidt and Thunnissen (1985) and Thunnissen and Nieuwenhuis (1989) proposed some modifications to the method of Jackson et al. With the modified method, differences in radiation temperature of a certain crop, as derived from thermal images, can be transformed directly into differences in relative transpiration. The developed linear relationship between crop temperature and daily relative transpiration is dependent on crop type and height. Therefore, mapping of crop transpiration from thermal images must be combined with crop classification. Crop classification is performed using reflection images obtained with a multi-spectral scanner.

In this paper the experience with remote sensing and hydrological modelling at the Winand Staring Centre will be discussed. The remote sensing approach is presented in detail while the applied hydrological model is only briefly described. Especially, the integration of both methods will be treated. With this integrated approach the description of hydrological conditions in agriculture can be improved.

A method has been developed to map crop transpiration from digital reflection and thermal images obtained from aircraft. Using remote sensing in this way, detailed information on the regional distribution of transpiration on flight days can be obtained. Crop water status can also be analysed by applying agro-hydrological simulation models. Models have the advantage that evolution in time can be accurately described. However, they have the limitation that actual field conditions, used as input, have to be generalized. For the flight day, remote sensing gives information in terms of patterns. It was found that, in general, an important improvement of the hydrological description af an area can be achieved by combining remote sensing data with hydrological model simulations.

With geographical information systems one-dimensional model simulations can be located using existing maps, resulting in simulation maps. The transpiration map obtained through remote sensing can be used to verify simulated transpiration for a test-area. If the simulation results deviate systematically from the remote sensing transpiration map, the applied input parameters are adjusted before further simulation takes place.

Two methods to determine actual crop transpiration

Determination of crop transpiration from remotely sensed imagery

The temperature of objects at the earth's surface is determined by the instantaneous equilibrium between gains and losses of energy. At the earth's surface, to a first approximation, net radiation $\mathbf{R_n}$ equals the sum of latent heat flux into the air $\mathbf{\lambda E}$, the sensible heat flux into the air $\mathbf{H}$ and the heat flux into the soil $\mathbf{G}$:

$$\mathbf{R_n} = \mathbf{\lambda E} + \mathbf{H} + \mathbf{G}. \tag{6.1}$$

In *Equation 6.1* the units of all terms are in W m^{-2}. Where $\mathbf{\lambda}$ is the latent heat of vaporization (J kg^{-1}) and $\mathbf{E}$ the evapotranspiration flux (kg m^{-2}s^{-1}). The term $\mathbf{R_n}$ can be split up into net short wave and net long wave radiation terms:

$$\mathbf{R_n} = (1 - \boldsymbol{\alpha})\mathbf{R_s} + \epsilon(\mathbf{R_l} - \boldsymbol{\sigma} \boldsymbol{T}_c^4), \tag{6.2}$$

where $\mathbf{\lambda}$ is the surface reflection coefficient, $\mathbf{R_s}$ the incoming short wave solar radiation flux (W m^{-2}), ϵ the emission coefficient, $\mathbf{R_l}$ the long wave sky radiation flux (W m^{-2}), σ the constant of Stefan Boltzmann (56.7 × 10^{-9} W m^{-2} K^{-4}) and T_c is the crop surface temperature (K).

When the crop is well-supplied with water, the net radiation energy is mainly used as latent heat for vaporization. If the latent heat flux decreases, the surface temperature increases, resulting in a rise of the sensible heat flux $\mathbf{H}$. Considering the transport of heat from the crop surface with temperature T_c to a certain height $\mathbf{z_{ref}}$ (m) with air temperature T_a (K), the transport equation can be expressed as:

$$\mathbf{H} = -\rho \mathbf{c_p}(T_a - T_c)/\mathbf{r_{ah}}, \tag{6.3}$$

where ρ is the density of moist air (kg m^{-3}), $\mathbf{c_p}$ the specific heat of moist air (J kg^{-1} K^{-1}) and $\mathbf{r_{ah}}$ the turbulent diffusion resistance for heat transport (s m^{-1}) from the crop surface to $\mathbf{z} = \mathbf{z_{ref}}$.

Combining *Equations 6.1, 6.2* and *6.3*, the relation between latent heat flux λE and surface temperature T_c can be found (Brown and Rosenberg, 1973; Stone and Horton, 1974):

$$\lambda E = \rho c_p (T_a - T_c)/r_{ah} + (1 - \alpha)R_s + \epsilon(R_l - \sigma T_c^4) - G. \quad (6.4)$$

From *Equation 6.4* it can be seen that λE depends on a number of meteorological and crop surface parameters. For a certain area, T_c can be remotely sensed by thermal infrared line scanning. When T_a, r_{ah}, α, R_s, ϵ, R_l and G are known (or estimated) λE can be computed. The resistance r_{ah} depends on wind velocity u, the roughness of the crop surface z_o and atmospheric stability (Dyer, 1967; Webb, 1970).

For clear sky conditions T_a, R_s, R_l and u can be taken as constant over a certain area. This means that standard meteorological measurements can be used. The parameter α has to be determined from field measurements or from reflection images. With more indirect procedures ϵ and z_o can be estimated by combining field observations with a vegetation-index derived from reflection images.

Jackson et al., (1977) related midday surface-to-air temperature differences linearly to 24-hour evapotranspiration and net radiation values. To estimate the slope of this relationship a crop-dependent analytical expression has been derived by Seguin and Itier (1983). Nieuwenhuis et al., (1985) and Thunnissen and Nieuwenhuis (1989) proposed to replace the surface-air temperature difference by the temperature difference ($T_c - T_c^*$), that exists between the crop transpiring under the restriction of the actual soil moisture condition and the crop transpiring under optimal soil moisture condition. The net radiation term was replaced by the 24-hour potential evapotranspiration rate of the crop. With these adjustments they obtained:

$$\lambda E_{24}/\lambda E_{p24} = 1 - B_r(T_c - T_c^*), \quad (6.5)$$

where λE_{24} and λE_{p24} are respectively the actual and potential 24-hour evapotranspiration rate (mm day^{-1}) and B_r (K^{-1}) is a calibration constant. By means of *Equation 6.5*, differences in radiation temperature of a certain crop, derived from thermal images, can be transformed directly into reductions in transpiration.

From TERGRA-model calculations (Soer, 1977), Thunnissen (1984) found that B_r can be described by a linear function of the wind velocity (u) at a height of 2.0 m above groundsurface:

$$B_r = a + bu. \quad (6.6)$$

Values for the regression coefficients a and b are given in *Table 6.1* for different types of crops and crop heights. For agro-hydrological purposes thermal images are usually recorded on clear days in the summer period. It was found that for such days *Equations 6.5* and *6.6* can be applied for the meteorological conditions prevailing in the Netherlands.

Determination of crop transpiration with a hydrological simulation model

Remotely sensed images characterize crop conditions at one time. For several agro-hydrological applications, however, determination of cumulative effects in time on the total crop yield is required. As an example, one can think of the effects of groundwater extraction for domestic purposes on the growing conditions of grassland and arable crops. The amount of water available for transpiration strongly influences dry matter production.

TABLE 6.1. Values for the coefficients a and b in *Equation 6.6* for a number of crops with crop height.

Crop	Crop Height(cm)	a (K^{-1})	b $(K^{-1}\ m\ s^{-1})$
Grass	<15	0.050	0.010
Grass	>15	0.050	0.017
Potatoes	60	0.050	0.023
Sugar beet	60	0.050	0.023
Cereals	100	0.090	0.030
Maize	200	0.100	0.047

With an agro-hydrological simulation model such as SWATRE (Feddes, Kowalik and Zaradny, 1978; Belmans, Wesseling and Feddes, 1983), the use of water by crops can be simulated during the entire growing season. SWATRE is a transient one-dimensional finite-difference soil-water-root uptake model, that applies a simple sink term and different types of boundary conditions at the bottom of the system. If the soil system remains unsaturated, one of the following three bottom boundary conditions can be used: pressure head, zero flux or free drainage. When the lower part of the system remains saturated, one can give either the groundwater level or the flux through the bottom of the system as input. In the latter case the groundwater level is computed. At the top of the system, 24-hour data on rainfall, potential soil evaporation and potential transpiration are needed.

SWATRE needs the following inputs:

- 24-hour meteorological data, including precipitation figures and data necessary to calculate potential evapotranspiration. Usually meteorological data from routine measurements are used. However, precipitation data, especially, may vary considerably within larger study areas;

- crop characteristics such as rooting depth. The variety of crops included in simulations is usually reduced to the main crop types. Without the use of remote sensing images no topographical location can be assigned to the crops. In that case the relative area of both crops per sub-region is known from agricultural statistics, which enables the simulation of overall transpiration for a sub-region for example;

- soil physical parameters including (un)saturated conductivity and water retention characteristics for each soil layer. Every soil type represented on the standard 1:50000 soil map includes a range of soil physical parameters. Numerous soil samples from all over the country were analysed to determine the average soil physical properties per distinguished soil layer. On the basis of the observed soil layers and average soil physical properties per distinguished soil layer, the soil classes for which simulations are performed are defined;

- groundwater table regime, which is schematized into drainage classes.

Model simulations are performed for every so-called simulation-unit. Simulation-units are units for which the water supply of crops is comparable. In practice, simulation-units are unique combinations of crop type, soil class and drainage condition for which model simulations are carried out. Performing model calculations for each simulation unit, crop transpiration can be simulated for the entire growing season. The obtained results are presented in tables but can also be assigned to simulation maps in a GIS, as discussed later.

Integration of remotely sensed imagery and SWATRE simulations in a geographical information system

General description of the geographical information system

As SWATRE is one-dimensional it gives output for one location only, without spatial overview. Until recently the integration of remote sensing data, maps and simulation results was carried out by hand. Therefore a GIS was developed based on an integrated use of remote sensing data and SWATRE simulations so that remote sensing techniques become operational for larger areas and practical use in water management. *Figure 6.1* shows a flow diagram of the system.

The geographical information system uses as input appropriate remote sensing images, ground truth, a digital soil/drainage map and the simulation model input parameters. In this system transpiration maps obtained from remotely sensed images are used to verify model simulations for a representative test-area on one or two days during the growing season. As SWATRE is well calibrated with accurately measured soil physical, hydrological, meteorological and crop parameters, the model itself needs no verification. But actually measured input parameters are usually not available so readily available standard parameters and generalized soil/drainage maps are used. Because of this input, model results must be verified. Especially, the accuracy of soil physical parameters is important, because they strongly influence simulations. Changing of the model input only takes place after careful evaluation of the differences between the simulated transpiration map and remote sensing transpiration map. If model results are in agreement with the remote sensing transpiration map, simulations can be performed for larger areas.

Remote sensing image processing to map crop transpiration

REQUIRED REMOTE SENSING IMAGERY

As the agricultural fields in the Netherlands are relatively small only earth observation satellite systems, such as SPOT and Landsat-TM, and multi-spectral scanning techniques from aeroplanes are applicable. To derive a transpiration map from thermal infrared (TIR) images it is necessary to obtain the following imagery for the test-area:

- a TIR image. To detect crop stress conditions from thermal images they must be recorded on clear summer days around midday, after a relatively dry period. Because of the acquisition time of about 10.30 am and the poor spatial resolution of 120 m, Landsat-TM thermal images are not useful. Moreover, the poor temporal resolution of 16 days is a severe constraint. Consequently, one is dependent on thermography from aeroplanes. To relate digital numbers of the TIR image to crop temperature, field measurements are necessary;

- reflection images of a date in the same growing season to classify the different crop types. Because of TM's spectral band in the middle-infrared range, an accurate

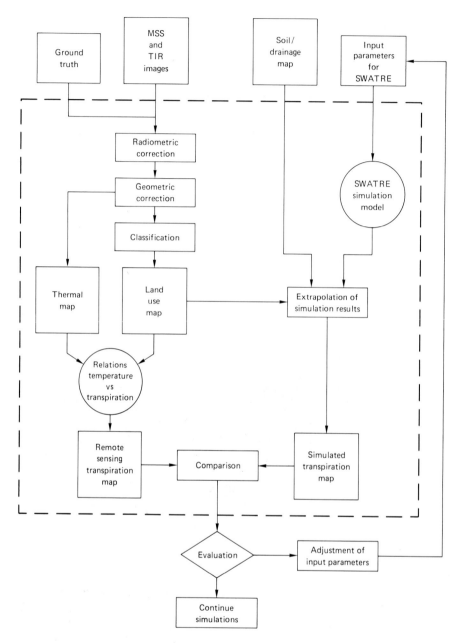

FIGURE 6.1. Flow diagram of the geographical information system in which model simulations and remotely sensed imagery are combined.

RAINFALL ESTIMATE MAP

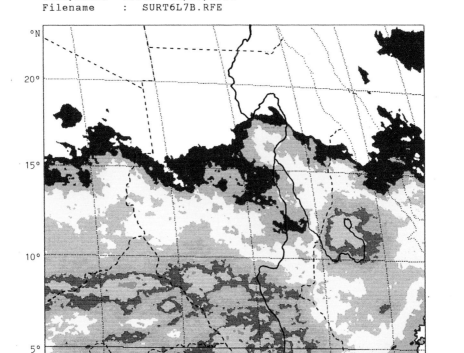

Plate 5.1. Rainfall estimate map for Sudan and surrounding area for 11 - 21 July, 1988. Cold cloud durations have been derived from hourly Meteosat TIR images and converted to rainfall with linear regressions. Data over Ethiopia and south of 9°N has not been calibrated and must not be interpreted quantitatively. The key is shown in the colour strip below the map, estimated rainfall ranging from zero up to 250 mm.

Plate 6.1. Transpiration map of part of a study area near a pumping station (P) situated in the eastern part of the Netherlands, as derived from reflection and thermal images of 30 July, 1982. crop relative transpiration decreases from blue (> 90%), green (70 - 90%), yellow (50 - 70%), red (30 - 50%), to magenta (< 30%). Black indicates non-agricultural land use. The isoline indicates a 10 cm drawdown according to the calculations reported by de Laat and Awater (1978).

Plate 6.2. Transpiration maps of a study area situated in the southern part of the Netherlands, as composed from reflection and thermal images of 22 July, 1983, top, and obtained by combining SWATRE simulations with a digitised soil map 1:50000, below. Crop relative transpiration decreases from dark blue (> 90%), cyan (70 - 90%), green (50 - 70%), yellow (30 - 50%) to orange (< 30%). Black indicates non-agricultural land use.

Plate 10.1. A subset of the digital field (DFBD) used as a reference, map comprising both training and test areas from four strata. The farm field boundaries show segments of the DFBD system in raster form.

Plate 10.2. Crop map for a 16000 Ha area. Per field classification (method B) was used, supported by DFBD.

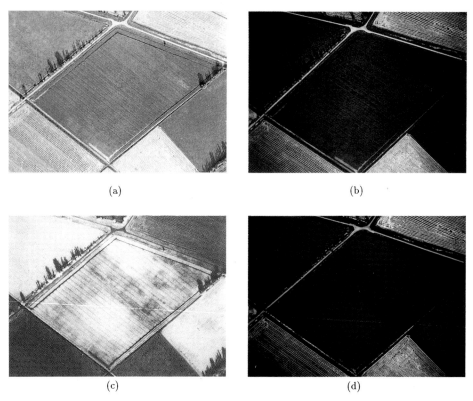

Plate 15.1. Appearance of wilting and senescing potato crop on four film types exposed simultaneously. (a) panchromatic black and white, (b) true colour, (c) Infrared black and white, (d) infrared false colour.

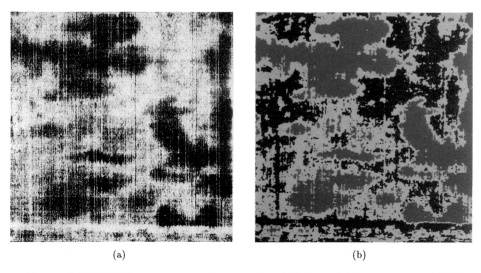

Plate 15.2. Herbicide damage on part of a sugar beet field shown (a) in grey tone and (b) in false colour based on the computed density slice indicating, green – unaffected crop, yellow – partial establishment, red – bare soil.

crop map can be obtained from one Landsat-TM multi-spectral scanner image recorded in the growing season. Both the Daedalus airborne scanner usually used in the Netherlands and SPOT lack the possibility to record reflection in the middle infrared range. If only SPOT or Daedalus images are available an accurate crop map can be obtained by applying a multi-temporal approach or by interactive updating of the classification results.

- reflection images of the same date as the TIR image to map crop height. Crop height can be derived from the Daedalus reflection images recorded at the same moment as the TIR image.

IMAGE PROCESSING

Because of the large scanning angle and the instability of the sensor platform, airborne scanner images must be corrected both radiometrically and geometrically (e.g. by combination with existing maps or satellite images) before further processing of the data is possible. Standard image processing routines, which are used to process satellite imagery, were found to be not powerful enough to correct airborne imagery. Therefore, we developed methods to correct Daedalus scanner imagery, as will be discussed.

RADIOMETRIC CORRECTION OF AIRBORNE SCANNER IMAGERY

Airborne scanner images often show a clear gradient in radiation intensity perpendicular to the flight direction. This is caused by the large viewing angle of airborne scanner systems in combination with a non-optimal flight situation (Barnsley, 1984). In the optimal situation the sun is at its highest point and the flight direction has the same azimuth as the sun. In practice this is usually not the case. A radiometric correction of the reflection images is necessary to classify the crop types. This is achieved by taking a representative subset of the image; a strip of about 100 lines covering the land use of interest (in this case agricultural fields), calculating the mean pixel values per column and drawing a histogram of the mean pixel values v column numbers. A linear or second order regression function is fitted and a correction value estimated for every column. In the resulting image the agricultural fields have a quite uniform intensity.

GEOMETRIC CORRECTION OF AIRBORNE SCANNER IMAGERY

During the scanning process a number of geometric distortions are introduced into airborne scanner data. As most of the distortions are non-systematic and local, due to an unstable atmosphere, they can only be corrected using Ground Control Points (GCPs). In some cases the use of a higher order polynomial proved to be insufficient for the correction of badly distorted images and therefore a new method, the Facet Transformation Method, was developed. This method of geometric correction is not based on one higher order polynomial function but on a large number of first order functions. Between the GCPs triangles (facets) are formed. The corner points of each facet form the three coordinate pairs needed to fit a first order transformation function for the pixels within the triangle. Advantages of the facet transformation method are:

- local distortions can be corrected by adding new GCPs on troublesome places. Every new GCP adds three new triangles and three new functions, whereas adding extra GCPs to a large number of GCPs when applying a higher order polynomial, gives almost no improvement;

- the accuracy of correcting a small image is equal to the accuracy of correcting a large image, whereas using higher order polynomials the accuracy decreases as the size of the images increases.

Land use and transpiration maps

Classification of the radiometrically and geometrically corrected remote sensing images is usually performed using a maximum-likelihood algorithm. For this process some ground truth is necessary. As already mentioned, crop type classification can cause problems if only Daedalus imagery of one date are available. In this case long grass and maize can hardly be distinguished. On false colour photographs, however, these crops can be separated because of differences in texture, so that classification results can be improved interactively based on a visual interpretation of simultaneously taken false colour photographs. If satellite images of the same growing season are also available classification of crop types is less troublesome.

For an up to date land-use map, grassland has to be divided into three classes depending on grass height, by applying a Normalized Difference Vegetation Index (**NDVI**), which is defined as (Tucker, 1977):

$$\text{NDVI} = (\text{IR} - \text{R})/(\text{IR} + \text{R}).$$

For **IR** and **R** the radiation intensities in, respectively, wavelength bands 9 (near-infrared) and 7 (red) of the Daedalus scanner have been used. In previous research it was found that a relation exists between **NDVI** and grass height (Project team Remote Sensing Studieproject Oost-Gelderland, 1985). The exact relation has to be determined for every region and date. Therefore grass height is measured on several plots on flight days.

Combining the land use map and the thermal map and using *Equations 6.5* and *6.6* the temperatures are automatically converted into a map with estimates of relative 24-hour crop transpiration. For the determination of crop temperatures T_c^* plots with potentially transpiring crops are traced. For short grass no transpiration values are computed because short, recently mown, grass has an incomplete soil cover. In that case, observed crop temperatures are to a large extent determined by relatively warm bare soil.

Production of simulation maps using simulated transpiration values

For an extrapolation of one-dimensional model results, based on simulation units - unique combinations of a soil class, drainage class and crop type -, it is necessary to know the spatial distribution of the soil classes, drainage classes and crop types. The distribution of crop types, which differs from year to year, is extracted from remote sensing imagery and available on a land use map. Soil and drainage classes are available on digitized soil maps. Combination of the land use map and soil map results in a simulation map to which simulated transpiration values are assigned for every simulation unit.

This method of extrapolation is as accurate as the maps used. The accuracy of land use maps obtained with remote sensing is in practice restricted to 80-90%. The standard 1:50000 soil map of the Netherlands is a generalization of reality. Because of large spatial variability, the restricted number of borings and the necessity to neglect small areas with deviating soils, a single unit on this map is usually an association of soil types. The chance of actually finding a particular soil type in places where it should be according to the map is at least 70%. The drainage condition is indicated on the 1:50000 soil map as so-called groundwater table fluctuation classes, based on the mean highest and the mean lowest groundwater table. They represent the average winter and summer water table, respectively, in a year with an average precipitation and transpiration. Actual groundwater tables can deviate from this map, which is another source of inaccuracy in the simulation map. As in practice no other means for extrapolation are usually available, the spatial accuracy of the simulation map will not be optimal. Because of the mentioned inaccuracies, the remote sensing transpiration map and the simulation map are compared on the basis of mean transpiration values per simulation unit.

Experience with transpiration mapping from remotely sensed imagery in agricultural water management

Two case studies are considered here. The first study deals with simulation models used as an aid in the interpretation of transpiration maps obtained through remote sensing. In addition, it was investigated whether simulation results could be verified using transpiration maps derived from remote sensing. In the second example a more recent study is described, in which remote sensing transpiration maps are used to verify model simulations using a GIS.

Interpretation of transpiration maps derived from remotely sensed imagery

In 1982 and 1983 remote sensing flights were performed in the eastern part of the Netherlands. Reflection and thermal images were recorded with the Daedalus scanner. The images acquired after a dry period were especially relevant to investigate the possibilities of remote sensing in agricultural water management. The transpiration map shown in *Plate 6.1* was used to study the effects of groundwater extraction by pumping stations on crop water supply. As phreatic groundwater is extracted in this region of the Netherlands, the groundwater level, and therefore crop water supply, are influenced. Around the centre of the extraction a more or less conical depression of the groundwater table occurs (see *Plate 6.1*). Because of the conical depression of the groundwater table circular drought patterns were expected around the site of the pumping station. This was, however, not observed. One of the reasons was the common practice of irrigating grassland using sprinklers. Because of this, numerous plots are present with crops well supplied with water and which are more or less potentially transpiring (blue in *Plate 6.1*).

Under natural conditions crop transpiration depends on the moisture availability in the root zone, which is determined by:

- the depth of the root zone;
- the moisture retention capacity of the root zone;
- the hydraulic conductivity of the subsoil;
- the groundwater level during the growing season.

To explain the occurrences of reduction in transpiration, calculations with SWATRE were performed. For Typic Haplaquod soils, moisture retention curves were determined on the basis of measurements of matric pressure head in-situ and volumetric moisture content in the laboratory (Thunnissen and Nieuwenhuis, 1989). For Plaggept and Typic Humaquept soils moisture retention curves were derived from the literature (Krabbenborg, Poelman and Van Zuilen, 1983). The hydraulic conductivity curves for the different soil layers were calculated from their soil texture and organic matter content according to the method described by Bloemen (1980). Rooting depths were determined *in-situ*. Measured groundwater levels were used as bottom boundary conditions. To validate the SWATRE simulations, calculated values of the hydraulic potential were compared with measured values for some plots on Haplaquod and Humaquept soils. In general they were in good agreement (Thunnissen and Nieuwenhuis, 1989). For the flight day, transpiration values derived from the remote sensing transpiration map were compared with the simulated values for different locations within the study area. Both results are in good agreement (*Figure 6.2*). Only for Plaggept soils is transpiration probably overestimated by SWATRE.

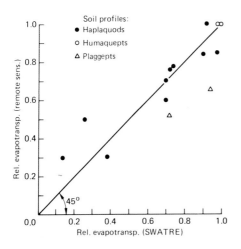

FIGURE 6.2. Relative 24-hour transpiration rates as obtained by the remote sensing approach and simulated with SWATRE for fourteen grassland plots on three different soil profiles.

The effects of groundwater extraction could not be determined directly from a visual interpretation of drought patterns on the transpiration map, as derived from remote sensing images, because of the application of sprinkling irrigation and the variability in soil physical and hydrological properties. However, information on effects of groundwater extraction could be obtained from a systematic analysis of transpiration by crop type, soil and drainage class. This systematic analysis was performed by hand. Nowadays such an analysis could be carried out faster and more accurately using GIS techniques.

Verifying simulated transpiration with remote sensing transpiration maps using a GIS

In the previous paragraph it was shown that, in general, remote sensing results are in good agreement with SWATRE calculations if soil physical parameters, groundwater levels and rooting depths are measured accurately. This implies that remote sensing images could be applied to verify simulation model results. Usually, no extensive fieldwork and laboratory analyses are carried out to test simulation of the water balance of an area. Soil and drainage classes are available on the standard soil map and crop parameters are taken from literature. Due to the inaccuracy and generalization of the input data, model results are not optimal.

With the information system presented in *Figure 6.1*, one of the constituents of the water balance - crop transpiration - can be verified on a single day. If systematic deviations occur between simulated and remote sensing transpiration, input parameters for the simulation model can be adjusted before water balance simulations for larger areas take place. It is also possible to include in the comparison between a remote sensing based transpiration map and simulation results, simulation maps of a few days before and after the flight day. This gives extra information on the accuracy of the simulation results.

The GIS is being used in a water management research project currently being carried out. The aim of this project is the development of water inlet schemes in the Province of Noord-Brabant, using model simulations. For a test-area, a simulation map and a transpiration map based on remote sensing were composed as described earlier. In the southern part of the test-area, crops are transpiring well (see *Plate 6.2*). Also, in the surroundings of the indicated canal, crop water supply is sufficient. Especially around forested areas, soils are sensitive to drought (red colour). This is not surprising as forested areas were not brought under cultivation because of their sensitivity to drought. *Plate 6.2* shows that transpiration maps derived from remote sensing images provide relatively detailed information, while on the basis of a generalized soil map more global information is obtained.

Conclusions and discussion

Remote sensing offers the possibility to obtain detailed information on the regional distribution of crop transpiration. The first case study showed, however, that for the hydrological interpretation of transpiration maps based on remote sensing, additional information is indispensable. Thanks to geographical information systems, results obtained with an one-dimensional simulation model can be combined with digitized maps presenting soil physical and hydrological features. In this way simulation maps can be composed. A second case-study demonstrated the possibilities of GIS and remote sensing techniques. For each simulation unit the mean simulated transpiration value has been compared with the corresponding mean remote sensing value. Mean transpiration values per drainage class and crop type were calculated for the two methods. Evaluation of the comparison showed that, in general, for Typic Haplaquod soils the results of both methods are in agreement, whereas the simulation model overestimated transpiration for Plaggept soils. This is probably due to an overestimation of capillary rise in Plaggept soils. It was found that if the remote sensing results deviate systematically from the simulation map, the applied soil physical input parameters have to be adjusted.

With the development of information systems based on the integration of remote sensing methods and hydrological modelling, an important improvement of the hydrological description of an area can be achieved.

References

BARNSLEY, M.J. (1984). Integration of multispectral data obtained at different view angles for vegetation analysis. In *Proceedings Earsel/ESA Symposium on integrative approaches in remote sensing*. ESA SP-214, Guildford, UK

BELMANS, C., WESSELING, J.G. and FEDDES, R.A. (1983). Simulation model of the water balance of a cropped soil: SWATRE. *Journal of Hydrology*, **63**, 271-286. (Also published as: Technical Bulletin No 21. ICW, Wageningen, The Netherlands)

BLOEMEN, G.W. (1980). Calculation of hydraulic conductivity of soils from texture and organic matter content. *Zeitschrift fur Pflanzenernahnung und Bodenkunde*, **143**, 581-605. (Also published as: Technical Bulletin No 120. ICW, Wageningen, The Netherlands)

BROWN, K.W. and ROSENBERG, N.J. (1973). A resistance model to predict evapotranspiration and its application to a sugar beet field. *Agronomy Journal*, **65**, 341-347

DYER, A.J. (1967). The turbulent transport of heat and water vapour in an unstable atmosphere. *Quarterly Journal Royal Meteorological Society*, **93**, 501-508

FEDDES, R.A., KOWALIK, P.J. and ZARADNY, H. (1978). *Simulation of field water use and crop yield*. Simulation Monographs, Pudoc, Wageningen, The Netherlands

HATFIELD, J.L., REGINATO, R.J. and IDSO, S.B. (1984). Evaluation of canopy temperature-evapotranspiration models over various crops. *Agricultural and Forest Meteorology*, **32**, 41-53

JACKSON, R.D., REGINATO, R.J. and IDSO, S.B. (1977). Wheat canopy temperature: a practical tool for evaluating water requirements. *Water Resources Research*, **13**, 651-656

KRABBENBORG, A.J., POELMAN, J.N.B. and VAN ZUILEN, E.J. (1983). Standaard vochtkarakteristieken van zandgronden en veenkoloniale gronden, deel 1 en 2. Rapport 1680, Soil Survey Institute, Wageningen, The Netherlands (in dutch)

LAAT, P.J.M. de and AWATER, R.H.C.M. (1978). Groundwater flow and evapotranspiration; a simulation model. Part 1: Theory. Basisrapport Commissie Bestudering Waterhuishouding Gelderland, Provinciale Waterstaat van Gelderland, Arnhem, The Netherlands

NIEUWENHUIS, G.J.A., SMIDT, E.H. and THUNNISSEN, H.A.M. (1985). Estimation of regional evapotranspiration of arable crops from thermal infrared images. *International Journal of Remote Sensing*, **6**, 1319-1334. (Also published as: Technical Bulletin No 37. ICW, Wageningen, The Netherlands

Projectteam Remote Sensing Studie Project Oost-Gelderland. (1985). Onderzoek naar de operationele toepassing van remote sensing technieken in de landbouw en natuurbeheer. ICW Rapport 17, Wageningen (in dutch)

SEGUIN, B. and ITIER, B. (1983). Using midday surface temperature to estimate daily evapotranspiration from satellite thermal IR data. *International Journal of Remote Sensing*, **4**, 371-383

SOER, G.J.R. (1977). The TERGRA model - a mathematical model for the simulation of the daily behaviour of crop surface temperature and actual evapotranspiration. Nota 1014. ICW, Wageningen, The Netherlands

SOER, G.J.R. (1980). Estimation of regional evapotranspiration and soil moisture conditions using remotely sensed crop surface temperatures. *Remote Sensing of the Environment*, **9**, 27-45. (Also published as: Technical Bulletin No 116. ICW, Wageningen, The Netherlands)

STONE, L.R. and HORTON, M.L. (1974). Estimating evapotranspiration using canopy temperatures: field evaluation. *Agronomy Journal*, **66**, 450-454

THUNNISSEN, H.A.M. (1984). Eenvoudige methode voor de bepaling van de regionale dagverdamping van een gewas met remote sensing. Note 1580 ICW, Wageningen, The Netherlands (in dutch)

THUNNISSEN, H.A.M. and NIEUWENHUIS, G.J.A. (1989). An application of remote sensing and soil water balance simulation models to determine the effect of groundwater extraction on crop evapotranspiration. *Agricultural Water Management*, **15**, 315-332

TUCKER, C.J. (1977). Use of near infrared/red radiance ratios for estimating vegetation biomass and physiological status. In *Proceedings 11th International Symposium of Remote Sensing of Environment*, **1**, 493-494

WEBB, E.K. (1970). Profile relationships: the log-linear range, and extension to strong stability. *Quarterly Journal Royal Meteorological Society*, **96**, 67-90

III.

LAND CLASSIFICATION AND CROP INVENTORIES

7
THEORETICAL PROBLEMS IN IMAGE CLASSIFICATION

P.M. MATHER

*Department of Geography, University of Nottingham,
Nottingham, NG7 2RD, U.K.*

Introduction

A remotely-sensed image is a discrete representation of the spatial variation in the magnitude of reflected or emitted energy over an area. The representation is discrete in three senses. First of all, the area is imaged in terms of individual ground units or pixels, the size of which is dependent on the instantaneous field of view of the detector system used. For example, images produced by airborne sensors may have a pixel size of 1 m, while images from the AVHRR sensor carried by the NOAA series of meteorological satellites have a pixel size of 1 km x 1 km at nadir. Secondly, the magnitude of the reflected or emitted energy is expressed on an integer scale, usually 0-255 or 0-1023. Thirdly, each image is sensed in each of a set of specific wavebands.

The magnitude of the reflected or emitted energy measured in each waveband for a single pixel is considered to be related to the characteristics of the material forming the surface cover over the ground area corresponding to that pixel. The purpose of classification is to establish rules whereby patterns of spectral reflection and emission in multiple wavebands can be identified in terms of land-cover types, such as particular crops or other kinds of vegetation.

In reality, there are always complicating factors and this area of study is no exception. Complications occur due to the effects of interactions between electromagnetic energy and the components of the atmosphere, due to the effects of the geometry of the imaging system, particularly when compared with topography, and due to the assumptions underlying the statistical techniques employed in the classification process. Some of these problems are considered in detail in other contributions to this conference. This paper is concerned with the use of classification procedures or decision rules applied to remotely-sensed data with the aim of deriving information concerning the nature and geographical distribution of agricultural crops. There is neither time nor space to provide an exhaustive review of the topic; an introduction is provided by Mather (1987a). For a more extensive treatment, the report by Drake, Settle, Hardy and Townshend (1987) is recommended.

Image classification: The per-pixel approach

Instruments carried by remote sensing satellites and aircraft detect and record electromagnetic radiation emanating from the ground surface. Ground-leaving radiance is ostensibly measured for unit areas called pixels, the dimensions of which are usually related to the field of view of the instrument and the sampling rate of the analogue-to-digital converter used to translate the signal received at the detector into a count on an integer scale (usually 0-255 or 0-1023). Measurements are taken in a number of wavebands; for example, the Landsat Multispectral Scanner (MSS) provides data in four wavebands between 400 and 1100 nm for pixels measuring 79 x 57 m; Landsat!Thematic Mapper (TM) has seven wavebands (from 480 nm to 12.5 μm) and a pixel size of 30 x 30 m, while the French SPOT satellite carries a High Resolution Visible (HRV) sensor which in multi-spectral mode has three wavebands (500 - 890 nm) and a pixel size of 20 x 20 m. In the visible and near-infrared regions of the spectrum the material at the ground surface interacts with and modifies the spectral distribution of the incident radiation, so that it is in theory possible to identify the ground surface cover type by studying the distribution of spectral radiance values recorded at the sensor. This is the principle upon which the interpretation of all remotely-sensed images, whether analogue or digital, is based.

The classification of digital multispectral remotely-sensed images begins with the assumption that variations in the multivariate pattern of radiances (or, more precisely, quantised counts) over an area are directly related to the nature of the ground surface cover of that area. It is further assumed that similar cover types have similar spectral reflectance properties, so that statistical characteristics of an assemblage of image pixels, thought to represent a particular cover type, can be used to define a decision rule which is capable of discriminating between that cover type and all others. The most commonly-used decision rule is based on the further assumption that the quantised radiances in the k wavebands for a particular cover type represent independent observations and that the population from which they are drawn follows a multivariate-normal distribution. This is the Maximum Likelihood rule, which is in widespread used and will be used here to demonstrate the principles involved.

On the basis of prior knowledge, derived from experience, field observation or the study of maps and aerial photographs, the investigator decides upon a number of crop vegetation types that are present in the area to be classified. Call this number m. The first step is to identify areas on the image that are representative of each type and to extract the multi-band pixel values for each such area. These m datasets form training samples, and there is one training sample per class. The Maximum Likelihood classifier requires that for each class the multivariate sample mean vector and inter-band variance-covariance matrix are calculated. The sample mean vector is denoted as $\bar{x}$ and the inter-band variance-covariance matrix as S.

Before proceeding further it is prudent to ask whether the m classes can be distinguished on the basis of these statistics. Usually this is achieved by a non-rigorous procedure which is valid only if the training data sets for each of the m classes can be assumed to be samples drawn from underlying multivariate-normal distributions. The statistic used is called the Divergence, J. It is calculated for classes a and b from:

$$J(a,b) = 0.5\text{tr}\{(\mathbf{S_a} - \mathbf{S_b})(\mathbf{S_b^{-1}} - \mathbf{S_a^{-1}})\} + 0.5\text{tr}\{(\mathbf{S_a^{-1}} + \mathbf{S_b^{-1}})(\bar{x}_a - \bar{x}_b)(\bar{x}_a - \bar{x}_b)'\}, \quad (7.1)$$

where tr refers to the "trace" of the matrix, i.e. the sum of its diagonal elements.

The result of this calculation is generally transformed to give the Transformed Divergence, J_T:

$$J_T(a, b) = 100(1 - \exp(-(J(a, b)/8))). \tag{7.2}$$

which has the effect of mapping the range of values onto the scale 0-100, though other constants (for example 2000) are used. If the 0-100 scale is employed then a value of J_T less than 75-80 implies that classes a and b are not statistically separable. Hence, it is important to distinguish between spectral classes (which are statistically separable) and target classes, in this instance representing agricultural crops. One spectral class may well represent two or more non-separable target classes, the combination of target classes resulting from a study of the results of the Transformed Divergence values.

Transformed Divergence values can be averaged for all m classes to give the Average Transformed Divergence. If this is done for all possible combinations of the k wavebands, then some reduction in the required Maximum Likelihood calculations can be achieved by selecting that subset of the k wavebands. It has generally been found, for example, that the number of Thematic Mapper wavebands can be reduced from seven to four without any significant reduction in separability as measured by the Average Transformed Divergence. Unfortunately, the number of possible combinations of $1, 2, \ldots, k$ wavebands is very large. For subsets of size c taken from the full set of k wavebands there are

$$\begin{pmatrix} c \\ \vdots \\ k \end{pmatrix} = \frac{k}{c!(k-c)!} \tag{7.3}$$

possible combinations. Thus, for $c=3$ and $k=7$ the number of combinations is 35, while for $c=4$ and $m=12$ the number rises to 495. However, a simple forward selection procedure can be used to locate a near-optimal solution. First, compute J_T for each of the k wavebands separately. Select the highest J_T value corresponding to waveband S_1. Now take combinations of band S_1 plus each of the (k - 1) remaining bands in turn, and again select the highest J_T which will correspond to band S_1 plus a second band S_2. Repeat this procedure for (S_1, S_2) plus each of the remaining wavebands until all k wavebands have been selected. The results will indicate the single best waveband (S_1), the best pair (S_1, S_2), triplet (S_1, S_2, S_3) and so on. The corresponding J_T values can be graphed and the plot inspected visually to determine the point at which there is little increase in separability for the extra waveband.

At this stage the Divergence analysis has indicated which of the m target classes must be combined to give r spectral classes, and the forward selection procedure has identified p of the k wavebands as the best discriminators. The Maximum Likelihood calculations proceed on the basis of this reduced number of classes and spectral bands as follows. If each of the r training samples can reasonably be considered to be drawn from a multivariate-normal distribution with mean $\bar{x}$ and variance-covariance matrix S, then the probability that a given pixel z is a member of class a is given by:

$$P(z) = 2\pi^{0.5p}|S_a|^{-0.5}\exp[-0.5(y' \, S_a^{-1} \, y)], \tag{7.4}$$

where $y = (z - \bar{x})$. Taking natural logarithms, eliminating constants and multiplying by -1 to make the expression tidier we get:

$$-2\ln(P(z)) = \ln|S_a| + y' \, S_a^{-1} \, y. \tag{7.5}$$

That is, the probability that pixel z is a member of class a is proportional to the logarithm of the determinant of the sample variance-covariance matrix for class a plus an expression known as the Mahalanobis Distance of z with reference to class a. Since the original expression of the probability has been multiplied by -1, then the class a with the largest probability of membership is that with the smallest value of $-2 \ln(P(z))$. The computations can be reduced further for a small number of wavebands (for example the four wavebands of Landsat MSS or the three wavebands of the SPOT HRV) by the use of hash-table methods as described by Mather (1987a). Each image pixel is processed by the method described above and is allocated to one of the r spectral classes, or if the largest probability of class membership falls below a specified threshold, then the pixel is labelled as "unknown". The ordinal identifiers ("labels") of the classes to which the pixels have been allocated are used to build an image which can be displayed in pseudo-colour by representing the label "1" by red, or some other suitable colour, the label "2" by green and so on.

Difficulties with the per-pixel approach

The "classical" approach experiences the following limitations:

1. The assumption that pixels are statistically independent is unrealistic; if geography has any laws, one of the first must be that things that are close together are more likely to be similar than things that are far apart. If contiguous groups of pixels are selected to form the training data sets, then the sample statistics derived from such datasets will be biased.

2. Even if independent pixels are selected to form the training data sets, classifier performance will be determined by the fact that sample statistics are used in calibration. Thus, for a heterogeneous class, considerably different decision boundaries may result if different sets of pixels are used to form the training data sets.

3. The relationship between the scale of the objects being classified and the size of the image pixel has not been taken into consideration.

4. Two of the key features used by the human visual recognition system, texture and context, are not used. Patterns are built up from below on a pixel by pixel basis on the assumption that each pixel is a separate entity. Texture, which is a measure of the local variation in pixel values, and context, the logical relationship between an object and its neighbours, are ignored.

5. Unless the area being imaged is completely uniform, variations in terrain will produce variations in reflected or emitted radiance depending on the geometry of the sun-sensor-target system. Thus, it is unrealistic to pretend that areas of similar surface cover will produce similar levels of reflected or emitted radiance at a given time.

6. Knowledge of external factors, such as the existence of field boundaries or of variations in soil type over the area are not taken into consideration.

Problems (1) to (3) relate to the nature of the training data sets. Ideally, each training data set should be sufficient to provide adequate sample estimates of the mean vector and the variance-convariance matrix of the corresponding target class. Often, though, a rectangular set of pixels is selected from the image as being representative of a particular class without regard to the fact that, due to the presence of positive spatial autocorrelation, the number of independent samples is less than the number of pixels in the training data set. It can be shown that interband covariances are underestimated (Campbell, 1981). Drake et al. (1987) avoid the problem by selecting sample pixels at random within a test area. Their results indicate that at least 100 pixels must be selected from each class, though the sample size is related to the number of wavebands, p. Classification accuracy is reduced if the sample size is insufficient, because of the effect of outliers or "rogue" pixels, which may be mixed pixels at a class boundary or within a field where, for instance, the crop is less well developed or the pixel radiance is affected by the presence of patches of bare soil. Mather (1987b) describes a method of reducing the effect of such rogue pixels by the use of robust estimators in the calculation of the vector $\bar{x}$ and the matrix S.

Problems (2) and (3) refer to the relationship between the ground dimensions of the pixel and the size of the objects making up the target. A homogeneous class will be one in which the objects are smaller than the pixel size. Thus, a row crop consisting of individual plants with soil visible between the plants and between the rows will appear homogeneous if the pixel size is, say, 5 m x 5 m whereas it will be heterogeneous if the pixel size is 5 cm x 5 cm. High-resolution imagery, such as that obtained from aircraft-borne scanners with a pixel size of 5 m, may therefore "see" a forest as a heterogeneous target because the individual trees are larger than the pixel size. At the same scale a field of wheat would present a homogeneous target. Training data samples drawn from heterogeneous classes must be larger than those derived from homogeneous targets because of their inherently greater variability, and there is likely to be more variability in the results obtained from a classification if the target classes are heterogeneous. Woodcock and Strahler (1987) present a discussion of the relationship between scale and heterogeneity.

The classical Maximum-Likelihood approach presented earlier assumes that each pixel to be classified is independent of its neighbours, which define the context of that pixel. Context can be thought of as the nature of the land cover surrounding a particular pixel. The human visual system uses context in the process of identifying an object, so that incongruous decisions are not made. The rationale underlying the incorporation of contextual data into the classification process is again the presence of positive spatial autocorrelation. The decision to label a particular pixel as wheat, for example, may well be influenced by the fact that the surrounding pixels are also labelled "wheat". If the neighbouring pixels were labelled "upland unimproved pasture" then the correctness of the decision to label the pixel as "wheat" may be called into question. The problem is, of course, how to incorporate contextual information. Since the classification process begins with a blank sheet, so to speak, the decision rule incorporating contextual information must be a recursive one, in the sense that an initial classification is made and the labels or names given to the pixels at iteration 1 are used as contextual information at iteration 2, and so on. Drake et al., (1987) describe various approaches to the problem, including their "contextual priors" algorithm which was developed from the work of Strahler (1980). These authors find that classification accuracy improves if contextual information is used, though the parameters of the relationships defining context are difficult to estimate and the computational cost is considerable. Nevertheless, they demonstrate the benefits of the use of a contextual classifier, claiming that "A remarkable improvement in the accuracy of all classes can be seen, particularly in the high variance classes ... " (Drake et al., 1987).

A second visual cue used by photo-interpreters is texture, which refers to the local variability in the reflectance values in the neighbourhood of the pixel to be classified. Texture can therefore be measured in each image waveband. A number of texture measures have been proposed, some based upon the image pixel values themselves and others on the Fourier transform of the pixel values in the local neighbourhood (Weszka, Dyer and Rosenfeld, 1976; Haralick, 1979). None of these measures has achieved any considerable increase in classification accuracy, except in specific cases, and at the moment there seems to be little point in including a texture measure feature in a classification exercise carried out for an agricultural area.

The Maximum Likelihood classifier uses class membership probabilities that are based on the assumption that the pixel values in a particular class follow a p-variate normal distribution, where p is the number of features (spectral bands, texture measures, etc.) used in the analysis. This assumption is difficult to satisfy, so investigators are often content to see that the distributions are uni-modal. However, because of differing terrain illumination conditions, caused primarily by variations in topography, a logically-uniform class (such as "wheat") may have a bi- or multi-modal distribution in feature space, for example if some of the wheat is growing on a horizontal surface, and other fields are located on the illuminated or shaded sides of a ridge or hill. There are consequently at least three modes in the frequency distribution of pixel values corresponding to the class "wheat". One way around the problem is to use three spectral classes - wheat, shaded wheat and illuminated wheat, but this is dodging the issue, for there is a continuum of possibilities. In a study of the classification of SPOT HRV data in a hilly area of North-West Wales, Jones, Settle and Wyatt (1988) used a first-order model of radiant transfer to adjust the pixel values in each band to take account of the illumination geometry, that is, the angular relationship between sun, sensor and target. A digital elevation model was used to determine slope and aspect in this procedure, which represents an attempt to resolve the problem rather than to avoid it.

The sixth problem identified here is also related to the use of external data, that is data other than that provided by the sensors themselves. Remotely-sensed data can provide spectral reflectance and emittance measurements at pixel locations in digital form. Nowadays an increasing amount of cartographic data are becoming available in a compatible format. Such data might consist of measurements of ground elevation at a series of regularly spaced points forming the digital elevation model discussed in the preceding paragraph. Such data can be used to correct for terrain effects on radiance. Other cartographic data include the boundaries of fields (though it is appreciated that field boundaries do not necessarily coincide with crop boundaries). Since the agricultural scientist is often interested in the identification of objects such as fields rather than in the individual pixels making up such fields, it is possible to use the field boundaries derived from digital maps to segment the image into a set of objects of interest and then to classify the objects, using the statistical properties of the pixels making up these objects. This approach was followed by Pedley (1986; 1987) in the classification of crops in the Bawtry area of South Yorkshire and by Csornai, Dalia, Farkasfalvy and Nadór, in Hungary (Chapter 10, these proceedings).

A more ambitious approach has been followed by a UK Alvey Project team (based at NUTIS, University of Reading, University of Sussex, the Institute of Terrestrial Ecology and Systems Designers Scientific Ltd). Their work is also based upon the concept of image segmentation rather than on building-up a set of regions from the individual pixels. A knowledge-based approach was used which took into account the size and shape of, and the topological and logical relationships among, the objects to be classified. Thus, if an object is described as "field" then the system assigns a field-specific label, such as wheat or barley to it. Field boundaries may be derived from digital maps or through the application to the image to be classified of edge-detection algorithms. This on-going work represents a new development in the classification of remotely-sensed images for agricultural purposes. It is described in a number of papers, including those by Mason, Corr, Cross, Hogg, Lawrence, Petrou and Tailor (1988) and Tailor, Corr, Casolini, Cross, Hogg, Lawrence, Mason, Petrou and Vango (1988).

Conclusions

Someone once remarked "the more I learn, the more I realise the extent of my ignorance". This statement is certainly true as far as the classification of agricultural crops from remotely-sensed imagery is concerned. In the 1970s a state of optimism prevailed; it was thought that repetitive multi-spectral digital imagery would provide sufficient information to allow the automatic allocation of pixels to their correct classes, and provide a stand-alone method of monitoring the Earth's vegetation cover. It was soon realised that this was a pipe-dream, and that the problem was far more complex than many early proponents of remote sensing had realised, or dared to state publicly. The difficulties include:

- the complexity of the algorithms used and their generally-unattainable requirements;
- the problem of relating the scale of the phenomenon being observed on the ground to the resolution of the imagery;
- the need for external data in the form of ground elevation and field boundary information.

These difficulties have been discussed above.

Two trends seem to be emerging; one relates to the perception of the objective of the classification process, which aims to identify objects on the ground such as fields. If knowledge of the boundaries of these objects is available, either from maps or from image processing operations, then rules can be applied based on simple logic using expert systems. The second trend is seen in the growing realisation that remotely-sensed data represent only a part of the relevant information available to an investigator. If other, conventional, data such as terrain elevation and field boundary data are integrated with the remotely-sensed data, then more acceptable results are likely to be achieved (e.g. see the discussion of work in China by Liu Guoxiang and Zheng Dawei, Chapter 8, and by Li Yuzhu, Chapter 9 of these proceedings). Both of these trends are discernible in the work of Mason et al., (1988), which shows how recent developments in knowledge-based systems can be combined with equivalent developments in the combination and analysis of spatial data within the context of a Geographical Information System.

Nevertheless, significant problems remain. The difficulty of obtaining remotely-sensed data for a particular date, due to cloud-cover problems, and the delays in receiving data after it has been collected are well known and cannot be resolved by improvements in computer processing methods. The weather-independence of data collected by active microwave instruments, such as that to be carried soon on board ERS-1, will eliminate one of these two difficulties. However, far less is known about the response of natural vegetation, crops and soil in the microwave region compared with the optical region of the spectrum.

References

CAMPBELL, G.C. (1981). Spatial correlation effects upon accuracy of supervised classification of land cover. *Photogrammetric Engineering and Remote Sensing*, **47**, 355-357

DRAKE, N.A., SETTLE, J.J., HARDY, J.R. and TOWNSHEND, J.R.G. (1987). The development of improved algorithms for image processing and classification. Final Report, Special Topic GST/02/129, Natural Environment Research Council, Swindon, UK

HARALICK, R.M. (1979). Statistical and structural approaches to texture. *Proceedings IEEE*, **67**, 786-804

JONES, A.R., SETTLE, J.J. and Wyatt, B.K. (1988). Use of digital terrain data in the interpretation of SPOT-1 HRV multispectral imagery. *International Journal of Remote Sensing*, **9**, 669-682

MASON, D.C., CORR, D.G., CROSS, A., HOGG, D.C., LAWRENCE, D.H., PETROU, M. and TAILOR, A.M. (1988). The use of digital map data in the segmentation and classification of remotely-sensed images. *International Journal of Geographical Information Systems*, **2**, 195-215

MATHER, P.M. (1987a). *Computer processing of remotely-sensed images: an introduction.* John Wiley and Son Ltd., Chichester, UK

MATHER, P.M. (1987b). Preprocessing of training data for multispectral image classification. In *Advances in digital image processing.* Proceedings 13th Annual Technical Conference of the Remote Sensing Society, Nottingham, UK

PEDLEY, M.I. (1986). Combined remotely sensed and map data as an aid to image interpretation and analysis. *International Journal of Remote Sensing*, **7**, 305-307

PEDLEY, M.I. (1987). Digital image classification of SPOT HRV data using a field based approach. In *Advances in digital image processing.* Proceedings 13th Annual Technical Conference of the Remote Sensing Society, Nottingham, UK

STRAHLER, A.M. (1980). The use of prior probabilities in maximum likelihood classification of remotely sensed data. *Remote Sensing of Environment*, **10**, 135-163

TAILOR, A.M., CORR, D.G., CASOLINI, P., CROSS, A., HOGG, D.C., LAWRENCE, D.H., MASON, D.C., PETROU, M. and VANGO, R.D. (1988). Development of a knowledge-based segmenter for remotely-sensed images. *Philosophical Transactions of the Royal Society, Series A*, **324**, 437-466. (Also in *Exploiting remotely-sensed imagery.* Ed. by Browning, et al., The Royal Society, London)

WEZKA, J., DYER, C.R. and ROSENFELD, A. (1976). A comparative study of texture measures for terrain classification. *IEEE Transactions. Systems, Man and Cybernetics*, **SMC-6**, 269-285

WOODCOCK, C.E. and STRAHLER, A.H. (1987). The factor of scale in remote sensing. *Remote Sensing of Environment*, **21**, 311-332

8

ESTIMATING PRODUCTION OF WINTER WHEAT BY REMOTE SENSING AND UNIFIED GROUND NETWORK. I. SYSTEM VERIFICATION

LIU GUOXIANG and ZHENG DAWEI

Beijing Municipal Academy of Agricultural and Forestry Sciences,[1]
Beijing, China

Introduction

This chapter describes the first successful project for estimating production of crops in a large area across provinces in China by remote sensing combined with a ground network. The accuracy reached 90 to 95% for an area of 2.6 million ha for 3 years. The flow chart of the program for estimating production is shown in this paper.

It is of special importance to China, a developing country with a population of 1.1 billion, to estimate cereal production accurately. After disintegration of the commune system in rural areas, the old administrative statistical system of cereal yield reporting has been questioned. Since 1982 the National Committee of Economy has supported a research project to estimate cereal output with the new technique of Remote Sensing. This project was first carried out in Beijing Province to estimate production of winter wheat, the most important crop in North China (Liu, Tian, Xiao, Zheng, Men, Yan and Lei, 1986). In 1984 it was extended to the Beijing-Tianjin-Hebei area. From 1985 the project was further extended to the whole of North China including 11 provinces. At the same time, a project for estimating rice production by remote sensing in South China has been executed, led by the Jiangshu Academy of Agriculture. In this paper the progress during the years 1983 to 1986 in Beijing-Tianjin-Hebei is introduced and further progress from 1986 is described in Dr. Li's paper in these proceedings. In 1983 and 1984 we studied the experiences and techniques of Area Sampling Frame Construction Using Satellite Imagery (ASFCUSI) (Wigton and Bormann, 1977), Large Area Crop Inventory Estimate (LACIE) (NASA, 1979), Agriculture and Resources Inventory Surveys Through Aerospace Remote Sensing (AgRISTARS) (NASA, 1981), and of some Chinese institutes (Zhang and Wang, 1982). The project was supported by the State Commission for the Economy. The main techniques and methods developed for estimating production of winter wheat include:

- making a stratified map by satellite imagery in order to establish a ground monitoring network;

---

[1]Xiao Shuzao, Men Xianyue (Institute of Meteorology, Tianjin) and Yan Yilin (Institute of Meteorology, Hebei) contributed to the writing and discussion.

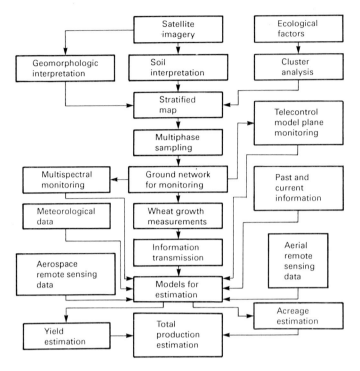

FIGURE 8.1. Flow chart of combined estimation of winter wheat by remote sensing and ground network

- a system for collecting, transferring and storing crop information;
- estimation of growing area using aerospace remote sensing of data and aerial infrared imagery as well as ground sampling;
- estimation of yield and total output by a model based on remote sensing and other data;
- dynamic monitoring of the growth and development of crops by remote sensing;
- techniques and methods of support research.

The flow chart of synthetic estimation is shown in *Figure 8.1*.

Results

Use of a stratified map by satellite imagery to establish ground monitoring network

The stratified map for area sampling frame utilizing satellite imagery was developed by Wigton and Bormann (1977) and has been introduced to many countries by FAO. The principle is to stratify an area using satellite imagery and take crop samples and assess

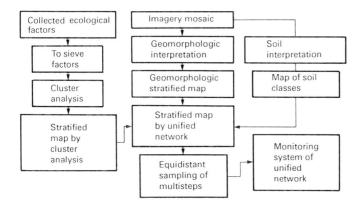

FIGURE 8.2. Flow chart of establishment of a unified network for estimating production by remote sensing

their relation to the statification by statistics. The technique was introduced by Dr. W.H. Wigton at a training class in 1981 in Beijing and was improved for the project described. The flow chart for estimating a unified sampling network is shown in *Figure 8.2*. The whole area was divided into 7 strata based on combined analysis of the stratified maps of soil, geomorphology and a clustering analysis of 5 factors which were selected from 15 ecological factors, including suitability of soil for growing wheat (weight 0.5), proportion of irrigated fields (weight 0.35), total rainfall from March to May (weight 0.1), average minimum temperature of January (weight 0.025) and mean temperature from September 1 to November 30 (weight 0.025). As a result the areas with similar characters of soil, geomorphology and ecological factors were incorporated into one stratum. Multi-temporal sampling of crops was carried out in each stratum on the map within 34 chosen sampling counties.

System for collecting, transferring and storing crop information

Crop information plays an important part in the estimation of output. It is the base for the interpretation of satellite imagery and data processing, and is the essential factor in the model of yield estimation.

Agronomic data

Based on the publication "Observation Methods of Agrometeorology" (National Bureau of Meteorology) and other sources, methods of estimation for agronomic elements for the unified network were formulated. Observations made included: sowing date, varieties, seeding quantity, fertilizer, number of seedlings and tillers before winter and in early spring, number of tillers with more than 3 leaves and of dead tillers, tiller or ear number and leaf area index at jointing and heading stages, assessment of filling state, grain weight and the final harvest yield. For storage of crop information, a data bank was established using an IBM-PC which includes all data from the ground network observations in 1983-

FIGURE 8.3. Stratified map of Beijing-Tianjin-Hebei Area for combined estimation of production of winter wheat by remote sensing and ground network

1984 and 1984-1985, vegetation indices based on AVHRR data from the NOAA satellite (Gatlin, Tucker and Schneider, 1981), data of sowing area and yield of different years, and the historical meteorological data. The crop information was transferred to the databank by telegraph in a unified format.

There are 3 levels of management of the ground monitoring system: province, county and segment. In total there are 102 segments in 34 sampling counties dispersed in 3 provinces and divided into 7 strata (*Table 8.1*). The necessary steps of segment management include engaging technicians, training them, guiding and inspecting them at work, examining and verifying the data and, finally, the establishment of the data bank. The flow chart of crop information is shown in *Figure 8.4*.

TABLE 8.1. Sampling error of growing area of winter wheat in different strata.

Strata Number	Growing area (ha)	Number of counties	Sampling counties	Yield in 1984 (kg ha^{-1})	Area error (%)
I	409200	11	5	4215	-1.5
II	164600	8	2	3570	1.8
III	522500	25	6	3173	-0.6
IV	362400	18	4	2640	-1.9
V	235100	16	3	3023	1.8
VI	559700	24	7	2415	0.7
VII	311500	16	4	1845	1.4

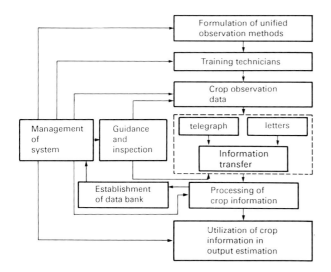

FIGURE 8.4. Flow chart of collection, transmission, processing, utilization and storage of crop information

Estimation of growing area of winter wheat

The growing area of winter wheat in Shunyi County, Beijing was calculated in 1983 using aerospace remote sensing data from a Landsat 3 image of the Beijing Area on March 16, 1981. It was found that the estimated area was 8.1% more than that published by the County Planning Committee. In order to distinguish wheat from the area of other crops, 20 segments were selected and the proportion of incorrectly classified area of other crops was determined (*Table 8.1*). Since 1985, with the guidance and help of the National Center of Meteosat, the growing area of winter wheat has been estimated utilizing data of TIROS-N imagery with an error about 5 to 10% (see part II of this paper by Dr. Li Yuzhu). Four methods were utilized for estimation of crop areas based on aerial remote sensing; i.e. graph paper, planimeter, Multicolor Data System Model 4200F and density slicing of infrared aerial photographs. The difference between the results analysed and those based on data of aerospace remote sensing was only 1%. The flow chart for the estimation of winter wheat is shown in *Figure 8.5*.

Independent estimates of the area of winter wheat were also obtained using methods without remote sensing. Based on the principles of sampling theory in statistics, homogeneously distributed blocks were selected on a topographic map and the proportion of winter wheat in each block was investigated. This is easy to carry out but needs much labour. Another method was to choose some family farms as samples and to determine the area of winter wheat in each, then multiply by the total number of farms to work out the growing area of sample segments. The results of both methods were similar.

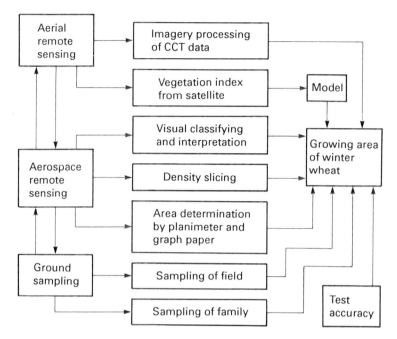

FIGURE 8.5. Flow chart of estimation of growing area of winter wheat

Estimation of yield and total output

The combined model for estimating production is established with the basic unit as a stratum in which conditions are similar. The yield model is mainly based on a sub-model of remote sensing combined with information from the ground sampling network. In the early stages of growth, when the method of remote sensing is not suitable, methods of agronomy, agrometeorology and economics are also used to supplement the data. The flow chart of production estimation is shown in *Figure 8.6*.

With guidance and help of the National Center for Meteosat (NCM), models were established for the estimation of yield from remote sensing data for different strata and good results were obtained using Meteosat data (Li, 1990).

By visual interpretation of aerial infrared imagery, it was found that crop growth and vigour can be distinguished from false colour photographs. Bright red indicates overgrown seedlings, with clear lines of furrows and even texture. For vigorous seedlings photographs show red and regular patterns as well as even texture. Regular pattern and a slightly rough texture, dark red and green indicates normal seedlings. A field with weak seedlings is usually of yellow-red or blue-red in colour and has a texture like fog. A model airplane was also used to monitor growth and development for comparison with data of Meteosat from the sampling segments.

Several spectral indices of vegetation were related with crop parameters including: Normalised Difference Vegetation Index (**NDVI**), Agricultural Vegetation Index (**AVI**), Relative Vegetation Index (**RVI**), Perpendicular Vegetation Index (**PVI**) and synthetic Green Range (**GR** or **GRw**). The formulae giving the indices are:

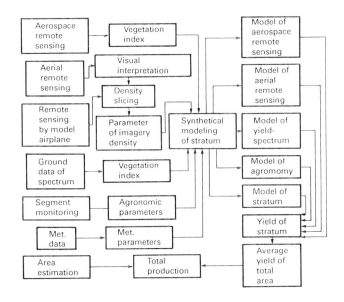

FIGURE 8.6. Flow chart of estimation of total production

- NDVI = $(MSS_7 - MSS_5)/(MSS_7 + MSS_5)$
- AVI = $(MSS_7 - MSS_5)$
- RVI = (MSS_7 / MSS_5)
- PVI = $0.939 MSS_7 - 0.344 MSS_5 + 0.09$
- GR = $-0.283 MSS_4 - 0.660 MSS_5 + 0.577 MSS_6 + 0.388 MSS_7$
- GRw = $-0.020 MSS_4 - 0.667 MSS_5 + 0.523 MSS_6 + 0.488 MSS_7$

where MSS_4 to MSS_7 are the albedos in different bands of the crop spectrum detected by the Landsat multi-spectral scanner, i.e. 500-600, 600-700, 700-800 and 800-1100 nm, respectively. Models of the relations between crop parameters and the vegetation indices in different stages of crop growth were developed as shown in *Table 8.2*.

Agronomic models of crop parameters for segments in each stratum were also developed for estimation of yield, as seen in *Table 8.2*. In the table Y is yield in kg ha^{-1}, LAI is leaf area index, DM is dry matter in kg ha^{-1}, BM is fresh weight in kg ha^{-1}, SJ is the number of stems in million ha^{-1}, H is the number of heads (in the same unit as SJ). The models presented in *Table 8.2* are those suitable for estimation of crop yields; others need to be improved.

Meteorological conditions are the main factors affecting yield fluctuation. In general the relation between crop yield and climate can be expressed by an equation of the form:

$$Y = Y_t + Y_m \tag{8.1}$$

TABLE 8.2. Models of relations between yield spectral indices and other factors.

Stage	Models	r	p
reviving	Y = -216.03 + 2016.58 LAI	0.85	0.05
	Y = 582.66 + 1.22 DM	0.93	0.05
jointing	Y = 1446.75 + 0.02386 SJ	0.68	0.05
	Y = 2032.01 + 3.97 DM	0.80	0.05
heading	Y = 1759.75 + 1.20 BM	0.86	0.05
	Y = 1963.94 + 1162.92 LAI	0.95	0.05
	Y = 941.79 + 0.06 H	0.93	0.05
	Y = -27.44 + 10.62 RVI	0.91	0.01
	Y = -1351.94 + 167.54 NDVI	0.91	0.05

where Y is the yield in kg ha^{-1}, Y_t is the trend yield, which can be worked out by multiple regression, and Y_m is called climatic yield,

$$Y_m = f(X_1, X_2, \ldots X_i, \ldots X_n),$$

where $X_1, X_2, \ldots X_i, \ldots X_n$ are different climatic factors affecting the yield. The form of the function can be worked out by successive regression and other methods. Some important models are listed in *Table 8.3* for the strata one to seven. For each class, the symbols in the relation represent the quantities listed below.

Dynamic monitoring of crop growth and development

Crop parameters reflecting the responses of growth and development are the key to agronomic models and the base for estimating production by remote sensing. For dynamic monitoring of crop growth and development, a portable radiometer was designed which can be operated conveniently using a combined mini-recorder and printer. On clear days at the crop stages of reviving, jointing and heading, respectively, measurements of crop spectral reflectance were carried out in the field at times from 9 a.m. to 4 p.m, local time. Measurements of growth include stem and head number, BM, DM, and LAI. Based on analysis of the relationship between spectral data and crop parameters, vegetation indices were found to be closely related with yield and crop growth (*Table 8.4*) Some correlations may be improved by non-linear regression; e.g. at the heading stage:

LAI = 4.7808 × 3748.37^{AVI} + 1with r = 0.792 and p = 0.01,

where the correlation coefficient is much higher than that of LAI in *Table 8.4*. It was also found that there are good relations between harvest yield (Y, kg ha^{-1}) and vegetation index at the heading stage:

Y = -27.4369+10.6206^{RVI}, with r = 0.914 and p = 0.01;
Y = -1351.94+167.54^{NDVI}, with r = 0.909 and p = 0.01.

At the 'reviving' stage there are good linear relations between LAI or DM and six vegetation indices, and all correlation coefficients are more than 0.7 and significant. At the jointing stage there are 3 vegetation indices significantly related with stem numbers or dry matter. But at the heading stage there are only 2 vegetation indices significantly

TABLE 8.3. Meteorological models.

Class	Relation
I	$Y = 1095.8 - 1854X_1 - 11.475X_2 - 12.015X_3 + 89.25X_4$
II	$Y = -289.5 + 0.69X_1 - 1.05X_2 + 119.355X_3 + 34.69X_4$
III	$Y = 2514 + 268.95X_1 + 36.78X_2 - 32.37X_3 + 18.24X_4$
IV	Y value is the same as for I
V	$Y = -1494.03 - 2.925X_2 + 13.569X_4 + 5.61X_5$
VI	Y value is the same as for V
VII	Y value is the same as for III

where I: X_1 - max depth of frozen soil in cm;
X_2 - sunshine from Oct.1-10, in hr;
X_3 - mean temperature from March 21-31, in °C;
X_4 - mean minimum temperature from Oct. 21-31, in °C;

II: X_1 - sunshine from April 11 to May 10, in hr;
X_2 - accumulated temperature from Dec. 21 to Jan. 20, in °C d;
X_3 - accumulated temperature from Jan. 21 to Feb. 10, in °C d;
X_4 - sunshine from Feb. 1-20, in hr;

III: X_1 - mean January temperature, in °C;
X_2 - mean November temperature, in °C;
X_3 - max depth of frozen soil, in cm;
X_4 - precipitation in November, in mm;

IV : X_2 - accumulated temperature below 0 °C, in °C d;
X_4 - mean temperature for Oct. and Nov. in °C;
X_5 - April rainfall, in mm;

related with BM, DM, LAI or Head number. In contrast, the relations between the albedos of single spectral bands as vegetation indices and crop parameters are usually not good. The correlation coefficients are less than those given or not significant.

Support research techniques and methods

With guidance and help of Prof. Lei Wanqun, it was found that there are good linear relationships between crop parameters or yield and the modified ratio or difference of spectral density based on imagery obtained by a model aeroplane from a height of 200 to 500 m. Infrared film and optical and geometrical processing were needed. A model aeroplane with radio remote control is very adaptable and of low cost, and therefore it is potentially very useful. The main problem is that the height and attitude are still difficult to control and measure accurately. Also, the geometrical processing of the image and the calculation of scale is not easy.

TABLE 8.4. Correlation coefficients between (a) yield or (b) crop parameter and vegetation indices at heading stage. (*: highly significant at p=0.01, others are not significant).

(a)
Vegetation indices	NDVI	RVI	PVI	GR	GRw
Yield (kg ha^{-1})	0.9100*	0.9250*	0.7566*	0.7400*	0.8366*

(b)
Crop parameter		BM	DM	LAI	H
Vegetation indices	AVI	0.2810	0.110	0.417	0.230
	NDVI	0.839*	0.610*	0.640*	0.680*
	RVI	0.81*	0.72*	0.62*	0.69*

Conclusion

The traditional statistical system of production estimation needs a large input of labour, and cannot obtain yield information promptly. In contrast, the main advantage of remote sensing is that all areas can be monitored dynamically, but the relation between crop yield and imagery is not simple. There are clear relations between LAI or DM and vegetation indices. However, conditions after heading have important effects on final yield particularly on grain weight. Besides, the National Economics Committee, (NEC), has called for relatively accurate estimation during the project, not after the project. Our strategy is to combine methods of remote sensing with a ground monitoring network. The progress obtained showed that the strategy is feasible. The ground network still needs much labour, but less than the traditional one, while the accuracy is higher and information is earlier.

It is the world trend to estimate and monitor crop growth by remote sensing. The resolution of Landsat is quite high but the cost of imagery data is too high and dynamic monitoring is difficult. During this project we tried to use data from Meteosat. The resolution is not high enough for small fields but is satisfactory for a large area. The cost is very low and dynamic monitoring is possible. The detailed application of the system is introduced in Dr. Li's paper.

Remote sensing using Meteosat reduces the number of segments studied, but the ground network is still necessary both for obtaining crop information and for testing the yield of samples. It is possible to use the Meteosat data for this application because in North China during most of the period of growth of winter wheat, it is almost the only over-wintering crop in the field. Consequently, winter wheat is easy to distinguish from other crops. There will be more difficulties in estimating production of other crops such as maize or rice.

Acknowledgement

The State Commission for the Economy is the main financial supporter of the project. Prof. Li Lianjie, Prof. Pei Xinde, Prof. Mei Nan, Prof. Lin Pei, Prof. Liu Zhonghe and Prof. Lei Wanqun have guided the research work. The National Center of Meteosat has guided the analysis of image data. Before the end of this project, research work was gradually combined with the further project of estimating production of winter wheat in 11 provinces of the whole of North China which is led by Dr. Li Yuzhu and Dr. Liu Guoxiang. We are also grateful to all people who contributed to this project during 1984-1986.

References

GATLIN, J.A., TUCKER, C.J. and SCHNEIDER, S.R. (1981). Use of NOAA-6 AVHRR Channel One and Two for monitoring vegetation. In Proceedings IGARSS 1981. IEEE, New York

LI YUZHU (1990). Synthetical Estimation of Winter Wheat in North China by Remote Sensing, (in press)

LIU GUOXIANG, TIAN FUSEN, XIAO SHUZHAO, ZHENG DAWEI, MEN XIANYUE, YAN YILIN, and LEI WANGUN (1986). *Research and Experiments of Production Estimation of Winter Wheat by Remote Sensing in Beijing-Tianjin-Hebei Area.* Beijing Press of Science and Technology, Beijing

NASA (1979). *Proceedings of the LACIE Symposium.* JSC-16015, NASA, Lyndon B. Johnson Space Center, Houston, Texas

NASA (1981). *Agriculture and Resources Inventory Surveys through Aerospace Remote Sensing.* Lyndon B. Johnson Space Center, Houston, Texas

WIGTON, W.H. and BORMANN, P. (1977). *A Guide to Area Sampling Frame Construction Utilizing Satellite Imagery.* UNOSAD, New York

ZHANG HONGMIN and WANG JIASEN (1982). Studies on spectral characters and model of output estimation of wheat. *Acta of Beijing University of Agriculture,* **8**, 8

9
ESTIMATING PRODUCTION OF WINTER WHEAT BY REMOTE SENSING AND UNIFIED GROUND NETWORK. II. NATIONWIDE ESTIMATION OF WHEAT YIELDS

LI YUZHU

Academy of Meteorological Science, Beijing 100081, China

Introduction

Large-scale crop estimation and vegetation monitoring has been one of the major applications of remote sensing techniques in agriculture, because it has counteracted the weaknesses of conventional methods of yield estimation which are generally short-period, small scale and need considerable resources. Moreover, artificial influences can occur in the conventional estimations of crop yields made by the agricultural and statistical administrations. As a result, the remote sensing of crop estimation has been studied for use in China. The development of the remote-sensing technique, especially the development of satellite remote-sensing since the 1970's, has provided a new scientific means for crop assessment. With the advantages of providing valuable information regarding the growing condition of crops, regular and continued large area coverage and daily, near real time, data acquisition, satellite remote-sensing has made possible assessment of the crop output in large areas.

The Large Area Crop Inventory Experiment (LACIE) which started operation in the middle seventies and its follow-up, Agriculture and Resources Inventory Surveys Through Aerospace Remote Sensing (AgRISTARS), have created the precedent in this field (Mergerson, Ozga, Holko, Miller, Winings, Cook and Hanuschak, 1982; NASA, 1978). In China, the research of crop estimation began at the beginning of the eighties and since then the research scope has been extending progressively. Most data used in current Chinese work described here are from the Landsat Multi-Spectral Scanner (MSS) images.

Initiated in 1985, the programme of "Research on Nationwide Synthetic Estimation of Winter Wheat Yield by Remote Sensing (NSEWWYRS)", which is supported by the State Commission for the Economy, sets the precedent in China of nationwide crop estimation by remote sensing. The programme covers over 90% of the winter wheat growing area in China, including Beijing, Tianjin, Hebei, Henan, Shandong, Shanxi, Shaanxi, Jiangsu, Anhui, Gansu and Xinjiang Provinces, totalling about 21 million hectares.

The NOAA satellite data offer the advantages of large area coverage, daily observations, more cloud-free images and comparatively lower cost of data processing (Schneider and McGinnis, 1982; Yates, Tarpley, Schneider, McGinnis and Scofield, 1984; Zhou, Xiao, Chen, Xhang and Dian, 1985) compared to other sources. The State Meteorological Administration (SMA) of China has the necessary ground receiving equipment, large data processing capability and advanced communication facilities for the operational work. By combined consideration of these advantages, we chose to use the data from the Advanced Very High Resolution Radiometer (AVHRR).

Data and processing

Satellite Data

The satellite data used in this research is mainly obtained from the Advanced Very High Resolution Radiometer and, secondly, from Landsat MSS and TM images. At the begining of the eighties, a series of studies have proved that the improved AVHRR of the third generation of the Operational Polar-orbiting Meteorological Satellite can be used effectively in vegetation monitoring (Hock, 1984; Malingreau, 1986; Schneider and McGinnis, 1982). Since the spectral sensitivities of the AVHRR channels 1 and 2 are similar to the Landsat MSS bands 5 and 7, respectively, we can catch most information provided by the MSS 5 and 7 through the multispectral synthetic analysis of the data from the AVHRR channels 1 and 2. Various vegetation indices of linear and non-linear composition of visible and near-infrared reflectance values, based on the crop spectral reflectance characteristics, can be used for assessing the crop growth conditions and yield (Perry and Lautenschlager, 1984; Tucker and Holben, 1981; Tucker, Gatlin and Schneider, 1984; Yates *et al.* 1984). We use mainly ratios and normalized differences of reflectance values as the vegetation indices. The data collected by AVHRR Channel 1 (0.58-0.68 μm) is used as the visible radiance; the data collected by Channel 2 (0.72-1.10 μm) as the near-infrared radiance. The two vegetation indices, **G_1** and **G_3** are calculated based on the following equations:

$$\mathbf{G_1} = (CH2/CH1) \times 10 \qquad (9.1)$$

$$\mathbf{G_3} = ((CH2 - CH1)/(CH2 + CH1)) \times 20 \qquad (9.2)$$

The **G_1** and **G_3** are also termed the greenness indices. The Satellite Meteorological Center (SMC) of the State Meteorological Administration (SMA) of China receives AVHRR data from the NOAA satellites and processes them into greenness digital images with 2 km resolution of the latitude-longitude grid (Chen Weiyung, in press). The digital images are calculated statistically, based on county areas, on a Bull DPS 7/717 Computer. The digital results are provided on computer compatible tape for analysis and interpretation.

Ground monitoring data

The ground monitoring data provide both the base of remote sensing synthetic estimation of winter wheat yield and important ground information for remote sensing of crop identification and growing condition assessment. In our study, the ground data come from the Ground Monitoring Network of NSEWWYRS, which consists of more than 400 observation sites in 140 counties. The observations include agronomic information, such as phenological stage, height and density of crop, observation area, irrigation and fertilizer inputs, and spectro-radiometric measurements at some sites. The observational results are sent via post or through meteorological organisation cables to the Forecast Centre of the Institute of Agrometeorology.

Geographic location

Although the digital image data provided by the SMA have been pre-processed, some differences from the real geographic locations still exist. Therefore, when the data are used correction must be made based on the distinct ground targets on the images (e.g. water bodies such as lakes, reservoirs and large rivers, and cities and islands) according to the real latitude and longitude.

Sun angle effects

Previous studies have shown that the reflectance values of vegetation indices are related to the sun angle. For the time being the NOAA satellite data are used without the sun angle correction in this work. In order to decrease the analysis error caused by the sun angle, we use the NOAA images at about 2:30 p.m. local time, when the satellite is passing the Equator and the local sun time is close to noon. Moreover, the image data at this time are easy to contrast with data from weather observations and ground spectro-radiometric observations.

The screening of G1 and G3 values

The daily G_1 and G_3 values provided for a particular period are not corrected for the atmospheric effects. In order to eliminate atmospheric influences, the maximum values are used in the data screening (Malingreau, 1986). In normal cases of crop growth, the biomass of winter wheat between the turning green and heading stages increases with time, as do the G_1 and G_3 values. In cases when the values decrease in this period, a check of the cloud and haze effects must be made against the ground agronomic data and the synchronous weather observations. In such cases, the G_1 and G_3 values of such days must be discarded (*Figure 9.1*).

Selection of monitoring phase

The vegetation information from the satellite images indicate all green vegetation on the ground. In order to identify winter wheat and reduce the interference of the green information from other plants, we have to select the optimal monitoring phases according to the crop and plant calendars.

Studies of the phenophases of major plants and crops other than winter wheat in the eleven provinces have indicated that the green crop in the cultivated fields is mainly winter wheat in the period from the last ten days of November to the first twenty days of April of the following year. Furthermore, the winter and spring is in a key period of growth and development when the growth and vigour of winter wheat will affect its yield directly. Therefore, this is the best season for the assessment of winter wheat by satellite remote sensing. Within this period, the subphase from complete cover to jointing stage provides the best information for area evaluation. The area of forest and other plants must be deduced from the digital images.

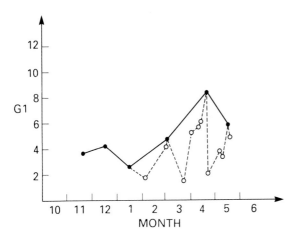

FIGURE 9.1. Diagram showing effects of procedure for data screening values of greenness index G_1 to reduce the atmospheric effects

Methods

Ground monitoring

Preliminary classification of the study area for winter wheat has been based upon meteorological satellite images of various phases and Landsat images, combined with cluster analysis of data on geomorphology, soil, production level and agrometeorological condition. (Liu, 1987; Liu and Zheng, these proceedings). The number of counties to be sampled was defined by the limit sampling error equation. The sampling county number in various regions was decided by the optimum allocation equation. The approach of multi-step sampling was used to determine the counties and observation sites in various regions to be sampled. By two phase sampling, more than 400 observation sites were selected from over 800 observation sites in 11 provinces and formed into a nationwide ground monitoring network. According to the observation procedure, the results were reported to the Forecast Centre of the Institute of Agrometeorology of SMA.

Winter wheat growth differs in various areas due to different geomorphologies, soil types, climates and cultivation conditions such as technical level, water and soil conditions, cultivation habits and varieties. This is inconvenient for the temporal and spatial analysis of the satellite data and causes misjudgement of the growth vigour of winter wheat. As a result, we set up a substratification for growth vigour monitoring of winter wheat according to the timing of the regional development of winter wheat. The results indicate that the growth vigour of winter wheat changed little in relation to G_1 values during periods of several days. Therefore, the average values for the major development periods of winter wheat (sowing, emergence, turning green, jointing, heading and maturity) in various counties are taken as a criteria for the region (with an error less than or equal to 5 days).

The G_1 profiles for winter wheat in various monitoring regions and sampled counties were then plotted. The combined analysis of G_1 values and the agronomic observation data (leaf area index, total stalk number and seedling height) indicated a high correlation. The years with high G_1 values have not only high growth vigour but also high grain output (*Figure 9.2*). The integrated G_1 value ($\sum G_1$) of the whole growth period or of the key growth period of winter wheat indicates that there exists a relationship between the integrated value and winter wheat output. In normal cases, the higher the integrated value the higher the grain output. An exception occurs in the cases of 'spindling' (the areas with 'spindling' winter wheat can be observed on the satellite images).

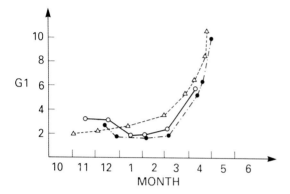

FIGURE 9.2. Time courses of G_1 values and wheat yields in Xinxiang County, Henan Province. The following key indicates years of the study, the final yields being shown against the corresponding lines in the figure in units of kg mu^{-1}, where 1 mu $\equiv$ 1/15 ha. Key: ——— 1985-1986; — — — 1986-1987; —·—·— 1987-1988. In these years the yields were 339.0 kg mu^{-1}, 343.5 kg mu^{-1} and 329.5 kg mu^{-1}, respectively

On the basis of dynamic monitoring, the estimates of yield can be made by comparison between similar years. In the year when a forecast is made, the G_1 profiles since emergence must be plotted and the interpretation of the satellite information can be based on the ground observation data in the sampled counties. Later, the growth assessment of winter wheat in the monitored area can be made. Finally, by comparison of the G_1 value with that in previous years, the estimation of the growth tendency of winter wheat in the current year can be concluded according to the analysis of the similarities between the years.

Yield estimation

The yield of winter wheat is estimated from the statistics of the remote sensing data so as to avoid the weaknesses of low resolution and large pixel area of the NOAA images. In the nationwide analysis, a county is taken as a sample unit. The analysis reveals a good correlation of yield with vegetation indices in the jointing-heading period, which is taken as the most suitable period for the estimation of winter wheat yield. The analysis also shows that greenness correlates better to the total yield than to the yield per unit area.

Prediction models have been built based on the historical yield and greeness data.

The basic form is:
$$y = a + bG \qquad (9.3)$$

where y = yield (unit area or total), G = greenness (average or integrated) and a, b are coefficients with appropriate units.

As it is not possible to obtain the same greeness conditions (date, vegetation stage, etc) which were used for building prediction models, a greenness correction is needed for the predicted year. One of the corrections is made for accumulated temperature (Xiao, Zhou, Chen, Dian, Zhang, Xiao, Meng, Zhao and Zhang, 1986). The greenness value of the predicted year is corrected by increasing (or decreasing) the G value per 100 degrees of accumulated temperature in the jointing-heading period of winter wheat in the previous years, or by interpolating the greeness values from the G profile of the predicted year directly to obtain the G value of the same development stage of winter wheat.

Area evaluation

Area values are the elements required for the estimation of production. Unfortunately, no confident conclusion can be made from the area data reported by various administration offices. For solving this problem, we have to use the satellite data. However, high resolution images such as TM or SPOT images are too expensive for large area estimation. Moreover, they may not provide real-time data. Therefore, we try to use AVHRR images to solve this problem. The steps in area evaluation are:

1. Selection of the greenness model for area evaluation with AVHRR. It can be seen from the comparison of the dependence of the various vegetation indices (greenness) on chlorophyll content of winter wheat, that the G_3 index is changing smoothly in the period from the heading stage to the milk stage. It can be supposed that in this period the greenness values reflect mainly area rather than changes of crop chlorophyll content (Xiao et al., 1986), i.e. the sum of the greenness values on certain land is related to the growing area of winter wheat on the same piece of land. In the NOAA digital images, every pixel is expressed by a digital number. Within a pixel, winter wheat occupies a certain area and the bare soil occupies the others, and this difference is referred to as the wheat-soil ratio. Therefore, it is possible to draw area information from G_3.

2. Based on the stratifications, the sub-strata are divided into categories according to the historical changes of the proportion of winter wheat grown and growth system; i.e. sub-strata with approximately the same wheat-soil ratio values.

3. Calculation of wheat-soil ratio $R_{ws} = A_w/A_s$, where A_w = wheat area, and A_s = soil area. The wheat-soil ratio can be calculated either by density slicing or percentage sampling of TM images in a certain region or by statistical sampling (Ran, Tian, Yan and Cao, 1987).

4. A correlation model between wheat-soil ratio and G_3 is built, based on historical data, of the form;
$$R_{ws} = a + bG_3 \qquad (9.4)$$
a and b are appropriate constants, and G_3 is the corrected greeness value.

According to the corrected $\mathbf{G_3}$ values of the predicted year and soil areas, the winter wheat area of the sub-strata are evaluated. The winter wheat areas of 11 Provinces are determined by accumulating winter wheat areas of the substrata.

Prediction experiments

Based on the studies of methodology of prediction, a quasi-operational prediction experiment for synthetic winter wheat yield estimation by remote sensing was started in 1986. Subsequently, the yield tendency, production tendency and production prediction have been issued to the relevant decision makers in March, April and May every year. In addition, the assessments of growth of winter wheat based on the analysis of remote sensing information have been broadcast on radio and TV.

The NSEWWYRS operational system consists of four sub-systems for information collection, transmission, processing and service *Figure 9.3*. The first sub-system is responsible for information and data collection, including real-time and non-real-time remote sensing, ground monitoring (agronomic and agricultural, etc) and also information from other sources.

The second sub-system is responsible for information and data transmission through various media. For the purposes of data standardization, uniform communication codes and formats have been designed. The data processing is realized on computers (mainly micro-computers). A data base including various data sources has been built and relevant necessary software has been written.

The third sub-system for data processing in the NSEWWYRS system applies remote sensing and other (agronomic, meteorological and economical) data for the production of prediction models. A series of integrated models have been built based mainly on remote sensing models.

The fourth sub-system supplies services by various routes and transmission media. At present the services include written reports, radio and TV broadcasting.

Accuracy and testing

According to the technical scheme, the accuracy of the NSEWWYRS system must be at least 90%, with respect to both production and frequency. Namely the error of the predicted yield must be less than 10% of the actual yield and 9 out of ten times predictions must be 'accurate' (i.e. error is less than 10%). *Table 9.1* shows the test results of the prediction in the period from 1986 to 1988.

TABLE 9.1. The accuracy of the NSEWWYRS wheat yield predictor.

Year	Error (%)
1986	-3.7
1987	+4.5
1988	-2.3

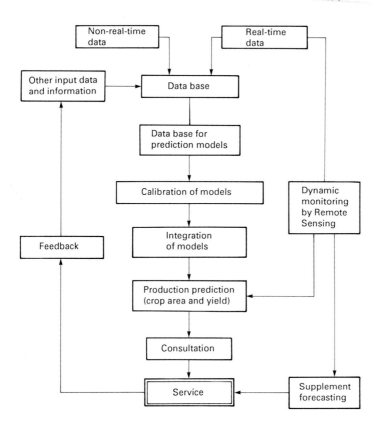

FIGURE 9.3. Operational flow diagram for the NSEWWYRS system for estimation of winter wheat yields

Discussion

Due to the limited period of the study, it is still necessary to improve our understanding of the mechanism of some aspects of the winter wheat yield estimation by remote sensing. Meanwhile, the quality of the meteorological satellite data should be improved, and it is expected that the prediction models will also be developed and optimized in the future.

References

CHEN WEIYUNG. Processing of NOAA's satellites data for crop estimation (in Chinese), *Journal of Meteorology.* in press

HOCK, J.C. (1984). Monitoring environmental resources through NOAA's polar orbiting satellites, *ITC Journal,* **4**, 263-268

LI YUZHU and QIAN SHUAN (1988). A preliminary study on dynamic monitoring winter wheat condition through NOAA's polar orbiting satellites (in Chinese), in press

LIU GUOXIANG (1987). Summation of the technology and methodology researches on remotely-sensed winter wheat estimation in the Beijing-Tianjin-Hebei District. In *Collected Works of The Technology and Methodology Researches on Remotely-Sensed Winter Wheat Estimation in the Beijing-Tianjin-Hebei District.* (in Chinese)

MERGERSON, J.M., OZGA, M., HOLKO, M., MILLER, C., WININGS, S.S., COOK, P. and HANUSCHAK, G. (1982). AgRISTARS DCLC Four State Project. In *Proceedings Machine Processing of Remotely Sensed Data Symposium.* USDA, Washington

MALINGREAU, J.P. (1986). Global vegetation dynamics satellite observations over Asia. *International Journal Remote Sensing*, **9**, 1121-1146

NASA (1978). Proceedings of the LACIE Symposium, Johnson Space Center, Houston, Texas.

PERRY, C.R.jr. and LAUTENSCHLAGER, L.F. (1984). Functional equivalence of spectral vegetation indices. *Remote Sensing of Environment*, **14**, 169-182

RAN ZHONGSHEN, TIAN FUSHENG, YAN YILING and CAO GUANGBIN (1987). Evaluation of winter wheat area through statistical sampling. In *Collected Works of The Technology and Methodology Researches on Remotely-Sensed Winter Wheat Estimation in the Beijing-Tianjin-Hebei District.* (in Chinese),

SCHNEIDER, S.R. and McGINNIS, D.F. (1982). The NOAA/AVHRR: A new satellite sensor for monitoring crop growth. In *Proceedings Machine Processing of Remotely Sensed Data.* USDA, Washington

TUCKER, C.J. and HOLBEN, B.N. (1981). Remote Sensing of Total Dry-Matter Accumulation in Winter Wheat. *Remote Sensing of Environment*, **11**, 171-189

TUCKER, C.J., GATLIN, J.A. and SCHNEIDER, S.R. (1984). Monitoring vegetation in the Nile delta with NOAA-6 and NOAA-7 AVHRR imagery. *Photogrammetric Engineering and Remote Sensing*, **1**, 53-61

XIAO QIANGUANG, ZHOU CISONG, CHEN WEIYING, DIAN CHONGGONG, ZHANG LIXIA, XIAO SHUZHAO, MENG XIANYUE, ZHAO ZHONGKAI and ZHANG GUIZONG (1986). Productivity estimate using meteorological satellite (in Chinese) *Remote Sensing of Environment*, **1**, 260-269

YATES, H.W., TARPLEY, J.D., SCHNEIDER, S.R., McGINNIS, D.F.and SCOFIELD, R.A. (1984). The role of meteorological satellites in agricultural remote sensing. *Remote Sensing of Environment*, **14**, 219-233

ZHOU CISONG, XIAO QIANGUANG, CHEN WEIYING, ZHANG LIXIA and DIAN CHONGGONG (1985). Application of AVHRR data on monitoring crop growth condition. *Research of Agronomy Modernisation* (in Chinese), **6**, 51-53

10

CROP INVENTORY STUDIES USING LANDSAT DATA ON A LARGE AREA IN HUNGARY

G. CSORNAI, O. DALIA, J. FARKASFALVY and G. NÁDOR

FÖMI Remote Sensing Centre,
H-1149 Budapest, Bosnyák tér 5. Hungary

Introduction

Agriculture is very important in the Hungarian economy. The major environmental factors such as soils, climate and terrain are suitable for intensive agriculture, almost thirty percent of the total land area is cropland, and the production is very high. Therefore, a comprehensive National Crop Information System (NCIS) is necessary. The one that has been operational for some three decades now, is based on reports from farms. The need for a more efficient, remotely sensed data based NCIS arose some years ago.

The average 60-80 hectare fields on the major crop growing regions provide a good opportunity to use even Landsat MSS data in crop monitoring. Since 1981 efforts have been made to develop the basic components of a satellite data based Crop Monitoring System (CMS). During this period a number of problems have been addressed, such as: crop mapping, inventory, crop development assessment and yield forecasting. The inventory segment of our satellite based CMS is closest to operational, though a lot of development is still needed.

A data processing system has been developed by our staff to support application studies. This system comprises sub-systems for image processing, reference data handling and geographic data modelling (Csornai, 1988). The system is now operational on a microVax compatible processor.

At first the potential of the methods and processing system was studied in different projects using, basically, per-point classification approaches to the mapping and inventory of the major crops in Hungary. Recently, a set of Geographical Information System (GIS) supported image analysis methods have been developed and tested for the same purposes.

Classification for crop mapping

Results of feasibility studies

The earlier studies were mainly intended to demonstrate the feasibility of using Landsat MSS data with the image processing methods we had developed. The set of image analysis methods and programs developed were tested on a land system (Csornai, Dalia, Gothar and Vámosi, 1983) for the first time. An ISODATA type clustering method was used - enhanced by built in intercluster distances (e.g. Jeffreys-Matusita, divergence, transformed divergence) together with look-up table maximum likelihood methods. Crop categories were estimated from the weighted mixtures of spectral components of the clusters. The accuracy assessment of the pixel based classification used ideal digital reference maps that did not account for the existence of the inhomogeneities in the field. Therefore, the term percent correct classification (PCC) is inadequate for use when so many inhomogeneities occur in the fields, because in most of the cases seemingly erroneous spots on the classification map may correspond to real field anomalies rather than spectral or data classification errors: these are rather classification errors of the user's classes or of the area estimates of them. Consequently, a clear distinction should be made when using the term PCC in this paper, keeping in mind that it does not qualify the spectral/data classification methods but rather the in-field heterogeneities and thus the raw area estimators.

The observed percent correct classification (PCC) values were in the range 85-95% for the major crops and a few of their development stage subclasses when single date Landsat MSS data were used. This held true for some farm areas and for a 10^5 hectare sample within the land system. A four date multi-temporal data set utilizing either raw data or multi-vegetation indices gave a 3-6 percent increase in classification accuracy. Demonstrating the potential of the Bayesian classifier for controlling the error distribution among the classes, the PCCs for single date Landsat MSS data increased in a similar way to the multi-temporal data set but using a maximum likelihood classifier.

A later (thorough) study (Csornai, Vámosi, Dalia and Gothar, 1983) clearly proved that the real performances of the classification procedures were even better when the existing in-field inhomogeneities were taken into account in the computation of a misclassification table. The 'pure' PCCs were in the range of 92-97% for the major crops: winter wheat, maize, sugar beet and sunflower.

All these investigations showed that Landsat MSS data and the processing methodology were adequate for crop mapping and inventory tasks within a land system. This success depends particularly on the large fields and the similarity in the weather and production patterns throughout large regions in Hungary.

Crop mapping and inventory on a complex county

Since 1982 Hajdu-Bihar County has been an area for testing different crop monitoring methods. It is located on the plain area of East Hungary (*Figure 10.1*). Of its 0.6 million hectare area, half is arable comprising a 10 percent sample of that of the whole country. Its soils are alluvial, varying from salt affected ones to very productive chernozems. With its intensive agricultural areas, scattered woods, and protected national parks, the area is complex and fairly representative of the whole crop growing area in Hungary, except for the average field size, which is bigger here by 10-20 percent than elsewhere in the

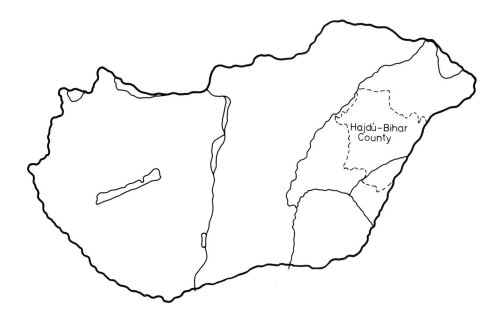

FIGURE 10.1. Hungary, indicating the 0.6 million hectare test area: Hajdu-Bihar County

country.

In a pilot project in 1987 (Csornai, Dalia, Farkasfalvy, Nádor and Vámosi, 1988), parallel to some new methods described below, a 'wall to wall' crop survey was done. Our aim was to compare the performances of satellite based methods with that of the traditional methods of crop mapping run by the top agricultural management.

Having chosen the constraints of low cost, complete and cloud free coverage of Landsat MSS data, two suitable dates of April 19 and September 2nd, 1984 were selected from our archive. None of the dates matched the calendar of all the major crops. Based on our experience it is believed that a single Landsat scene in the period June 1 - July 10 would result in the same or better mapping and area estimation accuracies.

Four strata were selected using soils maps and Landsat colour composites. The training and test areas (2-7 % of the total) were also chosen by stratification from a comprehensive digital reference data base.

The per-point approach was used as in the earlier studies (see previous subsection) for the four strata separately. The crop map appeared satisfactory, having good matches of categories along the strata boundaries. However, a detailed error analysis showed dependence of mis-classification values on stratum. The overall per-point PCCs were lower than previously, being in the 70-95 percent range. This was partly because of false ground data, due to human errors in the analysis of the large (not sampled), data set, and partly because of the unsuitable image dates. A direct derivation of area estimates for the county was compared to that of the Central Statistical Office of Hungary (CSOH), that is, the traditional data, and is shown in *Table 10.1.*

The use of integrated classes was necessary because of the incompatibility of Landsat

TABLE 10.1. Comparison of area estimates of Hungarian Central Statistical Office and Remote Sensing Centre using field based area sampling and bitemporal Landsat MSS data respectively.

Categories	Hung. Central Stat. Office (Ha)	FÖMI RSC (Ha)	Difference %
Winter Cereals	111124	114290	+2.8
Maize	117372	114218	-2.7
Alfalfa and pasture	168852	176860	+4.5
Total	297384	405368	+3.6

derived and CSOH categories. Despite the definition problems of classes such as 'maize' and 'maize for silage' the correspondence between CSOH and Landsat derived estimates is good. There are a number of problems still to be solved. However, the overall performance of Landsat MSS data analysis methods seems to be promising from the viewpoint of a National Crop Information System.

GIS supported image classification in crop inventories

For a NCIS the classification and the derived area estimation should be stable, that is, the estimation should have high accuracy and confidence values. A means to assure high performance is to integrate *a priori* information into the classification procedure. A more common way to enhance a thematic map is to use the available ancillary data and *a priori* information after the Landsat classification.

Newer approaches (e.g. Strahler, 1980) tried to incorporate the available *a priori* specific knowledge into the classification scheme. Others (e.g. Richards, Landgrebe and Swain, 1982; Swain, Richards and Lee, 1985) focused on establishing a conceptually unified basis for the potential support of a geographic information system, within the classification procedure. The use of GIS in the classification is a useful intermediate step before the application of dedicated expert systems.

As in Hungary the change rate of the position of the individual field boundaries is estimated as 2 to 5 % annually, classifiers making use of this fact seemed straightforward. For modelling purposes a digital field boundary data system (DFBD) was created for 80 farms out of 100 in Hajdu-Bihar County. The starting point was an updated set of farm field maps at scale 1:25000. This update was accomplished using topographic maps, farm field sketches and Landsat TM colour composites. The field maps were digitized, and geocode plus thematic data were coupled to the individual fields (see *Plate 10.1*). Thematic data comprised records of cultivar, fertilization used, yield, etc. from the period 1984-87, taken from existing data bases. Thus, a comprehensive digital field data base (DFBD) was set up within 4 man-months, for multipurpose utilization.

Classification methods

Prior to application the DFBD has to be updated. The first step of the update procedure is to apply a derivative operator to the actual satellite image. The Roberts edge detector has been selected based on its performance. A technique using fast Fourier transforms has also been studied for use in removing non-boundary contrast contributions. Until the complete automation of the whole check and update procedure, which is being developed, help from an interpreter is used via the display unit. The changes of DFBD tend to be of specific nature: splitting or merging parcels. The basic networks of roads and streams/rivers are steady features on the ground which aid location of field boundaries.

Once the DFBD is updated, either of the following two classification methods can be used. The first method (method A, e.g. Csornai et al., 1988) comprises a per-point classification plus a reclassification of the image using the DFBD. The pixels of a particular field are assigned to the most probable class. The reassignment is then based on the class membership statistics of the field pixels. If the variation is high, no reassignment takes place. The original version performed re-classification based on a two parameters majority decision. In the case of two farms of 2500 hectare each the average PCC increased from 72 to 90.7 % and from 93 to 99.6% respectively (Csornai, 1988) when Landsat MSS data were used.

The second method (method B, Farkasfalvy, 1987) is a real per-field method. Though good PCC arose from an ECHO type (Extraction and Classification of Homogeneous Objects) classifier (Kettig and Landgrebe, 1976; Csornai; Dalia et al., 1983) a more adequate way used agricultural fields instead of homogeneous segments as objects. Method B computes second order statistics for the individual fields defined by the DFBD. After a variance check for field homogeneity, those passing the check are clustered and classified per field similarly to ECHO. Inhomogeneous fields are classified per pixel.

Results

Landsat MSS results showed the definite advantage of the methods A and B over the per-point approach. Further indications were obtained using a very complex Landsat TM subscene to compare these methods on a common data base (Farkasfalvy, 1989). The subscene used was 512 by 512 pixel, from a July 8, 1987 scene over Hajdu-Bihar County in Hungary.

The underlying soil pattern has a big variance that is partially reflected in *Plate 10.2*, which shows the thematic output using method B in the classification. The extraordinary inhomogeneity of the area was particularly useful for a test. Many of the fields, that had markedly inhomogeneous vegetation cover, appeared on the classification map as homogeneous fields after using method B. Some others that did not meet the variance criteria, and therefore which method B left unprocessed, were classified per pixel. The corresponding PCCs (*Table 10.2*) clearly reveal the advantage of the new per-field classifiers. The values were computed from direct comparison of an ideal digital reference map (*Plate 10.1*) and the classification map; they therefore reflect both the real inhomogeneities of the fields plus the mis-classification. The increase of PCC in each class show the significantly better performance of the per-field classifiers (A, B) over the per-pixel one. The importance of a PCC is given by the number of pixels in the category (first column in *Table 10.2*). The low PCC value of method B for the soils can be traced back to the spectral and definition confusion between a maize subclass and bare soils: the fields where maize is grown for seed and for forage are bare at different times.

From more studies, it seems that the per-field method (B) is at least as good as method A, while both are significantly better than the per-pixel maximum likelihood or Bayesian classifiers.

TABLE 10.2. Comparison of the classification accuracies of the three methods on a small area. See text for explanation of methods

Categories	#pixels	Per-point class	Reclass. (A)	Per-field class. (B)
Wheat	41150	87%	91%	98%
Maize	37442	73%	78%	90%
Sugar-beet	12485	65%	83%	81%
Potato	4186	67%	83%	77%
Alfalfa	3687	65%	85%	96%
Soil	2026	87%	91%	45%
Water	3568	71%	100%	76%
Average PCC		80.1%	86.8%	91.6%

Conclusion

A remote sensing based National Crop Information System (NCIS) requires stable and high correctness and confidence values in crop identification and inventories. The comparison in Hungary of per-point and the new GIS supported classification based inventory studies showed a definite advantage of the latter. The benefits of increased correctness seem to exceed the cost of digital databases which can be used in many other problems. It has been observed that with the greater spectral and radiometric resolution of Landsat TM, the within-class variability of pixels is greater, causing a drop in classification accuracy compared to Landsat MSS. With the increased spectral capability of TM combined with these per-field methods, the overall performance may be remarkably better. These or similar methods can be used in different environments and countries. The methods need further improvement but seem promising in a national crop monitoring system. The general idea of integrating more *a priori* information from a GIS into the classification procedure, as in this work and that described for China elsewhere in these proceedings, is worth further use.

Acknowledgement

The results reported in this paper were achieved in a R and D project having been supported by the State Board for Technical Development, the Ministry of Agriculture and Food and the Hungarian Academy of Sciences. The authors are grateful to Dr. Judit Vámosi and Peter Zabó for their assistance in the countywide survey study.

References

CSORNAI, G. (1988). Agricultural land use mapping using geographical information system. In *Proceedings of the 5th Symposium ISSS Working Group of Remote Sensing*. Hungarian Society of Agricultural Sciences, Budapest, Hungary

CSORNAI, G., DALIA, O., FARKASFALVY, J., NÁDOR, G. and VÁMOSI, J. (1988). Regional vegetation assessment using Landsat data and digital image analysis. In *Proceedings* of the *5th Symposium International Society for Soil Sciences Working Group of Remote Sensing*. Hungarian Society of Agricultural Sciences, Budapest, Hungary

CSORNAI, G., DALIA, O., GOTHAR, A. and VÁMOSI, J. (1983). Classification method and automated result testing techniques for differentiating crop types. In *Proceedings of Symposium Machine Processing of Remotely Sensed Data*, Purdue University, West Lafayette, Indiana, USA

CSORNAI, G., VÁMOSI, J., DALIA, O. and GOTHAR, A. (1983). Vegetation status assessment and monitoring in agricultural areas by remote sensing. In *Proceedings of the XXXIVth International Astronautical Federation Congress*, Budapest, Hungary

FARKASFALVY, J. (1987). Comparison of the per-point and per-field classification methods in the processing of remotely sensed data. In *Proceedings of the 3rd Conference of Program Designers*, Eötvös Loránd University, Budapest, Hungary

FARKASFALVY, J. (1989). Comparison of different classification methods using Landsat TM data. In *Proceedings of the Conference of Artificial Intelligence*, Eötvös Loránd University, Visegrád, Hungary

KETTIG, R.L. and LANDGREBE, D.A. (1976). Classification of multispectral image data by extraction and classification of homogeneous objects. *IEEE Transactions Geoscience Electronics*, **GE-14**

RICHARDS, J.A., LANDGREBE, D.A. and SWAIN, P.H. (1982). A means for utilizing ancillary information in multispectral classification. *Remote Sensing of Environment*, **12**, 463-477

STRAHLER, A.H. (1980). The use of prior probabilities in maximum likelihood classification of remotely sensed data. *Remote Sensing of Environment*, **10**, 135-163

SWAIN P.H., RICHARDS, J.A. and LEE, T. (1985). Multisource data analysis in remote sensing and geographic information processing. In *Proceedings of Machine Processing of Remotely Sensed Data*, Purdue University, West Lafayette, Indiana, USA

IV.

PRODUCTIVITY

11
HIGH TEMPORAL FREQUENCY REMOTE SENSING OF PRIMARY PRODUCTION USING NOAA AVHRR

S.D. PRINCE

Geography Department, Room 1113, Lefrak Hall,
University of Maryland, College Park, Maryland 20742-8225, U.S.A.

Introduction

The potential of data from the NOAA spacecraft's Advanced Very High Resolution Radiometer (AVHRR) to monitor primary production in semi-arid vegetation is confirmed by an analysis of the results of 12 data sets from three Sahelian countries collected over a period of 8 years. Equations are presented to estimate production from multi-temporal sums of AVHRR vegetation indices, together with the associated errors.

The underlying remote sensing and primary production models are discussed and a physically based model is tested using a data set for Mali for which atmospherically corrected and sensor calibrated data are available. The possible interference of mixtures of surface brightnesses in the scene on satellite measurements of vegetation indices is discussed. The influence of variation in solar radiation on modelled primary production is shown to be much less than that of the vegetation index measurements. Application of the model to derive biological parameters such as the efficiency of conversion of solar radiation into plant material is demonstrated.

With the launch of the NOAA-6 polar-orbiting meteorological satellite in June 1979, which carried the first Advanced Very High Resolution Radiometer (AVHRR), a new era in the monitoring of primary production from space began. For the first time frequent measurements were available in the red and near-infrared regions of the spectrum so that, even after allowing for data losses due to cloud cover, observations could be obtained at intervals of 10-15 days (Prince, 1986). The first practical demonstrations of the potential of the AVHRR to monitor primary production were carried out in the semi-arid grassland savannas of Senegal (Tucker, Vanpraet, Boerwinkel and Gaston, 1983; Tucker, Vanpraet, Sharman and Van Ittersum, 1985) and Botswana (Prince and Tucker, 1986). The high temporal frequencies of measurements of the same site which are possible with the AVHRR are a consequence of its large field of view (1.1 km^2 resolution), however this advantage carries with it the problem of field data collection in such large areas. This problem seriously inhibits the experimental examination of the relationship between satellite measurements and production.

Because AVHRR data are available for the entire globe, there is great interest in the possibility of monitoring global primary production using these data. A number of studies (Goward, Tucker and Dye, 1985; Goward, Dye, Kerber and Kalb, 1987; Box, Holben and Kalb, 1989) have compared published primary production measurements for the world's major biomes with annual sequences of Global Vegetation Index data, a standard NOAA product derived from subsampled AVHRR data (NOAA, 1986). The primary production data were derived from many different field studies which used a variety of field techniques and sampling intensities and were generally not continued for more than one year. These production data have been applied to other areas, which were not themselves sampled, generally on the basis of a climate-based map of vegetation types or some other interpolation of production based on climatic variables. Clearly, the analysis can only relate mean annual production to the AVHRR data and cannot examine the ability of the satellite data to detect inter-annual differences, nor can the more subtle differences between vegetation types below the status of biomes be examined. At the continental and global scales it is unlikely that there can be much improvement in the quality of these production data for comparison with the satellite data. However, the quantity of samples from each biome could be increased and it is possible that the stratification of global vegetation could be improved.

Unresampled AVHRR data have a nominal resolution of 1.1 km for nadir views, but they can be used in conjunction with ground observations at a maximum resolution of between 10 and 100 km^2. When the data first became available, no adequate field production data existed at this scale and a new programme of data collection was established for this purpose in the Ferlo region of Senegal (Tucker *et al.*, 1983). The results of three years of monitoring in the Ferlo were reported by Tucker *et al.* (1985), and the good agreement obtained between satellite data and field observations encouraged a number of other programmes to be established. The results of these programmes enable the relationships between AVHRR data and field production, measured in specific seasons and in precisely defined vegetation types, to be tested for a considerably increased variety of conditions throughout the West African Sahel. They are analysed here, together with the data sets for Senegal between 1981 and 1983, giving, in all, 363 site records in 12 data sets from three countries covering eight years.

Theoretical analyses (Goudriaan, 1977; Kumar and Monteith, 1982; Asrar, Fuchs, Kanemasu and Hatfield, 1984; Sellers, 1985; Tucker and Sellers, 1986) and experimental studies on crops (Monteith and Elston, 1983; Tucker, Holben, Elgin and McMurtrey, 1980; 1981; Pinter, Jackson, Idso and Reginato, 1981; Walburg, Bauer, Daughtry and Housley, 1982; Hatfield, 1983) provide a basis to develop a model for satellite remote sensing of production founded on physical mechanisms. These analyses show that vegetation indices (VI's) may be expected to relate to the proportion of the incoming photosynthetically active radiation which is absorbed by the canopy (%APAR) at the time of measurement, and that these measurements should be used in conjunction with measurements of the actual global PAR incident on the vegetation during the same time period to determine the actual APAR. Temporal sums of APAR derived in this way are related to the gross primary production by means of a growth efficiency term, e with units of g MJ^{-1}. Following Steven, Biscoe and Jaggard, (1983),

$$\text{Gross production} = e\Sigma_t(a(\mathbf{NDVI}_t - \mathbf{r}_t)\mathbf{R}_{st}) \qquad (11.1)$$

Where:

- %APAR$_t$ = a(**NDVI**$_t$ − **r**$_t$).

- $NDVI_t$ is the mean Normalised Difference Vegetation Index measured at the surface of the vegetation during time interval t, and
  $NDVI = (IR - R)/(IR + R)$.

- r_t indicates the effect of the vegetation background on the $NDVI_t$.

- R_{st} is the global PAR incident during time interval t.

- a is a function relating **NDVI** to %APAR (%APAR = (**NDVI** × 1.055) - 0.0549) and is approximately equal to 1 (Goward and Dye, 1987).

The available evidence suggests that seasonal growth efficiency is relatively invariant (Monteith, 1977; Kumar and Monteith, 1982; Monteith and Elston, 1983; Asrar *et al.*, 1984; Cannell, Milne, Sheppard and Unsworth, 1987; Lindner, 1985; Palmer, 1988) although, in other cases (eg. Steven and Demetriades-Shah, 1987), variations due to disease, nitrogen nutrition and irrigation have been noted. Unfortunately, these efficiency values have mostly been derived in a way which is not comparable with e as defined here. Some report values for total solar radiation rather than PAR; some use incident radiation, others interception, and yet others measure or sometimes estimate reflection. The derivation of production values, in most cases, is equally inappropriate; some use above ground production, others total, most use net production but some calculate dark respiration, none incorporate photorespiration. Measurements of e on the basis of gross production and APAR are almost entirely lacking and would be of considerable interest. Until these are forthcoming, estimates of production based on the assumption of constant values of e are best regarded as estimates of the potential in the absence of any environmental constraint other than the incident solar radiation. An application of the model is illustrated here in which atmospherically corrected AVHRR data set for Mali in 1986 is analysed in conjunction with field production data to determine the value of e from satellite observations.

The analysis of temporal sums of VI's with respect to primary production in Senegal (Tucker *et al.*, 1985) in effect assumed e and R_s were constant and the satellite measurements of **NDVI** were directly proportional to the **NDVI** measured at the surface of the vegetation. Effects of soil and other non-vegetation components on the VI, represented by r_t in the comprehensive model, were not included.

$$\text{Gross production} = eR_s c \Sigma_t (NDVI_{t,satellite}) \times t_n. \tag{11.2}$$

Where:

- $NDVI_{t,satellite}$ is the **NDVI** calculated using AVHRR channels 1 and 2 without correction for atmospheric effects.

- c is a variable factor accounting for sensor calibration and for atmospheric effects on radiation transfer between the sun, earth and satellite.

- t_n is the number of days between each AVHRR **NDVI** composite. Thus, the slope of the relationship between summed vegetation index $\Sigma_t(NDVI_{t,satellite})$ and gross production was $eR_s c$ (Steven *et al.*, 1983).

Both the reduced and full models are considered below. First, the reduced model using uncorrected satellite data is applied to all the data, then the full model is applied using atmospherically corrected satellite measurements to calculate the **NDVI** for one data set.

Analysis of 1981-1988 data for Senegal, Mali and Niger

The data sets are listed in *Table 11.1* showing the countries, years and numbers of field samples which can be associated with independent satellite measurements. They are all in the Sahelian zone, where long-term mean annual rainfall is between 150 and 450 mm (Hiernaux, 1983), between 14° and 17° N and 16° W and 12° E. The Senegal study area consists of the Ferlo region in the northern part of Senegal (Tucker et al., 1983; Centre de Suivi Ecologique, 1989), the Mali study area consists of the Gourma (Hiernaux and Justice, 1986), and the Niger study area is in Tahoua, Maradi and Zinder provinces (Justice and Hiernaux, 1986; Wylie, Harrington, Pieper, Maman and Denda, 1988).

TABLE 11.1. Sahel rangeland production data sets

Country	Year	Number of sites	Source	Notes
Senegal	1981	42	Tucker et al., 1985	All on sand, herbs only.
	1982	49	ditto	42 on sand, 7 laterite.
	1983	95	ditto	75 on sand, 20 laterite.
	1987	16	Centre de Suivi Ecologique, Senegal	Sites on sand and laterite. (herb data only).
Mali	1984	11	International Livestock Centre for Africa, Mali	Large sites only (mostly sand) herb and tree data.
	1985	15	ditto	ditto
	1986	18	ditto	ditto
	1987	18	ditto	ditto
	1988	18	ditto	ditto
Niger	1986	22	New Mexico State University and Government of Niger	Herb data only
	1987	23	ditto	ditto
	1988	36	ditto	ditto

A harvest technique for measurement of net production (UNESCO, 1979; Beadle, Long, Imbamba, Holland and Olembo, 1985) has been used in all the studies because it is the only practicable procedure for measurement over large sample areas. The field procedures used in Niger are described by Wylie et al. (1988), those in Mali are described by Hiernaux and Justice (1986), and those in 1987 in Senegal by Centre de Suivi Ecologique (1989). In all of these programmes similar field sampling methods were adopted, using a large number of measurements in a sampling pattern designed to estimate the mean production in areas of between 10 and 100 km^2 in a few, fixed sample locations. In the Senegal study between 1981-1983 a large number of sites were sampled but with only one field measurement in each. In all programmes harvests were made of the above-ground standing crop of herbaceous vegetation. Below ground production is difficult to measure anywhere, but especially in the Sahel where the spatial heterogeneity of the vegetation demands large sample sizes, and so no comprehensive measurements are available. In Mali the annual leaf production of the woody vegetation has also been measured (Hiernaux and Justice, 1986).

The harvest method suffers from some well known errors when there is a significant amount of grazing, decay or exudation of metabolites from plant parts. Nevertheless, these problems are less likely to result in serious errors in estimation of seasonal production in the short life-cycle therophyte vegetation of the Sahel (UNESCO, 1979) where most herbivores are faced with a glut of forage during the short, but intense, growing season and decay is halted during dry periods (Penning de Vries and Djiteye, 1982).

AVHRR Local Area Coverage (LAC) data have been used for each study site, starting in late June and ending in early October of each year. Channels 1 and 2 were used to calculate **NDVI** and this single variable was mapped at full resolution in an equal longitude and latitude projection. Channel 5 was also mapped and used to create a cloud screen by setting all pixels with a temperature below 292K to zero. After adjustments in geographical registration by reference to ground control points, multidate maximum value composites were formed from short sequences of data throughout the growing season. Typically, about 30 LAC orbits were used to create approximately 9 composites covering the period 1 July - 30 September.

The Sahelian growing season generally starts in mid to late July after the first significant rains and continues until September. In some seasons growth can continue later in the year, especially in the deeper-rooted tree species. 1 July to 30 September has been adopted as a standard because of the movement of the Inter-Tropical Convergence Zone (ITCZ); dry air masses occur in the study areas before the beginning of June and again in early October and the increased transmission of near-infrared radiation under these conditions increases the AVHRR channel 2 signal and results in higher **NDVI** values which are not associated with any change in the surface vegetation. Until a routine water vapour correction is available it seems prudent to limit the satellite data use to the period of more uniform humidity, even at the expense of loss of measurements at the end of late growing seasons.

The results are given in *Figure 11.1*. Owing to the drought conditions which have occurred in the Sahel during the 1981-1988 period, many of the data points are near the origin, however wetter conditions in 1981 and 1988 together with the southern sites have provided a range of above ground herb production data of 0-3000 kg ha^{-1}. Because of the rather different vegetation sampling procedure used in Senegal between 1981 and 1983 these data have been omitted from the analysis. The restricted data set is plotted in *Figure 11.2*.

An analysis of covariance, using the herbaceous production as the dependent variable and the summed **NDVI** and data sets as independent variables, showed no significant interaction between data sets and **NDVI**, and only a marginally significant difference due to data sets alone (P=3.5%). The analysis was performed with summed **NDVI** as the independent variable because the variance of the satellite data is much less than that of the field data and the causal connection between herb production and the **NDVI** is not obviously in one direction. The results show that the entire data set can be represented by a narrow range of regression lines, all with the same slope (*Figure 11.2*). The herbaceous production (in kg ha^{-1}) can therefore be obtained from seasonal sums of AVHRR NDVI data using the equation:

$$\text{Herbaceous production} = \mathbf{a} + 108.29 * \mathbf{\Sigma_t}(\mathbf{NDVI_{t,satellite}} \times \mathbf{t_n}). \qquad (11.3)$$

The intercept **a** varies between -207 and 128, depending on site and year (*Figure 11.2*).

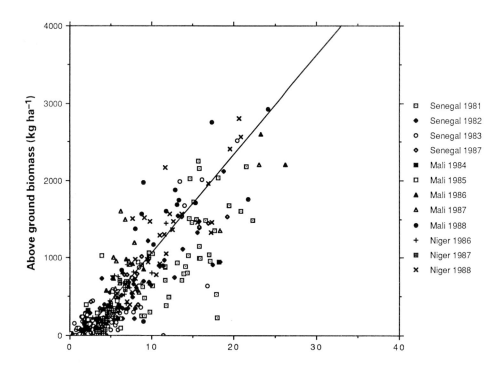

FIGURE 11.1. Above-ground herb production and seasonal sums of AVHRR **NDVI** for 363 Sahel sites in Senegal, Mali and Niger in eight years. Line fitted by regression model: Biomass = x + y $\Sigma_t(\mathbf{NDVI_{t,satellite}} \times \mathbf{t_n})$

This compares with a coefficient of 85.728 and a constant of -110.7 calculated for Senegal 1981-1983 using the data of Tucker *et al.*, (1985) alone. The confidence limits for prediction of the average production (Snedecor and Cochran, 1967) from observed values of $\Sigma_t(\mathbf{NDVI_{t,satellite}} \times \mathbf{t_n})$ using the present analysis are shown in *Figure 11.2* and of the same order as the range of constants for the different sites and years.

For the Mali data sets alone annual tree leaf production was also available. However, there was no evidence of major effects on the relationship with summed **NDVI** by using total production (trees+herbs), except for an increase in the production intercept. The year of observation was no longer significant in the total production analysis.

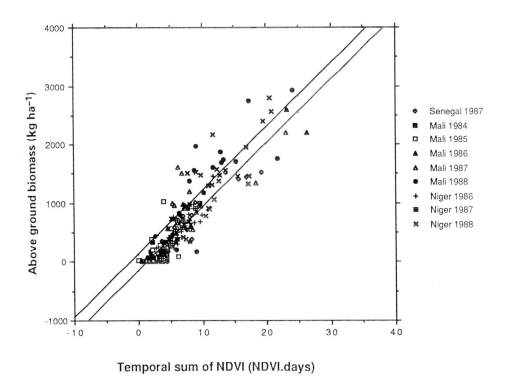

FIGURE 11.2. Above-ground herb production and seasonal sums of AVHRR NDVI for 172 Sahel sites for which there are detailed field measurements in Senegal, Mali and Niger in five years. Regression lines were calculated by an analysis of covariance in which the interaction of data set and $\Sigma_t(\mathbf{NDVI}_{t,\mathbf{satellite}})$ was shown to be not significant. Overall model $r^2 = 80\%$. Fitted lines are shown for the data sets with the maximum and minimum intercepts

Physically based production models

The Mali 1986 field data set was reanalysed in conjunction with atmospherically corrected **NDVI** data using the physically based model described in the introduction. For this, above and below ground gross production was needed. Since much of the vegetation is composed of herbaceous annuals, the end of season root/shoot ratio gives some indication of the below ground production (UNESCO, 1979). Measurements in three Mali sites in 1987 gave a mean value of 0.29 (Hanan, personal communication) which is used here to estimate the total standing crop from the above ground values measured in all the sites. In the IBP site at Fété Olé in Senegal (Bille and Poupon, 1972), at peak standing crop, root/shoot ratios varied between 2 and 1. However, Penning de Vries and Djiteye (1982) recorded root and shoot standing crops for Niono, Mali which result in ratios of 0.3 and 0.25.

The harvest method measures net primary production. For large areas of vegetation

TABLE 11.2. Sahel rangeland herb production; all sites and years.

Coefficient:	108	
Intercepts:	Senegal 1987	-149
	Mali 1984	-152
	Mali 1985	-35
	Mali 1986	-128
	Mali 1987	20
	Mali 1988	128
	Niger 1986	-130
	Niger 1987	-207
	Niger 1988	-8

Confidence limits for prediction of an individual biomass

Integrated NDVI (NDVI.days)	Confidence Limits (kg ha^{-1})
3	± 61
10	± 51
20	± 121
25	± 163

gross primary production can only be determined from the carbon dioxide flux above the canopy measured by profile techniques, by energy budgets which use the wind profile and the Bowen ratio, or by eddy correlation (Beadle et al., 1985). Owing to the technical difficulty and high cost of these methods, gross production data sets for the AVHRR scale of minimum ground sample do not exist for the Sahel. The respiratory loss of carbon dioxide, which accounts for the difference between net and gross production has been estimated using the relationship (Box, 1978):

$$\text{Respiration} = -4140 \log_e(1 - (\text{Net Production}/3000)). \qquad (11.4)$$

Pyranometer data were obtained from the Tombouctou and San stations in Mali, where Kipp CM5 integrators are installed. The two stations are 400 km apart and 500 km from the most distant field sites.

The satellite data consisted of 25 NOAA-9 AVHRR LAC orbits between 3 June and 3 October 1986. Only orbits having cloud-free views of a significant part of the Gourma region of Mali and a maximum view angle of 35° were used. The channel 1 and 2 data were calibrated using the provisional results of a new calibration procedure developed by Kaufman and Holben (personal communication), which uses two different techniques, viz. over the ocean, extreme nadir views for channel 1 and ocean glint for channel 2, and the radiance measured for an invariant desert target in the NE Sahara for both channels. The results show that the sensors have degraded by about 10% over the first four years of post-launch operation.

Atmospheric corrections to the satellite radiances (Chandrasekhar, 1960) were carried out using a procedure developed by B.N. Holben (personal communication). It is based on a modification of the atmospheric radiative transfer model of Ahmad and Fraser (1982) which allows for multiple scattering and includes the effects of polarization. In its modified form it attempts to account for the effects of ozone absorption, Rayleigh scattering, Mie scattering due to atmospheric dust and water vapour absorption to the appropriate atmospheric path length. Daily field measurements of atmospheric optical thickness were made using sun photometers at up to five stations no more than 100 km from the sites. Analyses of spatial variation in optical thickness suggest that 100 km is an acceptable range for interpolation (Holben and Eck, 1990). The corrections used are preliminary and, in some cases, changes in the values used are possible after further developments. The atmospheric correction accounts for the effect of variation in atmospheric path length which otherwise would reduce the **NDVI** of off-nadir views. However, vegetation rarely behaves as a true Lambertian reflector and as a result VI's vary with viewing angle (Pinter, Jackson, Idso and Reginato, 1983; Lee and Kaufman, 1986). Maximum value **NDVI** composites (Holben, 1986) were therefore used to select nadir views and also to reject pixels containing clouds which were not detected by the thermal mask.

In the Sahel there are often extensive patches of bare soil between areas of vegetation. The soil is generally brighter than the vegetation in all channels and this can significantly influence the VI measured for the pixel as a whole (Pech, Davis, Lamacraft and Graetz, 1986). Dead plant material also has well-known effects on the detection of the VI of the vegetation component of a scene (Asrar *et al.*, 1984; Tucker and Sellers, 1986; Huete and Jackson, 1987). No correction could be made for these effects, which must therefore account for an unknown amount of variation in the model results.

Few measurements of e are available which relate APAR to gross production and there are none for Sahelian grasslands, so it would not be justified to assume any specific value. e was therefore derived from the model. The results have been analyzed in a regression model in which the dependent variable is the gross production and the independent variable is the total absorbed photosynthetically active radiation in each of the 18 study sites. The slope of the regression is an estimate of e, the efficiency of gross production (Charles-Edwards, Doley and Rimmington, 1986). In this analysis a value of 1.18 g per MJ of absorbed photosynthetically active radiation was obtained (*Figure 11.3*).

Conclusions

Analyses of the greatly enlarged and improved data base of Sahelian herbaceous production and AVHRR data which is now available as a result of continuing research programmes confirms the existence of a strong linear relationship between production and seasonal sums of **NDVI** data. Nevertheless, there is considerable variation present and the confidence intervals for prediction of production vary between $\pm$ 51 kg ha^{-1} and $\pm$ 163 kg ha^{-1} in the range of production observed. The analyses presented here eliminate some potential causes of variation and point to others for future research. For example, little variation can be attributed to the different programmes and years of observation since as much, if not more, variation occurs within one year than between countries and years. The remaining possibilities can be considered in two categories: first the remote sensing model which seeks to measure **NDVI** at the surface of the plants from space and includes many factors such as atmospheric effects and sensor stability and the effects of mixtures of surface brightnesses in each pixel; and second, the primary production

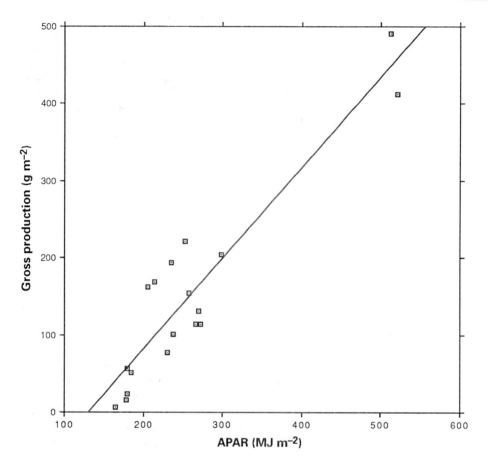

FIGURE 11.3. The relationship between absorbed photosynthetically active radiation and gross production for Mali field sites in 1986

model which attempts to account for production in terms of **NDVI**, solar radiation and an efficiency of energy fixation.

The results of the analysis of the atmospherically corrected and calibrated data for Mali in 1986 are encouraging but, in order to assess the effects of these factors, measurements are needed of the same site over several years when the satellite calibrations have changed; also when the primary production is considerably greater than in 1986 and larger effects of the atmosphere on the satellite measurements of **NDVI** may be expected; and in different climatic zones where the impact of cloudiness on solar radiation is more significant. The effect of the duration of the growing period on the period of summation of APAR needs further consideration. In some seasons in the southern part of the Gourma there is green vegetation in October and sometimes into early November. At this time the mean water vapour content of the atmosphere declines by as much as 80% and atmospheric correction for water vapour seems essential. Nevertheless, the results for one year in a limited part of Mali do support the view that the principal variable in the comprehensive model is the %APAR.

A possible cause of a low value for efficiency derived from the satellite data lies in the effect of the bright soil and dead vegetation background on the measurement of the **NDVI** of the vegetation component. Most soil backgrounds in the Sahel are slightly more reflective in the infrared than in the red with the result that they have an **NDVI** greater than zero. When summed over the growing season this effect is exacerbated and the sort of offset seen in *Figure 11.3* is to be expected. Furthermore, most Sahelian soils are more reflective in both red and infrared spectral bands than is the vegetation, often quite markedly so. The effect on the "scene **NDVI**" calculated from the reflectance of the entire mixed soil, live and dead vegetation scene, for the summed red components and the summed infrared components, is a reduction as compared with the "average **NDVI**" (sum of the **NDVI**'s for each component weighted by their relative areas) which is the index directly related to APAR. Thus, the function represented by r_t in the comprehensive model may have an important effect on the calculation of **e** and on the overall generality of the model.

The addition of solar radiation to the primary production had little effect in reducing the spread of points in the Mali data but this could be because there was little variation in solar radiation in the short growing season of the Sahel. However, consideration of an extreme case of variation in solar radiation throughout the growth cycle, for example a winter cereal in a temperate region, suggests that it is never a major variable. The solar radiation and **NDVI** for winter wheat grown in Beltsville, Maryland, U.S.A. calculated from data given in Tucker *et al.* (1980), shows that the radiation varies by a factor of 3 whereas the **NDVI** varies by a factor of 10. Thus, although theoretical objections to the simplified model (eg Steven *et al.*, 1983; Choudhury, 1987) are justified, radiation interception remains by far the most important variable (Monteith, 1977).

The 1986 Mali data gave a value of 1.18 g MJ^{-1} for the efficiency of conversion of PAR into *gross* production. Most of the published values of efficiencies are given for above-ground *net* production or for absorbed radiation expressed as the equivalent flux of total radiation. The 1986 Mali efficiency expressed as above-ground net production per unit PAR was 0.62 g MJ^{-1} and for above-ground net production per unit total solar radiation was 0.31 g MJ^{-1}, compared with published values which are in the 2-4 g MJ^{-1} range. Thus, no matter on what basis the values for e are compared, the value derived for Mali is below the reported values for crops. Whereas a considerably lower value for the efficiency of fixation of energy might be expected for a semi-arid region on an area or scene basis, there is no obvious physiological reason why there should be any difference on the basis of APAR; the mineral nutrient status of Sahelian soils is normally adequate, at least when compared with that of savanna soils further south (Penning de Vries and Djiteye, 1982), leaf area adjustments are very rapid in response to changes in soil moisture so the **NDVI** already accounts for the major effect of drought on production, and the increases in stomatal and mesophyll resistance and in respiration associated with drought are small (Legg, Day, Lawlor and Parkinson, 1979).

No doubt there are several causes of the low efficiency obtained by satellite measurements and the analyses point the way to improvements in the technique. However, the demonstration of the ability to determine biological parameters directly from satellite data is a challenging possibility and amply confirms the fundamental validity of the approach.

Acknowledgements

This work forms part of the Global Inventory, Monitoring and Modelling System (GIMMS) group study of remote sensing of grasslands primary production and uses methods and data collected in collaboration with members of the GIMMS group at NASA's Goddard Space Flight Center and the Remote Sensing Laboratory in the Geography Department at the University of Maryland. The field data were available through collaboration with the organisations listed in *Table 11.1*. Partial financial support was provided by NASA grant NCC 5-26.

References

AHMAD, Z. and FRASER, R.S. (1982). An iterative radiative transfer code for ocean-atmosphere system. *Journal of Atmospheric Science*, **39**, 656-665

ASRAR, G., FUCHS, M., KANEMASU, E.T. and HATFIELD, J.L. (1984). Estimating absorbed photosynthetic radiation and leaf area index from spectral reflectance of wheat. *Agronomy Journal*, **76**, 300-306

BEADLE, C.L., LONG, S.P., IMBAMBA, S.K., HOLLAND, D.O. and OLEMBO, R.J. (1985). *Photosynthesis in relation to plant production in terrestrial environments*. Tycooly Publishing/United Nations Environment Programme, Oxford

BILLE, J.C. and POUPON, H. (1972). Recherches écologiques sur une savane saheliénne: biomasse végétale et production primaire nette. *La Terre et la Vie*, **2**, 366-382

BOX, E.O. (1978). Geographical dimensions of terrestrial net and gross primary productivity. *Radiation and Environmental Biophysics*, **15**, 305-322

BOX, E.O., HOLBEN, B.N. and KALB, V. (1989). Accuracy of the AVHRR Vegetation Index as a predictor of biomass, primary productivity and net CO_2 flux. *Vegetatio*, **80**, 71-89

CANNELL, M.G.R., MILNE, R., SHEPPARD, L.J. and UNSWORTH, M.H. (1987). Radiation interception and productivity in willow. *Journal of Applied Ecology*, **24**, 261-278

CENTRE DE SUIVI ECOLOGIQUE (1989). Suivi de la production végétale au Sénégal pendant les hivernages 1987 et 1988. CSEDOC77:14/03/89. Dakar

CHANDRASEKHAR, S. (1960). *Radiative Transfer*. Dover, New York

CHARLES-EDWARDS, D.A., DOLEY, D. and RIMMINGTON, G.M. (1986). *Modelling Plant Growth and Development*. Academic Press, Sydney

CHOUDHURY, B.J. (1987). Relationships between vegetation indices, radiation absorption, and net photosynthesis evaluated by a sensitivity analysis. *Remote Sensing of Environment*, **22**, 209-233

GOUDRIAAN, J. (1977). *Crop micrometeorology: a simulation study*. Wageningen Centre for Agricultural Publishing and Documentation. Wageningen, Netherlands

GOWARD, S.N. and DYE, D.G. (1987). Evaluating North American net primary productivity with satellite data. *Advances in Space Research*, **7**, 165-174

GOWARD, S.N., DYE, D., KERBER, A. and KALB, V. (1987). Comparison of North and South American biomes from AVHRR observations. *Geocarto International*, **1**, 27-39

GOWARD, S.N., TUCKER, C.J. and DYE, D.G. (1985). North American vegetation patterns observed with the NOAA-t Advanced Very High Resolution Radiometer. *Vegetatio*, **64**, 3-14

HATFIELD, J.L. (1983). Remote sensing estimators of potential and actual crop yield. *Remote Sensing of Environment*, **13**, 301-311

HIERNAUX, P.H.Y. (1983). Recherches sur les systèmes des zones arides du Mali. CIPEA Rapport de Recherche No.5, Addis Ababa

HIERNAUX, P.H.Y. and JUSTICE, C.O. (1986). Suivi du développement végétal au cours de l'été 1984 dans le Sahel Malien. *International Journal of Remote Sensing*, **7**, 1515-1531

HOLBEN, B.N. (1986). Characteristics of maximum-value composite images from temporal AVHRR data. *International Journal of Remote Sensing*, **7**, 1417-1434

HOLBEN, B.N. and ECK, T. (1990). Temporal and spatial variability of aerosol optical depth in the Sahel region. *International Journal of Remote Sensing*. (in press)

HUETE, A.R. and JACKSON, R.D. (1987). Suitability of spectral indices for evaluating vegetation characteristics on arid rangelands. *Remote Sensing of Environment*, **23**, 213-232

JUSTICE, C.O. and HIERNAUX, P.H.Y. (1986). Monitoring the grasslands of the Sahel using NOAA AVHRR data: Niger 1983. *International Journal of Remote Sensing*, **7**, 1475-1497

KUMAR, M. and MONTEITH, J.L. (1982). Remote sensing of crop growth. In *Plants and the Daylight Spectrum*. Ed. by H. Smith. Academic Press, London

LEE, T.Y. and KAUFMAN, Y.J. (1986). Non-Lambertian effects on remote sensing of surface reflectance and vegetation index. *IEEE Transactions on Geoscience and Remote Sensing*, **GE-24**, 699-708

LEGG, B.J, DAY, W., LAWLOR, D.W. and PARKINSON, K.J. (1979). The effects of drought on barley growth: models and measurements showing the relative importance of leaf area and photosynthetic rate. *Journal of Agricultural Science, Cambridge*, **92**, 703-716

LINDNER, S. (1985). Potential and actual production in Australian forest stands. In *Research for Forest Management*. Ed. by J.J. Landsberg and W. Parsons. CSIRO, Melbourne

MONTEITH, J.L. (1977). Climate and the efficiency of crop production in Britain. *Philosophical Transactions of the Royal Society, London. Series B*, **281**, 277-294

MONTEITH, J.L. and ELSTON, J. (1983). Performance and productivity of foliage in the field. In *The Growth and Functioning of Leaves*. Ed. by J.E. Dale and F.I. Milthorpe. Cambridge University Press, Cambridge

NOAA (1986). *Global Vegetation Index Users' Guide*. Satellite Data Services Division, National Climatic Data Center, National Environmental Data and Information Service, National Oceanic and Atmospheric Administration, U.S. Department of Commerce. Washington D.C.

PALMER, J.W. (1988). Annual dry matter production and partitioning over the first 5 years of a bed system of Crispin/M.27 apple trees at four spacings. *Journal of Applied Ecology*, **25**, 569-578

PECH, R.P., DAVIS, A.W., LAMACRAFT, R.R. and GRAETZ, R.D. (1986). Calibration of Landsat data for sparsely vegetated semi-arid rangelands. *International Journal of Remote Sensing*, **7**, 1729-1750

PENNING DE VRIES, F.W.T. and DJITEYE, M.A. (1982). *La productivité des pâturages sahéliens*. Wageningen Centre for Agricultural Publishing and Documentation, Wageningen

PINTER, P.J. Jr., JACKSON, R.D., IDSO, S.B. and REGINATO, R.J. (1981). Multidate spectral reflectance as predictors of yield in water stressed wheat and barley. *International Journal of Remote Sensing*, **2**, 43-48

PINTER, P.J. Jr., JACKSON, R.D., IDSO, S.B. and REGINATO, R.J. (1983). Diurnal patterns of wheat spectral reflectances. *IEEE Transactions on Geoscience and Remote Sensing*, **GE-21**, 156-163

PRINCE, S.D. (1986). Monitoring the vegetation of semi-arid tropical rangelands with the NOAA-7 Advanced Very High Resolution Radiometer. In *Remote Sensing and Tropical Land Management*. Ed. by J.T. Parry and M.J. Eden. Wiley, London

PRINCE, S.D. and TUCKER, C.J. (1986). Satellite remote sensing of rangelands in Botswana. II. NOAA AVHRR and herbaceous vegetation. *International Journal of Remote Sensing*, **7**, 1555-1570

SELLERS, P.J. (1985). Canopy reflectance, photosynthesis and transpiration. *International Journal of Remote Sensing*, **6**, 1335-1372.

SNEDECOR, G.W. and COCHRAN, W.G. (1967). *Statistical Methods, 6th edition*. Iowa State, Ames

STEVEN, M.D., BISCOE, P.V. and JAGGARD, K.W. (1983). Estimation of sugar beet productivity from reflection in the red and infrared spectral bands. *International Journal of Remote Sensing*, **4**, 325-334

STEVEN, M.D. and DEMETRIADES-SHAH, T.H. (1987). Spectral indices of crop productivity under conditions of stress. In *Advances in Digital Image Processing*. Remote Sensing Society, Nottingham

TUCKER, C.J., HOLBEN, B.N., ELGIN, J.H. and McMURTREY, J.E. (1980). Relationship of spectral data to grain yield variation. *Photogrammetric Engineering and Remote Sensing*, **46**, 657-666

TUCKER, C.J., HOLBEN, B.N., ELGIN, J.H. and McMURTREY, J.E. (1981). Remote sensing of total dry-matter accumulation in winter wheat. *Remote Sensing of Environment*, **11**, 171-189

TUCKER, C.J., VANPRAET, C., BOERWINKEL, E. and GASTON, A. (1983). Satellite remote sensing of total dry matter production in the Senegalese Sahel. *Remote Sensing of Environment*, **13**, 461-474

TUCKER, C.J., VANPRAET, C.L., SHARMAN, M.J. and VAN ITTERSUM, G. (1985). Satellite remote sensing of total herbaceous biomass production in the Senegalese Sahel: 1980-1984. *Remote Sensing of Environment*, **17**, 1571-1581

TUCKER, C.J. and SELLERS, P.J. (1986). Satellite remote sensing of primary production. *International Journal of Remote Sensing*, **7**, 1395-1416

UNESCO (1979). *Tropical Grazing Lands Ecosystems*. United Nations Educational, Scientific and Cultural Organisation, Paris

WALBURG, G., BAUER, M.E., DAUGHTRY, C.S.T. and HOUSLEY, T.L. (1982). Effects of nitrogen nutrition on the growth, yield and reflectance characteristics of corn canopies. *Agronomy Journal*, **74**, 677-683

WYLIE, B., HARRINGTON, J., PIEPER, R., MAMAN, A. and DENDA, I. (1988). *1987 Pasture assessment early warning system. Research on satellite-based pasture assessment implementation techniques*. Niger Integrated Livestock Project, Ministry of Animal and Water Resources, Niamey, Niger

12

ESTIMATING GRASSLAND BIOMASS USING REMOTELY SENSED DATA

E.T. KANEMASU, T.H. DEMETRIADES-SHAH and H. SU

Evapotranspiration Laboratory, Waters Annex, Department of Agronomy, Kansas State University, Manhattan, Kansas, 66502, U.S.A.

and

A.R.G. LANG

CSIRO Center for Environmental Mechanics, GPO Box 821, Canberra, ACT 2601, Australia.

Introduction

The strong correlation between vegetation canopy near-infrared to red reflectance ratio and leaf area or biomass is well documented in the literature. A physical model developed at Kansas State University shows that two principal reasons why this empirical relationship is variant and non-unique are variations in canopy geometry and soil background reflectance (Shultis and Myneni, 1988). The normalized difference (spectral) vegetation index (**NDVI**) shows a strong near-linear relationship to the fraction of visible radiation intercepted by a vegetation canopy and appears to be almost insensitive to variations in canopy geometry. Ground-based **NDVI** measurements over prairie grassland sites were taken at regular intervals over the 1988 growing season. These measurements were used to compute canopy seasonal light interception using the Shultis and Myneni model. For a selected site the intercepted energy was regressed against above-ground dry matter production. The slope of this relationship is the light use efficiency (LUE). This estimate of LUE was combined with intercepted light - calculated from the **NDVI** measurements - to predict productivity at several independent test sites. Good agreement between predicted and measured values of dry matter production over the season was obtained. This approach was extended to predict grassland productivity using satellite spectral data.

Photosynthesis and transpiration are two basic processes of vegetative canopies that are strongly dependent upon the amount of green leaf area. Remote sensing has been seen as a potential tool for estimating canopy attributes, such as leaf area index and biomass, in plant growth and yield models (Kanemasu, Heilman, Bagley and Powers, 1977; Wiegand, Richardson and Kanemasu, 1979). The strong correlation between vegetation canopy near-infrared/red (**IR/R**) reflectance ratio and leaf area index (LAI) or biomass per unit ground area has been well documented in the literature (Kanemasu, Niblett, Manges, Lenhert and Newman, 1974). However, linear or quadratic relationships do not extrapolate to other sites and years (Weiser, Asrar, Miller and Kanemasu, 1986). These empirical relationships are dependent upon viewing geometry, canopy morphology, radiation geometry, background spectral characteristics and spectral characteristics of plant parts.

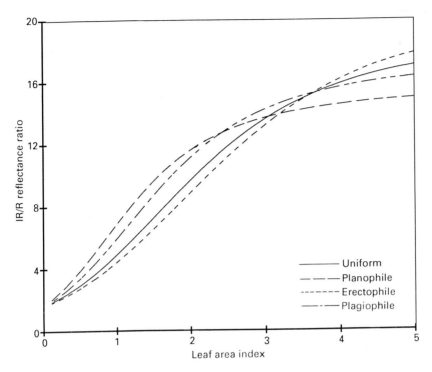

FIGURE 12.1.a The effect of leaf-angle distibution on the relationship between canopy **IR/R** reflectance ratio and leaf area index (modelled from Shultis and Myneni, 1988)

Radiation canopy models (e.g. SAIL, SUITS, CUPID) have been developed to describe the radiation streams within and above the canopy (Suits, 1972; Verhoef, 1984). A primary input to these models is leaf area. The radiation streams leaving the top of the canopy are the information captured by remote sensing systems whether carried by satellite, aircraft, or ground-based platforms. Radiation models have been inverted to predict canopy geometry and leaf area index from spectral reflectance measurements (Goel and Strebel, 1983).

Shultis and Myneni (1988) developed a physical model of the interaction of radiation with a vegetation canopy. *Figure 12.1a* shows the modelled relationship between the canopy **IR/R** reflectance ratio and LAI for different leaf angle distributions. Canopy geometry has a large effect on the relationship. Note the curvilinear relationship between the **IR/R** ratio and LAI. At low values of LAI (<3), however, this relationship is almost linear. *Figure 12.1b* shows the effect on the (modelled) canopy **IR/R** reflectance ratio versus LAI relationship of changing the soil spectral reflectance. There appears to be a significant effect due to changes in soil reflectance.

Another widely used spectral vegetation index is the Normalized Difference Vegetation Index (**NDVI**):

$$\mathbf{NDVI} = (\mathbf{IR} - \mathbf{R})/(\mathbf{IR} + \mathbf{R}).$$

This index is strongly correlated to the fraction of photosynthetically active radiation

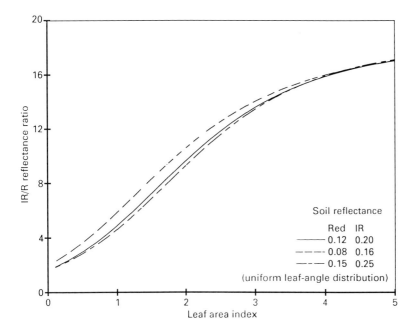

FIGURE 12.1.b The effect of soil background reflectance on the relationship between canopy **IR/R** reflectance ratio and leaf area index (modelled from Shultis and Myneni, 1988)

(i.e. radiation between 0.4 and 0.7 μm) intercepted by the canopy (Hatfield, Asrar and Kanemasu, 1984). *Figure 12.2a* shows the modelled relationship between **NDVI** and the fraction of photosythetically active radiation (PAR) intercepted for different leaf angle distributions. A near-linear relationship exists, and canopy geometry does not cause large variations in the relationship; however, changes in soil reflectance can cause some variations in this relationship as shown in *Figure 12.2b*.

Lapitan (1986) developed an empirical relationship between **NDVI** and light interception for wheat. Using this relationship and reflectance measurements, Garcia, Kanemasu, Blad, Bauer, Hatfield, Major, Reginato and Hubbard (1988) and Mojarro (1988) computed seasonal interception curves for different nitrogen (N) treatments of wheat. The treatments resulted in large differences in LAI and visible light interception. Measurements of above-ground dry matter and estimated cumulative light interception were regressed, and the slope is the dry matter yield of energy or light use efficiency (LUE). Values of light use efficiency were about 2.6 g MJ^{-1} of PAR for three years. There was not a significant difference in LUE among N treatments. Apparently, N increased the growth and permitted greater light absorption by the canopy. Presumably, the greater leaf photosynthesis promoted by N is not a major factor in accumulating above-ground biomass. Therefore, it would be possible to use the LUE value and the intercepted PAR estimated from remotely sensed **NDVI** measurements to predict above-ground biomass production.

The possibility of extending this methodology to estimate grain production is questionable. The strong relation between LAI and above-ground wheat biomass (until max-

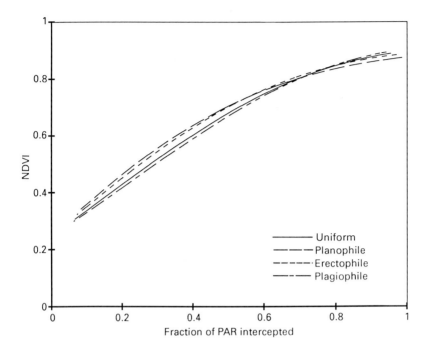

FIGURE 12.2.a The effect of leaf-angle distribution on the relationship between normalized difference vegetation index (**NDVI**) and fraction of PAR intercepted (modelled from Shultis and Myneni, 1988)

imum LAI) results in a similar relation between spectral indices (**IR/R** and **NDVI**) and biomass at anthesis (Kanemasu, Fuchs, Myneni, Blum, and Sears, 1988). Mojarro (1988) found a high correlation between wheat grain number (kernels per m^{-2}) and dry matter at anthesis. Therefore, we anticipate a good correlation between spectral reflectance at anthesis and grain number. Grain number is a major yield component. The other yield component is kernel weight. Mojarro (1988) found that kernel weight is negatively linearly related to senescence rate, and senescence rate is linearly related to the measured decline in the **IR/R** reflectance ratio. A question remains as to the uniqueness of these relationships across genotypes and environments. Kanemasu et al. (1988) examined the yield components of 25 genotypes over a two-year period and could not find a relationship of intercepted PAR to any of the yield components. This suggests a complex process model is required for predicting grain yield; however, remote sensing can play a role in assessing within-season yield estimates.

In this investigation we propose the use of **NDVI** to get a reliable estimate of the amount of PAR intercepted by grassland canopies. Growth is then predicted from a knowledge of the amount of PAR intercepted and the efficiency with which this energy is used to produce dry matter.

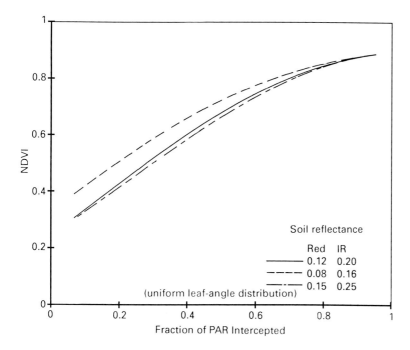

FIGURE 12.2.b The effect of soil background reflectance on the relationship between canopy normalized difference spectral vegetation index (**NDVI**) and fraction of PAR intercepted (modelled from Shultis and Myneni 1988)

Materials and methods

Ground-based measurements

This experiment was conducted on prairie grassland sites on and around the Konza Research Natural Area, just outside the town of Manhattan, Kansas and south of Kansas State University. The prairie is dominated by three species of grass, Big Bluestem (*Andropogon gerardii* Vitman), Little Bluestem (*Andropogon scoparius* Michx), and Indiangrass (*Sorghastrum nutan* L. Nash). While these are C4 grasses, there are numerous C3 grasses as well. The sites were part of the First ISLSCP[1] Field Experiment (FIFE) supported by the National Aeronautics and Space Administration (NASA). Ground-based spectral reflectance measurements were taken with an Exotech Model 100 AX radiometer; a 4-band radiometer with spectral sensitivity similar to the Landsat Multispectral Scanner (MSS). The wavelengths for the four bands are 0.5 to 0.6 μm, 0.6 to 0.7μm, 0.7 to 0.8μm, and 0.8 to 1.1μm. The radiometer has a 15° field of view. Measurements were nadir-viewing from a height of about 1.25m above the grass canopies. Sequential readings over a standard barium sulphate panel were used to calculate the grassland reflectance factors. Readings were taken on clear days within 2 hours of noon. Each site was designated as a circular

---

[1] International Satellite Land Surface Climatology Project

area of about 2800 m². Six points spaced approximately radially from the center of each site were randomly selected for the reflectance measurements. A 0.1 m² area at each of the points was sampled within a day or so after taking the reflectance measurements, for above-ground biomass and leaf area index measurements. The radiometric and biophysical data were collected from each site approximately every 10 days from May to October 1988. The fraction of photosynthetically active radiation (PAR) intercepted by the grassland sites was calculated from the radiometer **NDVI** measurements using the radiative transfer model of Shultis and Myneni (1988).

Other inputs to the model were the soil background reflectance, canopy leaf angle distribution, leaf spectral reflectance and transmittance, the sun angle, and the proportion of direct and diffuse illumination. The bare soil reflectance readings were obtained by taking reflectance measurements after removing all standing vegetation. A uniform leaf angle distribution was assumed. Typical values for the spectral properties of prairie grass leaves were provided by Dr. Betty Walter-Shea (Department of Agronomy, University of Nebraska) as part of FIFE. These were 0.08 for the red band reflectance and transmittance and 0.45 for the reflectance and transmittance in the near-infrared band. About 70% of the solar radiation was assumed to be in the direct beam, and the remaining 30% was assumed to be uniformly diffuse sky radiation. Direct measurements of the fraction of PAR intercepted by the grass canopies were also taken at each of the sampling points. A 1 m long line quantum sensor (LI-COR Inc.) was placed at the base of the canopy to estimate PAR transmitted to the base of the canopy. Simultaneous readings of the incident PAR were taken with another quantum sensor above the canopy. Interception was calculated as 1 minus transmittance. These direct measurements of interception were, however, considered problematic. The structure of the swards was highly variable and was frequently short and brush-like. The line quantum sensor, which was 2.5 cm by 2.5 cm in cross-section, tended to part the sward, or project above the canopy when the grass was short, so that interception was under-estimated. In other places, where there was standing dry vegetation, the fraction of PAR intercepted by actively growing foliage was over-estimated. Interception estimated from **NDVI** values does not involve disturbance of the sward, and provides an estimate of interception even if the grass is short. The reflectance method also samples over a larger area than the line quantum sensor. *Figure 12.3* shows **NDVI**-calculated, against directly measured fractional interception for a site in 1988. The points are scattered, due to the spatial variability of vegetation within the site. Also, the slope of the best-fitting line is less than the 1:1 relationship, because of the high intercept on the vertical scale. The line quantum sensor is unable to register correct values for low interception, which occur when the sward is short.

The fractional interceptions for the days in between the measurement days were calculated by linear interpolation. The amount of energy intercepted by the canopy for each day was calculated from the product of the interception fraction and the daily solar radiation, which was recorded at a nearby weather station. PAR was assumed to be 45% of the total solar radiation. For a selected site the seasonal values for above-ground green dry matter were regressed against accumulated intercepted PAR to determine the light use efficiency of the prairie grassland. This value of LUE and the interception measurements (calculated from canopy **NDVI** measurements) for other sites were used to predict their seasonal growth.

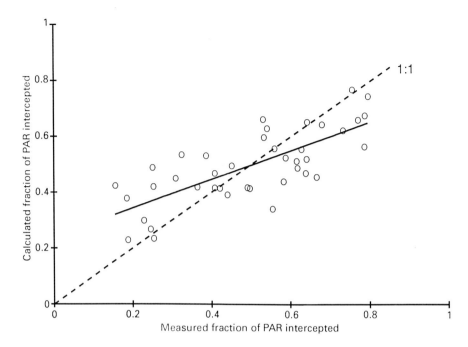

FIGURE 12.3. Fractional interception of incident PAR calculated from ground-based **NDVI** values (Shultis and Myneni, 1988) versus line quantum sensor measured values on FIFE site 7 in 1988

Satellite derived measurements

Twenty-six SPOT images were acquired from March to November of 1987, and 2 cloud-free images were acquired for the 1988 season. All the image data were obtained from FIFE Information System (Sellers, Hall, Asrar, Strebel and Murphy, 1988), and were prepared with level-1 calibration processing. An affine transformation (Jensen, 1986) was applied to the image geometric correction with error of about 0.5 pixels (20m resolution per pixel). A nearest-neighbour re-sampling technique was used to rectify the images without changing the original digital number (DN). DN data were extracted from the sites with each image band. A normalized difference vegetation index was calculated as follows:

$$\text{NDVI}_{\text{SPOT}} = (\text{ch3} - \text{ch2})/(\text{ch3} + \text{ch2})$$

where **ch2** and **ch3** are the DN values for channels 2 and 3 of SPOT, respectively. The wavelengths for the three SPOT channels are 0.5 to 0.59 μm, 0.61 to 0.68 μm, and 0.79 to 0.89 μm. Ground-based **NDVI** measurements for May 25 and June 20, 1988 were regressed against concurrent SPOT Normalised Difference (**NDVI**$_{\text{SPOT}}$) values to obtain a linear empirical relationship between ground-based **NDVI** measurements and SPOT values. The relationship between ground radiometer measurements and SPOT measurements (**NDVI**$_{\text{SPOT}}$) was given by:

$$\text{NDVI} = \text{NDVI}_{\text{SPOT}} * 1.16 + 0.16.$$

From this relationship the 1987 SPOT **NDVI** were used to infer what the ground-based values would have been for the 1987 season. These inferred ground-based **NDVI** values were used to calculate intercepted PAR for the 1987 season, as described earlier, using the model of Shultis and Myneni. It would have been preferable to have used an atmospheric radiation model to estimate the reflectance values at earth's surface in the MSS wavebands from the SPOT DN values; however, we did not have time for this analysis.

Results and discussion

Figure 12.4 shows the relationship between above ground dry matter produced for a selected grassland site (FIFE site 7) and the accumulated PAR interception over the 1988 growing season. Intercepted PAR values were calculated from the ground-based **NDVI** measurements using the model of Shultis and Myneni, as described earlier. The slope of the curve in *Figure 12.4* is the LUE and has a value of about 0.5 g per MJ of intercepted PAR. Weiser et al. (1986) reported average LUE values of about 1.4 g per MJ for similar sites. As the season progresses, growth rates decrease, because of shortages of water and nutrients, and due to developmental changes in the vegetation (personal communications from Dr. David Hartnett). Light use efficiency decreases gradually as the season progresses, and reaches a value of zero at the top of the growth curve. Our measurements were taken one month later into the growing season (when LUE was lower) than measurements taken by Weiser et al. This is the principal reason for the lower average value for LUE that we observed.

Using the value of 0.5 g MJ^{-1} for LUE and ground-based measurements of **NDVI** (which were used to calculate the PAR intercepted), growth at three other grassland sites was predicted. *Figures 12.5a-c* show the predicted and measured growth at these sites during the 1988 growing season. Each of the measured values in *Figure 12.5a-c* are the means of 6 samples taken around each site for a particular date. The error bars show the standard deviations about the mean values. The predicted values are generally well within sampling error and are in good agreement with measured values. The relatively large standard deviations are characteristic of the highly spatially variable groundcover. Only sites that were burned in the spring (around mid-April) and were not grazed were chosen. Burning the tall-grass prairie is a recommended practice for the rancher in this area. The main purpose of burning is to encourage new high-feeding quality growth. Unburned sites were excluded from the analysis because dry standing vegetation from the previous year was not distiguished from the dry biomass produced in the current growing season. On the unburned and ungrazed sites, dry vegetation frequently composed more than 50% of the total standing biomass. Serious errors in the analysis would have been introduced for these sites. Grazed sites were excluded because this form of analysis is invalid if biomass is removed without an estimate of the amount being removed.

Figures 12.6a-c show the measured and predicted values for prairie growth at three grassland sites (FIFE sites 8, 11, and 23) during the 1987 growing season. All three sites were burned in the spring and ungrazed, except for the site shown in *Figure 12.5c* which was burned and very lightly grazed. The predicted growth was calculated using SPOT satellite **NDVI** values to estimate intercepted PAR (as described in section 2.2) and a light use efficiency of 0.50 g MJ^{-1}. The predicted dry matter values and the measured values are in agreement, within the sampling error of the measured values.

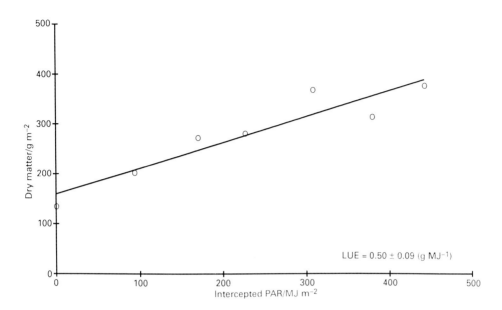

FIGURE 12.4. Above ground dry matter versus intercepted PAR on FIFE site 7 during 1988

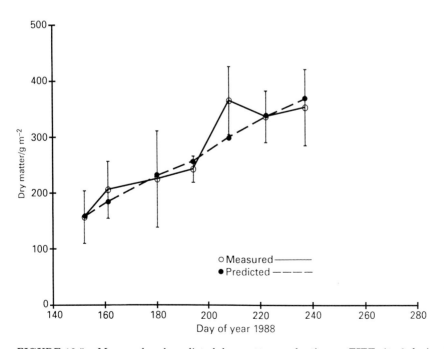

FIGURE 12.5.a Measured and predicted dry matter production on FIFE site 8 during 1988

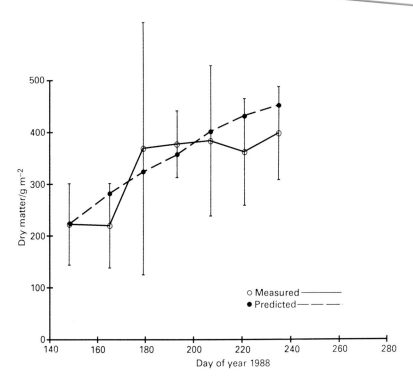

FIGURE 12.5.b Measured and predicted dry matter production on FIFE site 11 during 1988

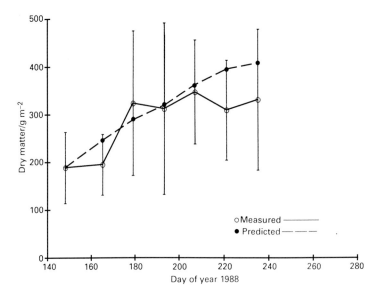

FIGURE 12.5.c Measured and predicted dry matter production on FIFE site 20 during 1988

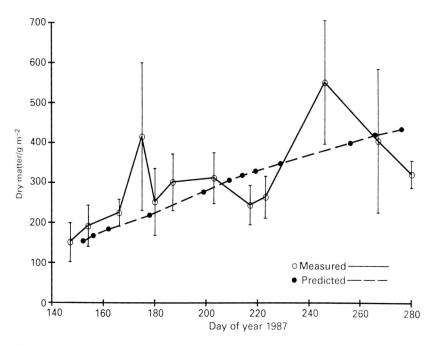

FIGURE 12.6.a Measured and predicted dry matter production on FIFE site 8 during 1987

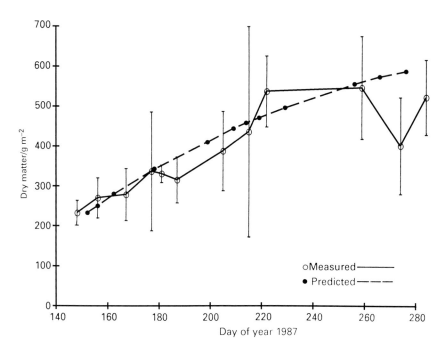

FIGURE 12.6.b Measured and predicted dry matter production on FIFE site 11 during 1987

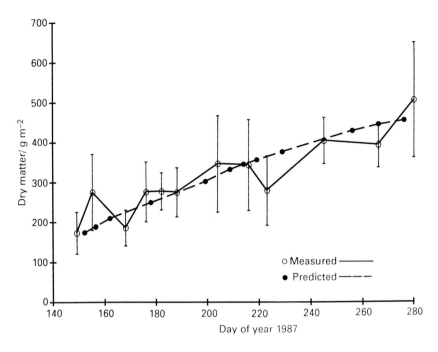

FIGURE 12.6.c Measured and predicted dry matter production on FIFE site 20 during 1987

An estimate of the efficiency with which intercepted PAR is used for dry matter production is of major importance to this method of predicting growth. If environmental or intrinsic plant factors significantly alter the efficiency with which intercepted energy is converted into dry matter, then a revised value for light use efficiency is required for estimating dry matter production. In this experiment biomass data were used from two seasons which were significantly different in terms of rainfall. *Figure 12.7* shows the 10-day summation rainfall for the 1987 and 1988 growing seasons. The 1988 season was drier than the 1987 season; there was not much rainfall until late into the growing season (days 180-200). In our experiment the light use efficiency did not appear to vary significantly between the years, and a value of 0.5 g MJ^{-1} determined for a selected site for the 1988 season (*Figure 12.4*) was used for all predictions. For both seasons the predicted values are within the range of biomass measurement error and the 0.5 g MJ^{-1} value for the light use efficiency appears to be valid, despite the differences in the rainfall. We would have anticipated higher LUE values. Others have, however, reported considerable variations for light use efficiency. Kiniry, Jones, O'Toole, Blanchet, Cabelguenne and Spanel, (1989), for example, found large within-species and between-species variations of light use efficiencies for maize, sorghum, rice, wheat and sunflower crops. Kiniry *et al.*, could not attribute the within-species variations to any known environmental factor.

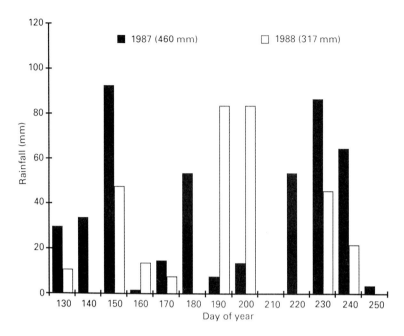

FIGURE 12.7. 10-day accumulated rainfall at Konza headquarters during the growing seasons of 1987 and 1988. Numbers in parentheses indicate the total amounts over the period shown

Conclusions

Variations in the canopy structure and soil background reflectances appear to be the principal factors causing empirical relationships between spectral vegetation indices and biomass or leaf area to be variant and non-unique. The relationship between the normalized difference spectral vegetation index and the fraction of solar energy intercepted by the crop is, however, near-linear and appears to be less sensitive to variations in canopy structure and soil background reflectance. If the efficiency with which intercepted energy is converted into dry matter is known, then growth can be predicted more reliably from reflectance-estimated interception than from direct empirical relationships between spectral vegetation indices and biomass. The results of our experiment to test the use of this procedure for estimating the growth of grassland, using ground and satellite spectral data, showed that predicted values of dry matter production agreed well with measured values of dry matter over different sites and seasons.

Acknowledgements

We thank Larry Ballou and his team for their hard work of collecting and processing the plant samples and Kerri Atwood for assistance with the collection of field spectral measurements. This work was financed by NASA under grant NAG 5-389 and the FIFE program. We appreciate the assistance of Dr. Ken Shultis in providing the output from his radiation model.

References

GARCIA, R., KANEMASU, E.T., BLAD, B.L., BAUER, A., HATFIELD, J.L., MAJOR, D.J., REGINATO, R.J. and HUBBARD, K.G. (1988). Interception and light use efficiency of winter wheat under different nitrogen regimes. *Agricultural and Forest Meteorology*, **44**, 175-186

GOEL, N.S. and STREBEL, D.E. (1983). Inversion of vegetation canopy reflectance models for estimating agronomic variables. I. Problem definition and initial results using the Suits model. *Remote Sensing of Environment*, **13**, 487-507

HATFIELD, J.L., ASRAR, G. and KANEMASU, E.T. (1984). Intercepted photosynthetically active radiation estimated by spectral reflectance. *Remote Sensing of Environment*, **14**, 65-75

JENSEN, J.R. (1986). *Introductory digital image processing, a remote sensing perspective.* Prentice-Hall, Englewood Cliffs, New Jersey

KANEMASU, E.T., NIBLETT, C.L., MANGES, H., LENHERT, D. and NEWMAN, M.A. (1974). Wheat: Its growth and disease severity as deduced from ERTS-1. *Remote Sensing of Environment*, **3**, 255-260

KANEMASU, E.T., HEILMAN, J.L., BAGLEY, J.O. and POWERS, W.L. (1977). Using Landsat data to estimate evapotranspiration of winter wheat. *Environmental Management*, **1**, 515-520

KANEMASU, E.T., FUCHS, M., MYNENI, R., BLUM, A. and SEARS, R. (1988). *Application of remote sensing technology to wheat breeding.* BARD final report. US. 658-83. Beltsville, MD

KINIRY, J.R., JONES, C.A., O'TOOLE, J.C., BLANCHET, R., CABELGUENNE, M. and SPANEL, D.A. (1989). Radiation use efficiency in biomass accumulation prior to grain filling for five grain-crop species. *Field Crops Research*, **20**, 51-64

LAPITAN, R. (1986). Spectral estimates of absorbed light and leaf area index: Effects of canopy geometry and water stress. Ph.D. dissertation. Dept. of Agronomy, Kansas State University, Manhattan, Kansas

MOJARRO, F. (1988). Analysis of the effect of water, nitrogen and weather on growth, grain yield, biomass production, and light use efficiency of winter wheat (*Triticum aestivum L.*). Ph.D. dissertation. Dept. of Agronomy, Kansas State University, Manhattan, Kansas

SELLERS, P.J., HALL, F.G., ASRAR, G., STREBEL, D.E. and MURPHY, R.E. (1988). The first ISLSCP field experiment (FIFE). *Bulletin of the American Meteorological Society*, **69**, 22-27

SHULTIS, J.K. and MYNENI, R.B. (1988). Radiative transfer in vegetation canopies with anisotropic scattering. *Journal Quantitative Spectroscopy and Radiation Transfer*, **39**, 115-129

SUITS, G.H. (1972). The calculation of the directional reflectance of a vegetation canopy. *Remote Sensing of Environment*, **2**, 117-125

VERHOEF, W. (1984). Light scattering by leaf layers with application to canopy reflectance modeling: the SAIL model. *Remote Sensing of Environment*, **16**, 125-141

WEISER, R.L., ASRAR, G., MILLER, G.P. and KANEMASU, E.T. (1986). Assessing grassland biophysical characteristics from spectral measurements. *Remote Sensing of Environment*, **20**, 141-152

WIEGAND, C.L., RICHARDSON, A.J. and KANEMASU, E.T. (1979). Leaf area index estimates for wheat from Landsat and their implications for evapotranspiration and crop modeling. *Agronomy Journal*, **71**, 336-342

13
REMOTE SENSING TO PREDICT THE YIELD OF SUGAR BEET IN ENGLAND

K.W. JAGGARD and C.J.A. CLARK

AFRC Institute of Arable Crops Research
Broom's Barn, Higham, Bury St. Edmunds, Suffolk IP28 6NP, U.K.

Introduction

In England all of the sugar-beet crop is grown on contract to British Sugar plc. The Agriculture Commission of the EEC has fixed the company's sugar production quota at 1.14 Mt per annum: this sugar, and the beet from which it is produced are subject to guaranteed prices, which are usually much higher than the world market price. Sugar produced in excess of quota must be sold outside the EEC at whatever price can be obtained on the world market. In addition to sugar, the company also produces approximately 0.75 Mt of dried animal feeds each year as a by-product from the sugar extraction process. These products have to be sold in markets whose prices can fluctuate greatly in response to demand and the supply of alternative feedstuff. With this type of marketing operation it is essential that the industry has accurate production forecasts.

After harvest, beet can only be stored for a few weeks before being processed in one of the twelve factories, where storage capacity for beet, sugar and animal feed is limited and expensive. The factories usually start to operate in late September and aim to finish the campaign by early January, depending upon the size of the crop. Therefore, early and accurate forecasts of beet yield are essential for drawing up reliable plans of factory operation such as delivery schedules, opening and closing dates, provision of storage capacity and the need for fuel and other raw materials.

In recent years the area of beet has been quite stable at 200,000 ± 5,000 ha., and the farmers accurately report this area to the sugar company. Therefore, accurate forecasts of total yield are almost wholly dependent on predictions of the yield per unit area. During the last 15 years these yields have varied by more than twofold, as shown in *Figure 13.1*.

Until a few years ago the yield forecasts were based on relationships between time and the weights of samples dug from a random selection of beet fields during the summer (Church and Gnanasakthy, 1983). However, these forecasts had several serious disadvantages:

- Collecting and processing the samples was labour intensive and hence expensive.

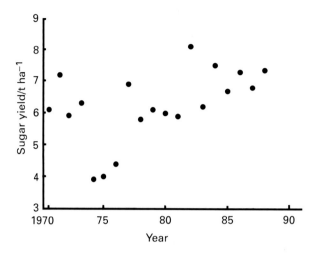

FIGURE 13.1. Variations in the sugar yield of the English sugar-beet crop since 1970

- Any alterations to the methods had to be used for several years before the data could provide reliable forecasts.

- The methods provided no objective way to allow for unusual conditions to crop growth after the samples were collected.

- Reliable forecasts could not be made before early September - much later than required for effective planning of production.

- Large variations in beet yield from spot to spot within a field, easily seen in aerial photographs of the crop (*Figure 13.2*), often resulted in samples being a poor representation of the population as a whole: consequently yield forecasts could be inaccurate.

The forecasting system described in this paper, based on a simple model of crop growth and remotely-sensed data about the crop canopy, was developed in response to the shortcomings outlined above.

Predicting growth of the beet crop

The productivity of beet crops is related directly to the amount of light intercepted by the foliage. This relationship can be described as

$$dw/dt = ef R_s \tag{13.1}$$

where productivity is denoted as dw/dt, e is the net coefficient of conversion of solar energy to plant material, and f is the fraction of solar irradiance, R_s, intercepted by the crop. One form of the integral of this equation is

$$w = e \int f R_s \, dt, \tag{13.2}$$

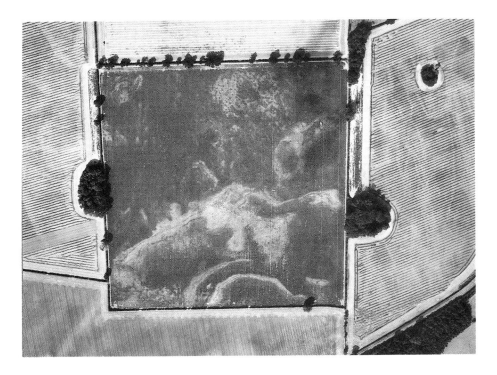

FIGURE 13.2. Photograph showing variations in beet growth, and hence yield, on a sandy soil. Such variations can cause large inaccuracies in yield forecasts which are based on the weights of samples of only a few plants. (Courtesy of ADAS Aerial Photography Unit)

where e is assumed to be constant.

In this form, with a constant (or at least a conservative) conversion coefficient, the equation suggests that yield can be predicted from knowledge of e and from measurements of f and R_s. Solar irradiance on a daily basis varies little from place to place over quite large distances and is easily measured. However, the fraction of light intercepted varies greatly according to the husbandry of the crop, and can also vary over short distances within a field, producing the variation evident in *Figure 13.2*. Thus, many measurements of f are needed if the canopy of the English beet crop is to be characterized.

The fraction of light, f, intercepted by a beet crop at any one time can be estimated from a near-infrared/red reflectance ratio (Steven, Biscoe and Jaggard, 1983), and many measurements can be obtained cheaply, rapidly and promptly by mounting a suitable spectrophotometer in an aircraft. In our case we used a helicopter to carry a photometer with an upward-looking, cosine-corrected, head and a downward-looking head with a 10° field of view. The instrument, made by Macam Photometrics, recorded the light intensity in two bands, 600-660 and 780-940 nm, once per second while passing over sugar-beet crops. The detected area had a diameter of approximately 35m. These measurements, made on three or four occasions during the summer, were used to estimate the fraction and thus the amount of light intercepted by beet crops, so that yields could be forecast.

As the season progressed, so the forecasts could be revised.

Predicting crop yield

Measurements were made from a helicopter throughout East Anglia in 1985 and 1986, and in each year surveyed 20 beet fields whose yields could be precisely determined at the end of the season. In 1985 the objective was to use regional weather data and measurements of the fraction of light intercepted to predict yield from each field. Development of the crop canopies was based on real and average temperatures (Milford, Pocock, Jaggard, Biscoe, Armstrong, Last and Goodman, 1985), and yields were forecast on the basis of *Equation 13.2*, using a value for e of 1.0 g stored sugar per MJ of light intercepted (Scott and Jaggard, 1985). This value is typical of healthy, unstressed beet crops. The forecasts used measured solar radiation and temperature data up to the date of making a forecast and average weather data thereafter. Irrespective of harvest date, it was assumed that all growth stopped on 31 October.

These yield forecasts were biased in favour of the growth of plants in the centres of the fields, and made no allowance for date and efficiency of harvest or losses during handling and storage of the beet prior to delivery to the factory. Therefore, these forecast yields were compared with those obtained by the farmer when he delivered his crop. The resulting relationship was used to adjust the primary predictions made for the 20 fields in the experiment in 1986, in an attempt to forecast the farmers' yields.

The comparison between predictions made in early August and the growers' yields is shown in *Figure 13.3*. On individual fields the accuracy was poor. In part at least, this must be because the dates and efficiency of harvest and storage of the crop differed between farmers. However, the aim of this study was to develop a technique to forecast yield on a regional or national scale, not on individual fields. This was simulated by calculating the yields for combinations of five fields which were grouped at random. The figure shows that the actual and predicted values were very similar.

National forecasts of sugar yield

The forecasting system outlined above was tried on a national scale in 1987 and 1988. The Macam spectrophotometer and its logger were mounted in a helicopter which was used to fly a zig-zag track of approximately 1000 km in the beet-growing areas. This length of track enabled us to make approximately 2000 reflectance measurements on 300-350 beet fields (about 1% of the national crop) in 2 days. The flight-path was plotted accurately and, except for deviations due to air traffic control strictures, was flown on four occasions between mid-June and late August. In both years forecasts were made in early July and the results (*Table 13.1*) are compared with the total amounts of sugar (on an area basis) purchased by British Sugar from farmers in those seasons. The agreement between July-predicted and purchased amounts was very good in 1987 and it did not improve as the predictions were revised throughout the summer. During the summer and autumn of that year the radiation receipts were similar to the long-term average and rainfall was sufficient to prevent any prolonged water stress (Jaggard, 1988).

In 1988 the agreement was reasonable, but poorer, and predictions made later in summer were for smaller yields. The inaccuracy was caused by widespread symptoms

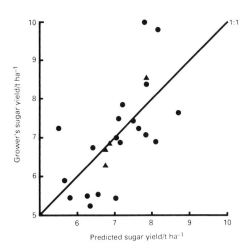

FIGURE 13.3. A comparison of predicted and delivered sugar yields from 20 fields (•) and random combinations of five fields (▲) in 1986. The line is not a fitted regression, but the 1:1 line of perfect equality

TABLE 13.1. The sugar yield (t ha^{-1}) of the beet crop in 1987 and 1988 as predicted in July from remotely-sensed data and as purchased by British Sugar plc.

	Predicted	Purchased	% Error
	Sugar (t ha^{-1})		
1987	6.8	6.75	+0.7
1988	7.7	7.31	+5.3

of virus yellows (Cooke, 1989), a complex of virus infections which cause the leaves to turn yellow and results in inefficient conversion of light to stored sugar. The early forecasts, made before virus symptoms were apparent, did not allow for the disease effect on the growth of the plants later in summer. Later forecasts, made when symptoms had appeared and were influencing the reflectance ratios, underestimated the delivered yields.

Despite the difficulty of dealing with crops whose foliage suddenly becomes chlorotic, this prediction system forecast the yield of crops of different size to within 5%. The system is now being used commercially and research is continuing in an attempt to discover ways to use remotely-sensed measurements to discriminate foliage amount from foliage condition. This will be the first requirement of a system which accurately and objectively forecasts the yields of crops affected by severe stress and disease.

References

CHURCH, B.M. and GNANASAKTHY, A. (1983). Estimating Sugar Production from preharvest samples. *British Sugar Beet Review*, **51**, 9-11

COOKE, D.A. (1989). Growth of the UK sugar-beet crop in 1988. AFRC Institute of Arable Crops Research, Report for 1988, 248

JAGGARD, K.W. (1988). 1987 and the growth of the crop. Rothamsted Report for 1987, 162-165.

MILFORD, G.F.J., POCOCK, T.O., JAGGARD, K.W., BISCOE, P.V., ARMSTRONG, M.J., LAST, P.J. and GOODMAN, P.J. (1985). An analysis of leaf growth in sugar beet. IV. The expansion of the leaf canopy in relation to temperature and nitrogen. *Annals of Applied Biology*, **107**, 335-347

SCOTT, R.K. and JAGGARD, K.W. (1985). The effects of pests and diseases on growth and yield of sugar-beet. *Proceedings of the 48th Winter Congress of the International Institute for Sugar Beet Research.* 153-169

STEVEN, M.D., BISCOE, P.V. and JAGGARD, K.W. (1983). Estimation of sugar beet productivity from reflection in the red and infrared spectral bands. *International Journal of Remote Sensing*, **4**, 325-334

V.

STRESS

14

HIGH-SPECTRAL RESOLUTION INDICES FOR CROP STRESS

M.D. STEVEN, T.J. MALTHUS, T.H. DEMETRIADES-SHAH[1], F.M. DANSON[2] and J.A. CLARK

University of Nottingham, U.K.

Vegetation Indices

Spectral indices have been used for some time for monitoring vegetation by remote sensing (e.g. Tucker, 1979; Curran, 1980). The original indices were based on combinations of visible and near-infrared bands, although other techniques have recently been proposed using microwave backscatter (Choudhury and Tucker, 1987; Bouman and Goudriaan 1989). Much developmental work at the moment centres around how vegetation indices should be constructed and interpreted.

Construction and interpretation of vegetation indices

The general principles behind the construction of vegetation indices (VIs) is that they should be sensitive to the presence of vegetation, more specifically to the amount and/or density of foliage, and insensitive to other environmental variables, such as soil background, atmospheric attenuation or solar angle. These requirements are difficult to achieve given that the environment is uncontrolled and in the general applications of remote sensing, unspecified or unknown.

The interpretation of vegetation indices is also problematical as there are different ways in which the amount or density of foliage may be specified. Much effort was spent in the past trying to establish relationships between the visible/near-infrared VIs and Leaf Area Index. More recently, a wide measure of acceptance has been gained for the interpretation of VIs in terms of intercepted or absorbed photosynthetically active radiation (PAR), (Steven, Biscoe and Jaggard, 1983; Asrar, Fuchs, Kanemasu and Hatfield, 1984; Goward, Tucker and Dye, 1985; Sellers, 1985; Kumar, 1988). This interpretation allows a direct relationship with the productivity of vegetation, mediated by a term representing the "efficiency" of conversion of absorbed radiation into biomass (Monteith, 1977). This "paradigm shift" in interpretation has probably come about not so much from the theoretical and empirical studies that established the validity of this interpretation but

---

[1]Current address. Department of Agronomy, Kansas State University, Manhattan, Kansas 66506, U.S.A.
[2]Current address. Department of Geography, University of Sheffield, Sheffield, U.K.

from a growing realisation that a direct measure of productivity is of greater practical value for biologists or agriculturalists than a measure of standing biomass at a particular date.

This history has a lesson for us: the relationships developed between spectral signatures of vegetation and the biophysical properties of canopies are almost always indirect. Interpretations that are put on spectral signatures or indices must be physically realistic to work, but to be useful they must also be suitably matched to the applications. We are fortunate that these criteria work in the same direction when visible/near-infrared VIs are interpreted in terms of productivity, but this may not be the case with indices for more subtle biophysical parameters.

In this paper, the applications under consideration are the detection of stress effects in vegetation and the estimation of their consequential effects on the productivity of agricultural systems. For this purpose, we broaden the scope of the term 'vegetation indices' from indices of canopy density only, to indices of the full range of biophysical parameters that respond to stress. In searching for suitable indices it is also worth bearing in mind that for the purpose of monitoring stress effects it may not be necessary to find a direct spectral response to the stress. Stress effects are very complex and, in general, not very well understood, but in practice some surrogate physiological marker which results in spectral change may be equally useful.

Limitations of vegetation indices

The conventional vegetation indices are based on ratios or other combinations of bands in the visible and near-infrared regions of the solar spectrum. A variety of indices have been devised with different constructions, but they are all functionally equivalent (Perry and Lautenschlager, 1984). In healthy green vegetation there are well established relationships of VIs with productivity but, when chlorosis occurs in stressed vegetation, these indices confound variations in vegetation cover with vegetation colour. The limitations of these indices were discussed by Steven (1985). For example, the relations between a vegetation index and light interception are shown in *Figure 14.1* for a series of Nitrogen application treatments in Spring Barley. The vegetation index is represented by the natural logarithm of the ratio of near-infrared (**IR**) and red (**R**) reflected radiation. Measurements were made using a Macam spectral ratio meter and normalised to a Kodak grey card (Steven, Biscoe and Jaggard, 1983). The curves are seasonal trajectories, and the downturn at the end of the season is characteristic of most crops. However, here there are also significant differences between treatments in mid-season, because leaf chlorophyll varied by almost a factor of two. Growth rates for these treatments were measured by the differences between successive harvests, taken every two weeks, and relations with productivity are shown in *Figure 14.2*, expressed per unit of incident solar radiation. There are large differences in the spectral ratio between treatments which do not correspond to any difference in productivity, and large temporal changes in the net productivity (about a factor of 2) are not detected by the spectral ratio. There were seasonal differences in productivity of 37% between these treatments, which were almost entirely accounted for by differences in intercepted light: the efficiency of light use did not differ significantly, in spite of chlorosis. In these circumstances the vegetation index therefore over-predicted differences in productivity, as shown in *Figure 14.2*.

On the other hand, in two trials of sugar beet deliberately infected with virus yellows disease, substantial differences in yield were accounted for mainly by lower efficiency

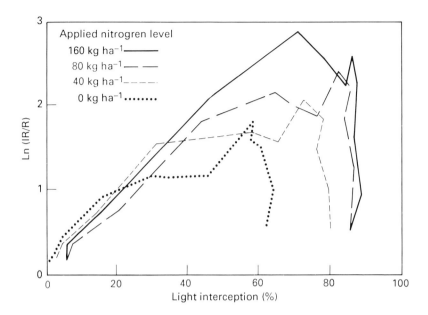

FIGURE 14.1. Vegetation index as a measure of the fraction of solar radiation intercepted by the canopy. Data for spring barley at Sutton Bonington in 1983, at four levels of applied Nitrogen. (Re-drawn from Steven, 1985)

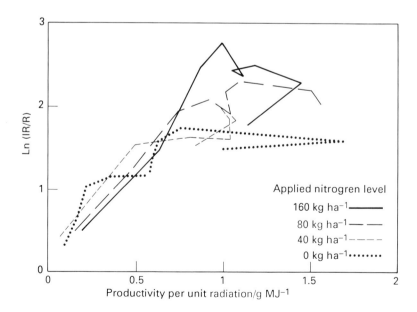

FIGURE 14.2. Vegetation index as a measure of productivity. Data from the same trial as in *Figure 14.1*

in the diseased crop. In this case, the reduction in VIs with chlorosis in the diseased treatments were consistent with the loss in efficiency (*Table 14.1*). However, it does not follow that the chlorosis was the cause of the decrease in efficiency and, as seen in the Nitrogen experiment, this kind of compensation does not always follow.

TABLE 14.1. Losses (%) in canopy production parameters due to virus yellows in sugar beet and corresponding differences in seasonally integrated vegetation indices. Key: **W** - Dry matter production; **fS** - intercepted solar radiation; **e** - dry matter energy quotient(or 'light use efficiency'); **IR/R** - integrated spectral ratio; ln(**IR/R**) integrated log transformed spectral ratio. (Note: the vegetation indices were weighted according to incident solar radiation.)

Year	δW	δfS	δe	$\delta(IR/R)$	$\delta(\ln(IR/R))$
1982	18	3	15	18	11
1983	30	13	20	22	14

In a drought experiment on field beans, carried out as part of the same programme, all treatments showed similar relations between ln(**IR/R**) and light interception (*Figure 14.3*). However, the droughted crops only achieved 59% of the optimum yield, which was accounted for by 17% less light intercepted and 29% lower efficiency. The vegetation index accounted accurately for differences in light interception, but with these large differences in efficiency would only account for about 1/3 of the loss in yield.

Effects of stress on vegetation

Stress may be defined as any factor that reduces the productivity of the canopy below its potential or optimum value. The response of vegetation productivity to stress (properly referred to as "strain") in terms of our accounting can be either an effect on the fraction of light intercepted and absorbed by the canopy or an effect on the efficiency with which that light is used to photosynthesise biomass. A large number of studies of crop growth under stress have found that many stresses have a larger effect on the fraction of light intercepted than on efficiency, at least in the long term. This observation accounts for the degree of success that conventional vegetation indices have had: essentially, they measure ground cover. Our principal concern is to extend the scope of vegetation monitoring to conditions where the strain is expressed in the efficiency, and to examine spectral signatures during the early stages of stress, before its effects on leaf area and light interception are significant.

Requirements for monitoring stressed vegetation

To monitor stressed vegetation, a set of indices is required that will meet the following requirements. It should enable us:

- to measure canopy density or light absorption for photosynthesis;
- to establish the presence of significant stress;
- to distinguish different classes of stress and

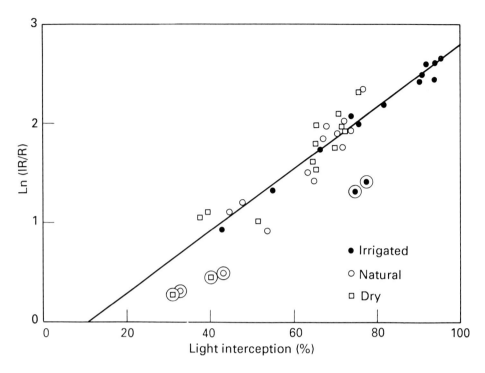

FIGURE 14.3. Vegetation index as a function of radiation intercepted by bean canopies under various water treatments. (Re-drawn from Steven, 1985)

- to measure the degree of stress and its effect on productivity.

The conventional two band VI such as **IR/R** used here or the Normalised Difference Vegetation Index (discussed elsewhere in these proceedings) can only meet the first of these requirements, subject to the limitations explained earlier. To fit the other requirements, we have been exploring the possibilities of using data with high spectral resolution across the visible and solar-infrared range of wavelengths.

We classify stresses, as shown in *Figure 14.4*, according to their effects on leaf area, leaf pigments and on vegetation metabolism and physiology. Stress effects which act on leaf area alone may be monitored with conventional VI's. Our current work is primarily concerned with stresses that induce chlorosis and with testing candidate indices to separate leaf colour from leaf amount. Other work is broadly concerned with forms of stress that affect productivity by closure of the stomata, using drought stress both as a model and as the most important example of this class of stress, and on the sensing of responses to disease.

This classification scheme is based on the premise that while stresses may take many forms, the responses (or forms of strain) exhibited by vegetation are limited in number and vegetation may respond to a range of stresses with similar patterns of strain. However, these strain responses are not exclusive and many stresses will give rise to a combination of responses.

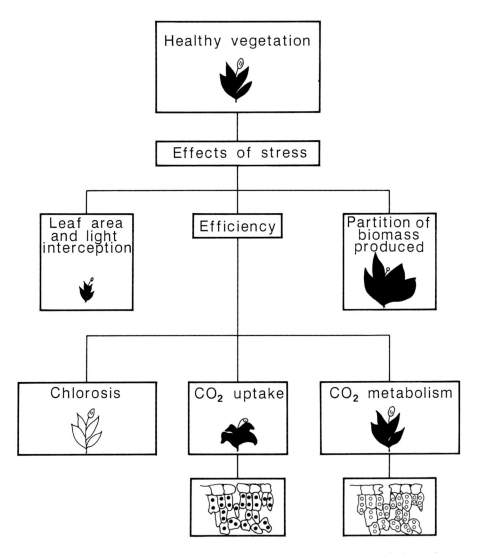

FIGURE 14.4. Schematic model of vegetation responses to stress. Counter-clockwise from top: Healthy vegetation; reduction in leaf area and light interception; loss of chlorophyll (represented by white leaves); reduction of CO_2 uptake by closure of stomata; restrictions to the uptake and metabolism of CO_2 within the cells; changes in the partition of biomass produced, leading to a reduction in the harvestable component

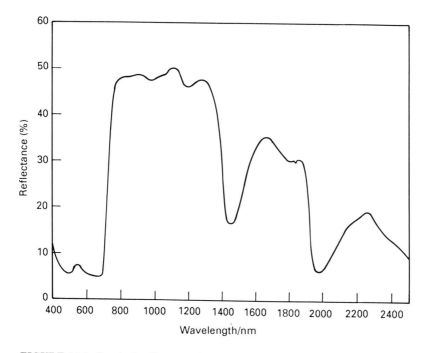

FIGURE 14.5. Spectral reflectance function of a sycamore leaf, relative to a barium sulphate panel. Measured in the laboratory with the IRIS spectro-radiometer

The need for high spectral resolution

Spectral resolution is defined as the discriminability of narrow spectral features (or loosely, the wavelength sampling rate). A number of features on the spectral signature of vegetation are too fine to discriminate using broad band sensors. *Figure 14.5* illustrates a typical leaf reflectance spectrum measured by a spectro-radiometer with a spectral resolution of a few nm. Many of the finer features, such as the weaker water absorption bands at 950 and 1170 nm, cannot be resolved by the broader band systems such as the Landsat Thematic Mapper. The use of fine spectral features reduces ambiguity of the spectral signal by helping to eliminate or reduce the effects of background variables. Much of the value of high spectral resolution lies in its ability to represent the slopes of absorption bands. These slopes correspond to regions of competing effects and are sensitive to small changes.

Technology

Developments in remote sensing have usually been technology led and applications of high spectral resolution are no exception. Imaging spectrometers that will measure the reflected radiance in several hundred narrow bands are planned for the next generation of satellites, and some have been tested already from airborne platforms or from the Space Shuttle (*Table 14.2*). The main impetus behind the development of imaging spectrometers has been to meet the needs of geological exploration, but the potential of this technology for vegetation monitoring is enormous. For example, the Moniteq PMI and AVIRIS imaging spectrometers each present more than two hundred channels of information for each pixel, and AVIRIS will be matched by HIRIS in the EOS satellite system (Goetz and Herring, 1989).

TABLE 14.2. Imaging spectrometers in operation or under development: PMI, Programmable Multispectral Imager; AIS, Airborne Imaging Spectrometer; AVIRIS, Airborne Visible Infrared Imaging Spectrometer; HIRIS, High Resolution Imaging Spectrometer (Wessman, Aber, Peterson and Melillo, 1988; Goetz and Herring, 1989)

Instrument	Spectral range	Number of bands	Resolution
Airborne systems			
Moniteq PMI	430 - 800 nm	288	2.6 nm
AIS tree scan mode	900 - 2100 nm	128	9.3 nm
rock scan mode	1200 - 2400 nm		
AVIRIS	400 - 2400 nm	224	9.6 nm
or	1200 - 2400 nm	32	36.8 nm
Spaceborne systems			
HIRIS	400 - 2400 nm	224	9.6 nm

History of high spectral resolution: the red-edge

The red-edge, that is the region 700-750 nm where the vegetation reflectance changes from very low in the chlorophyll red absorption to very high in the infrared, has long been recognised as a critical indicator of leaf vigour. Horler, Barber and Barringer (1980) studied a range of chlorotic and healthy leaves in the laboratory. They found that the wavelength of the red-edge was a good indicator of leaf chlorophyll and that when leaves were laid across backgrounds of different soils, this wavelength was invariant. However, they were unable to demonstrate the same effect conclusively for vegetation canopies, because when leaves overlap in canopies the spectral signal responds to the chlorophyll content of more than a single leaf. Similarly, Demetriades-Shah and Steven (1988) found that the red-edge for individual leaves was almost perfectly correlated with chlorophyll content, but that there was no correlation at all between the canopy red-edge and the red-edge wavelength (or chlorophyll content) of its component leaves. Indeed, *Figure 14.6* shows that the canopy red-edge in chlorotic and non-chlorotic canopies of sugar beet was almost invariant and occurred at a wavelength about 10 to 20 nm longer than the red-edge for individual leaves.

A number of studies (Collins, Chang, Raines, Canney and Ashley, 1983; Gauthier and Neville, 1985) have tried empirically to exploit variations in the red-edge for geobotanical applications. However, it has never been clear whether geobotanical anomalies become apparent in the red-edge because of spectral differences between tolerant and non-tolerant species in natural vegetation, because of spectral changes of specific species with stresses in general or because of spectral changes in vegetation due to the specific stresses occurring on metal-polluted soils.

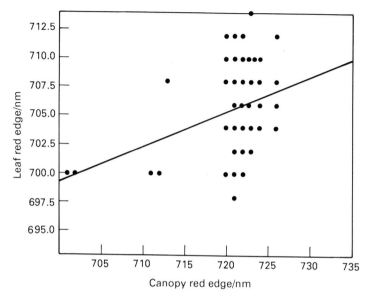

FIGURE 14.6. The wavelength of the red-edge measured over canopies of sugar beet and its relationship to the red edge for individual leaves from the same canopy. (Re-drawn from Demetriades-Shah and Steven, 1988)

Data processing

The implication of increased spectral resolution is that while the potential information contained in the data is much greater than in broader bands, the realisation of that potential is increasingly difficult because of the volume of data involved. Exploratory approaches, using arbitrary combinations of bands, have often been applied in the past to new forms of remotely-sensed data, but the burden of computation time to manipulate the data volumes produced by imaging spectrometers makes this approach unmanageable, Moreover, there is considerable redundancy in the data due to spectral autocorrelation. What is required is an *a priori* analytical scheme to reduce the data at each pixel to a small number of parameters, chosen to maximise the information with respect to vegetation from all the information available.

One method of data reduction is to model critical features of the spectrum by fitting a suitable shaped curve. Miller, Hare, Neville, Gauthier, McColl and Till, (1985) and Miller, Hare, Hollinger and Sturgeon, (1987) used an inverse-Gaussian curve to model the red-edge. This procedure reduces the slope to four parameters and these parameters can be tested for their suitability as vegetation strain indices. In these studies, some of the derived shape parameters were shown to be independent of illumination variation.

Noise is a particular problem with high spectral resolution data, as the instruments are operating close to their limits of sensitivity. Smoothing and filtering of the data are required in order to resolve the finer spectral features. A variety of smoothing techniques are available, including the curve fitting procedure described by Miller *et al.*, (1985; 1987), which smooths and extracts indices in one operation. Similarly, the derivative technique described by Demetriades-Shah and Steven (1988) involves fitting third-order polynomials to short segments of the curve. The fitted curve and its derivatives are then calculated from the fitted coefficients.

DERIVATIVE TECHNIQUES

Differentiation is an established technique in analytical chemistry to resolve the components of a spectrum and to reduce the effects of background spectral interference (O'Havers, 1982). Differentiation eliminates additive constants while reducing linear functions to constants, and is the basis of the red-edge technique. The derivative spectrum for a leaf (corresponding to the spectrum in *Figure 14.5*) is shown in *Figure 14.7* and that for a vegetation canopy in *Figure 14.8*. Both show highest values of the first derivative at the red-edge, as well as lower peaks either side of the green and at other wavelengths.

Demetriades-Shah and Steven (1988) tested a number of the major peaks on the derivative spectrum as indices of vegetation density and chlorosis, using the **IR/R** vegetation index as a standard for comparison. Although the red-edge proved not to be a satisfactory index of chlorophyll in canopies, there were several derivative features between 630 and 700 nm that showed reasonable promise. Other derivative features in the near-infrared were as good or better at estimating leaf area index or light interception than the conventional VI (**IR/R**) used for comparison.

Aims and methodology

The aims of our studies are to construct indices of vegetation strain that are as independent as possible of soil background, species, foliage density and other extraneous factors. We are particularly interested in determining how these measures of strain are related to variations in the efficiency of light use in vegetation. The approach that we have adopted involves:

- identification of candidate indices for stress monitoring and

- testing of these indices under various environmental conditions for their sensitivity to unwanted variables.

The work described earlier was concerned with crop canopies in the field. We are now re-examining some of the basic properties of leaves described by Guyot (these proceedings). Our interest is in the variations of the spectral properties with stress to assess which features may be exploitable. For this we need to consider both the sensitivity of such properties to the variable of interest and the sensitivity to other variables within their probable range in the environment. Measurements have been made with an IRIS Mk IV spectro-radiometer from the NERC[1] equipment pool. Experiments have studied the spectra of layers of leaves over a range of soil backgrounds and differences in the reflectance spectra between plant species measured over a constant background.

---

[1] Natural Environment Research Council

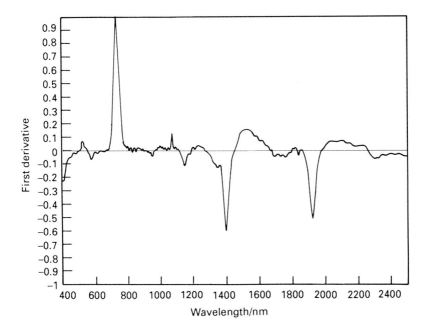

FIGURE 14.7. Normalised first derivative of the reflectance spectrum of a sycamore leaf (measured on the same leaf as in *Figure 14.5*)

Experimental design

Data from the leaf layering experiment have been used to test models of light transmission and reflection from leaves (Miller, Steven and Demetriades-Shah, 1989). However, our purpose in setting up the experiment was to establish indices of vegetation and of vegetation strain that are independent of the soil background. We also require indices that are sensitive to layering of leaves and others that are not or, in a less than ideal world, some that are more sensitive and some that are less. In experiments to examine stress effects it is important to de-couple as far as possible all the biophysical parameters involved. For example, in a field experiment it is very difficult to look at chlorosis and leaf cover independently because, in the longer term, almost all forms of stress tend to produce fewer or smaller leaves than a healthy canopy and the effects of chlorosis and leaf cover *per se* are inevitably confounded.

On one occasion we did, fortuitously, decouple leaf cover from chlorosis in a field trial of sugar beet (reported by Demetriades-Shah and Steven, 1988): The de-coupling occurred in a Nitrogen fertilizer trial, which would normally produce large dark leaves with High N and small pale leaves with Low N. In spite of our best efforts to prevent aphid infestation, aphids got in and caused a variation of colour at random across the trial due to sugar beet yellows virus. We therefore had variations in leaf amount due to N and an independent variation in leaf colour due to disease. This de-coupling, however, is very difficult to achieve on demand! If these parameters are not de-coupled strange

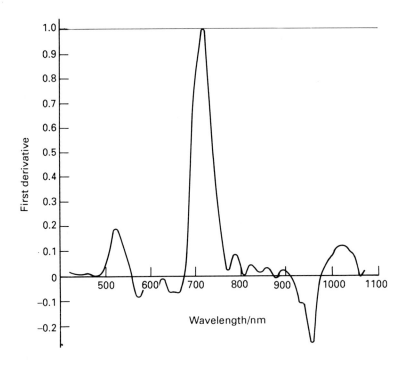

FIGURE 14.8. Normalized first derivative of the reflectance spectrum from 450 - 1100 nm for a closed canopy of sugar beet (Adapted from Demetriades-Shah and Steven, 1988)

statistical effects may occur, such as correlations of chlorophyll with the signal in the near-infrared (where we know *a priori* that there is no direct relationship of reflectance with the concentration of chlorophyll). In general, it is probably not possible (and certainly not practical) to try to de-couple all variables at once. Our approach to the analysis of spectral signature variations with stress is, therefore, through a series of partially de-coupled experiments.

Parameterization of biophysical properties

We have already discussed the issue of referring vegetation indices to leaf area index or to PAR interception and productivity. A second problem in relating spectral data to vegetation responses to stress is how the strain should be specified in terms of the biophysical parameters.

- How should we measure chlorosis? Is it sufficient to measure mean chlorophyll content per unit leaf area or do we also require an index of its variability and its distribution within the canopy?

- To specify water stress, should we use leaf water content, turgor, leaf water potential or some other measure?

We have not yet attempted to resolve these questions, indeed in many instances we have resorted to surrogate measures of these parameters for simplicity. For example, it is known that the reflectance of leaves in the near infrared is caused by multiple scattering between water-filled cells and air spaces. The actual magnitude of reflectance must depend on the detailed microphysical structure of leaves, which is difficult to measure. Instead we measured the thickness of leaves with a micrometer and the leaf water thickness by the difference between fresh and dry weights. We then used the air/water ratio as a surrogate parameter for the microphysical properties of leaves that determine reflectance. The correlation of this parameter with reflectance for a range of species, i.e. as a 'common index' of leaf water status, is shown in *Figure 14.9*. The correlation is significant from 750 - 1950 nm, although it only explains about 25% of the variance in reflectance. However, between 750 and 1350 nm the air/water ratio is a better correlate of reflectance than two alternative measures tested; percentage water content and leaf water thickness.

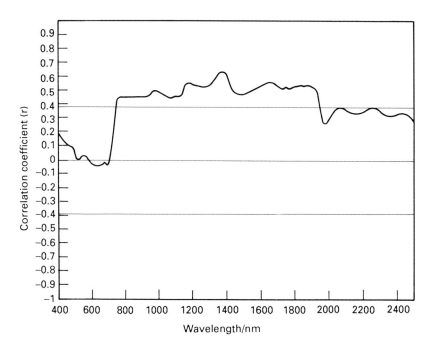

FIGURE 14.9. Spectrum of correlation coefficient between air/water ratio and reflectance for leaves from 26 plant species. (Re-drawn from Danson, Steven, Malthus and Clark, 1990)

A further point with establishing strain indices is that we are interested in the biophysical parameters only over part of their range. Large increases in reflectance with decreasing leaf water content were found by Gaussman (1985), who measured these changes by detaching a leaf from a plant and allowing it to dry out. In monitoring water stress in an agricultural environment, however, we are mainly concerned with that part of the range which a leaf might exhibit while still attached to a living plant.

Identification and testing of candidate indices

When considering candidate indices for biophysical parameters we are interested in the strength of response to the parameter in question, the immunity of the response against noise and the robustness of the response in different environments. (Differences between sites and environments introduce another form of noise that is not measured in an individual experiment). One approach is to use correlation spectra for a range of parameters. Robustness can be assessed by comparing results on different data sets. However, one index may work better than another in a particular environment due to the range and variability of the background noise parameters.

An example of the first stage of this approach is illustrated by our investigation of possible indices for water status. *Figure 14.10* shows the spectrum of correlation between the derivative reflectance spectrum and water thickness in leaves of various species, and *Figure 14.11* shows the corresponding correlation with the percentage water content by weight. In both figures, several maxima and minima exceed the threshold of significance and we might consider as candidate indices those that explain more than, say, 50% of the variance. We can argue that the weaker water absorption bands at the short wavelength end of the infrared spectrum should respond to greater thicknesses of water whereas the stronger absorption bands at longer wavelengths should respond only to the surface layers and may be a better indicator of relative rather than absolute water content. There is some evidence to support this argument in these data; certainly, the correlation peaks with percentage water become weaker at shorter wavelengths whereas this does not appear to occur with water thickness. However, when the same correlations are examined using a different data set from leaves of other species, only the derivative peaks at about 1050, 1150 and 1300 nm remain consistently correlated with water thickness, with another weaker possible candidate at 2300 nm. The maximum correlations with percentage water occur at different wavelengths from the previous data set and none of these features can be considered as reliable indices.

Once candidate indices have been identified by such procedures, the form of the correlation must be examined in a wide range of experimental conditions. An example of testing is illustrated for indices of vegetation cover. The ideal cover index would be independent of plant species, soil background, chlorophyll content and the degree of overlap of leaves in the canopy. To test for variations with soil and leaf overlap, the spectral reflectance of pea canopies was measured in the laboratory over backgrounds of sand and peat, representing extremes of soil colour. The tops of the plants were then cut back with scissors and the measurements repeated. These treatments effectively de-coupled leaf cover from leaf area index, because the leaves of the seedlings were clustered about the stems, so that pruning reduced leaf area proportionately more than the reduction in cover. *Figure 14.12* shows the relationship obtained in these measurements between a conventional vegetation index and cover. Two curves are apparent, corresponding to the different soil backgrounds, but there were no serious differences between the original and pruned canopies in the slopes of the relationships with cover. A number of candidate derivative indices from the infrared spectrum were also tested, some of them removing soil effects but exhibiting differences with pruning. The best candidate was the first derivative at 1520 nm, illustrated in *Figure 14.13*, which suppresses both soil and pruning effects while exhibiting a strong relationship with cover. Further testing with other data sets will be required to establish the consistency of this result and the effects of other forms of environmental noise.

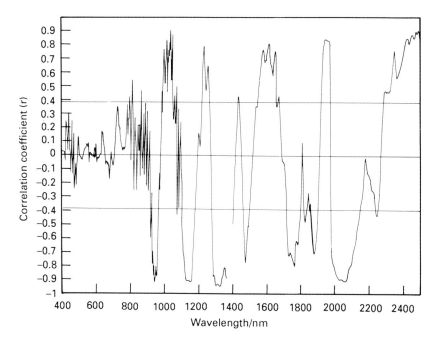

FIGURE 14.10. Spectrum of the correlation coefficient between leaf water thickness and the first derivative of reflectance for leaves from 26 species. (Re-drawn after Danson et al., 1990)

Future developments

There are a wide range of possibilities for the development of spectroscopic diagnosis of vegetation, and also a number of constraints on their practical implementation. A methodology which has been applied to extract information from other forms of multi-valence data is the technique of inversion (Campbell, these proceedings). For example, we have referred to the different sensitivities of the water absorption bands to different ranges of water depth. Inversion techniques applied to a series of measurements in these bands might allow a range of factors related to water status to be isolated. A related approach that might offer dividends is to "engineer" indices for specific biophysical parameters from algebraic combinations of simple extractable features of the spectrum. To apply such indices to the monitoring of the spectral consequences of strain in vegetation will, however, require parallel progress in the understanding of stress responses in vegetation, particularly as they effect productivity.

The most serious constraint to the application of new technology is, however, likely to be logistic. Imaging spectrometers may soon be deployed on satellites, but the sheer rate at which data is generated by these instruments means that we cannot expect sequential monitoring of agricultural crops through the growing season, at least for the foreseeable future, although they may have a contribution to make in geobotanical applications. One possible role for this technology in agriculture may be in commissioned airborne surveys to investigate specific problems. Another possibility is that ground-based research such

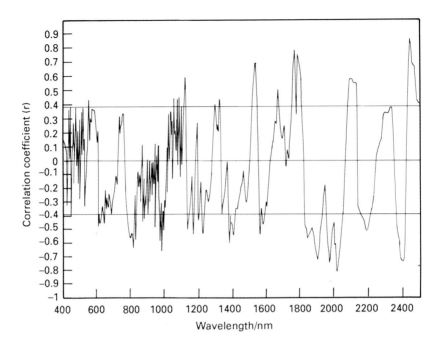

FIGURE 14.11. Spectrum of the correlation coefficient between relative water content and the first derivative of reflectance for leaves from 26 species. (Re-drawn after Danson et al., 1990)

as ours may lead to the development of data reduction algorithms, essentially transforming imaging spectrometer data to a few critical indices of vegetation condition. Rather, as Mather (these proceedings) described for the use of raster data in Geographical Information Systems, where areas of a uniform pixel class may be described by the values at the edge of the uniform field, the stored output from an instrument such as HIRIS might be reduced only to these critical indices or, at most significant spectral features. Alternatively, vegetation indices could be determined from a relatively small number of narrow-band measurements by a satellite-borne instrument specifically designed for that purpose.

A further attractive possibility which deserves consideration is to offer the technology to farmers. In an agricultural environment where there are increasing economic and legal constraints on the use of chemicals, a tractor-mounted device to diagnose stress in small areas has considerable potential advantages. Detection of variations in foliage density and/or conditions within a field, on the scales described by Blakeman and by Evans in these proceedings, can allow the possibility of immediate prophylactic treatment of affected plants. The result would be more effective use of chemicals and more efficient use of man and machine time, with benefit both to the farmer and the environment. The principle is being applied already to control the application of fertilizer. This is bringing Remote Sensing down to Earth.

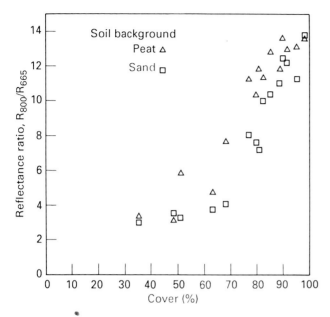

FIGURE 14.12. The relationship between the near-infrared:red reflectance ratio and vegetation cover for pruned and unpruned pea canopies on soil backgrounds of contrasting colour: peat, dark and sand, light

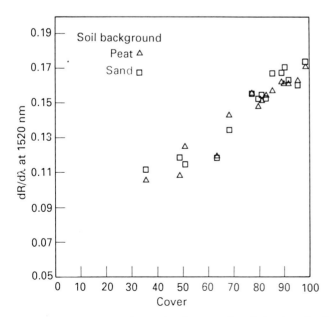

FIGURE 14.13. The relationship between first derivative of reflectance at 1520 nm and vegetation cover. Canopies as in *Figure 14.12*

Acknowledgements

The authors thank the Agricultural and Food Research Council and the Natural Environment Research Council of the United Kingdom for financial support of their past and current research on the spectral properties of vegetation.

References

ASRAR, G., FUCHS, M., KANEMASU, E.T. and HATFIELD, J.L. (1984). Estimating absorbed photosynthetic radiation and leaf area index from spectral reflectance in wheat. *Agronomy Journal*, **76**, 300-306

BOUMAN, B.A.M. and GOUDRIAAN, J. (1989). Estimation of crop growth from optical and microwave soil cover. *International Journal of Remote Sensing*, **10**, 1843-1855

COLLINS, W., CHANG, S-H., RAINES, G., CANNEY, F. and ASHLEY, R. (1983). Airborne biogeophysical mapping of hidden mineral deposits. *Economic Geology*, **78**, 737-749

CHOUDHURY, B.J. and TUCKER, C.J. (1987). Monitoring global vegetation using Nimbus 7 37 GH3 data. Some empirical relations. *International Journal of Remote Sensing*, **8**, 1085-1090

CURRAN, P.J. (1980). Multispectral remote sensing of vegetation amount. *Progress in Physical Geography*, **4**, 315

DANSON, F.M., STEVEN, M.D., MALTHUS, T.J. and CLARK, J.A. (1990). High spectral resolution data for determining leaf water content. *International Journal of Remote Sensing*, (in press)

DEMETRIADES-SHAH, T.H. and STEVEN, M.D. (1988). High resolution spectral indices for monitoring crop growth and chlorosis. In *Proceedings 4th International Colloquium on "Spectral Signatures of Objects in Remote Sensing"*, Aussois, France 18-22 January 1988, ESA SP-287, European Space Agency, Paris

GAUSSMAN, H.W. (1985). Plant leaf optical properties in visible and near-infrared light. Graduate Studies Texas Tech University No.29. Lubbock, Texas.

GAUTHIER, R.P. and NEVILLE, R.A. (1985). Narrow-band multispectral imagery of the vegetation red reflectance edge for use in geobotanical remote sensing. In *Proceedings 3rd International Colloqium on "Spectral Signatures of Objects in Remote Sensing"*. Les Arcs, France, 16-20 Dec. 1985. ESA SP-247. European Space Agency, Paris

GOETZ, A.F.H. and HERRING, M. (1989). The High Resolution Imaging Spectrometer (HIRIS) for EOS. *IEEE Transactions on Geoscience and Remote Sensing*, **27**, 136-144

GOWARD, S.N., TUCKER, C.J. and DYE, D.G. (1985). North-American vegetation patterns observed with NOAA-7 advanced very high resolution radiometer. *Vegetatio*, **64**, 3014

HORLER, D.N.H., BARBER, J. and BARRINGER, A.R. (1980). Effects of heavy metals on the absorbance and reflectance spectra of plants. *International Journal of Remote Sensing*, **1**, 121-136

KUMAR, M. (1988). Crop canopy spectral reflectance. *International Journal of Remote Sensing*, **9**, 285-294

Steven, M.D., Malthus, T.J., Demetriades-Shah, T.H., Danson, F.M. and Clark, J.A.227

MILLER, J.R., STEVEN, M.D. and DEMETRIADES-SHAH, T.H. (1989). Reflectance of layered bean leaves over different soil backgrounds: Measured and simulated spectra. (in preparation)

MILLER, J.R., HARE, E.W., HOLLINGER, A.B. and STURGEON, D.R. (1987). Imaging spectrometry as a tool for botanical mapping. *SPIE Proceedings*, **834**, San Diego, August 1987

MILLER, J.R., HARE, E.W., NEVILLE, R.A., GAUTHIER, R.P., McCOLL, W.D. and TILL, S.M. (1985). Correlation of metal concentration with anomolies in narrow band multispectral imagery of the vegetation red reflectance edge. In *Proceedings International Symposium on Remote Sensing of Environment, Fourth Thematic Conference - Remote Sensing for Exploration Geology*. San Francisco, April 1-4 1985

MONTEITH, J.L. (1977). Climate and the efficiency of crop production in Britain. *Philosophical Transactions of the Royal Society, London*, **B,281**, 277-294

O'HAVERS, T.C. (1982). Derivative spectroscopy and its applications in analysis. derivative spectroscopy: theoretical aspects. *Analysis Proceedings*, January 1982, 22-28

PERRY, C.R. and LAUTENSCHLAGER, L.F. (1984). Functional equivalence of spectral vegetation indices. *Remote Sensing of Environment*, **14**, 169-182

SELLERS, P.J. (1985). Canopy reflectance, photosynthesis and transpiration. *International Journal of Remote Sensing*, **6**, 1335-1372

STEVEN, M.D. (1985). The physical and physiological interpretation of vegetation spectral signatures. In *Proceedings 3rd International Colloqium on "Spectral Signatures of Objects in Remote Sensing*, Les Arcs, France, 16-20 Dec. 1985 ESA SP-247, European Space Agency, Paris

STEVEN, M.D., BISCOE, P.V. and JAGGARD, K.W. (1983). Estimation of Sugar Beet productivity from reflection in the red and near-infrared spectral bands. *International Journal of Remote Sensing*, **4**, 325-334

TUCKER, C.J. (1979). Red and photographic infrared combinations for monitoring vegetation. *Remote Sensing of Environment*, **8**, 127-150

WESSMAN, C.A., ABER, J.D., PETERSON, D.L. and MELILLO, I. (1988). Remote sensing of canopy chemistry and nitrogen cycling in temperate forest ecosystems. *Nature*, **335**, 154-156

15

THE IDENTIFICATION OF CROP DISEASE AND STRESS BY AERIAL PHOTOGRAPHY

R.H. BLAKEMAN

Aerial Photography Unit, ADAS, Block B, Government Buildings, Brooklands Avenue, Cambridge, CB2 2DR, U.K.

Introduction

In an age of high technology satellite sensors and computer based image analysis equipment, there is a tendency to overlook the role aerial photography has played, and continues to play, in agricultural remote sensing. Although aerial survey equipment and methods have changed little over the past forty years, the aerial photograph is still the most widely used source of remotely sensed data in crop studies (Curran, 1989; Lo, 1986; Simonett, 1983). The benefits of aerial photography in the study of crop diseases were recognised as early as the 1920's. Oblique black and white photographs were used to differentiate living and dead cotton plants and to pinpoint the foci of infection by root rot fungus (*Plymatotrichum omnivorum*) (Neblette, 1927). The current use of aerial photography for crop studies owes much to the research work into crop reflectance characteristics and film/filter combinations by Colwell in the USA (Colwell, 1956). In the UK, Brenchley pioneered the application of these findings in his study of potato blight (*Phytopthora infestans*), in commercial potato crops in the fens of East Anglia (Brenchley, 1966).

This paper looks at the advantages of aerial photography and illustrates its use in crop studies by the Agricultural Development and Advisory Service (ADAS), part of the U.K. Ministry of Agriculture, Fisheries and Food. The characteristics of different films and the factors affecting the choice of film are outlined. The reason for the failure of satellite sensors to meet the operational needs of agriculturists are also considered.

Advantages of Aerial photography

Although aerial photography cannot give as great a synoptic view as satellite imagery, it has proved reliable in detecting changes in crop condition at a farm and regional level (Curran, 1987; MAFF, 1986). Its advantages as an operational aid in agriculture can be summarised as:

- reliability and timeliness of acquisition;

- a good vantage point - photographs are usually obtained from heights between 150 m (500 ft) and 3000 m (10000 ft);

- fine resolution - dependent on film, contrast and scale, but typically 0.1 - 1.0 m;

- completeness of cover - a survey can be designed to cover a particular study area;

- ease of interpretation;

- the opportunity to extend observations to ground samples;

- ease of measurement;

- the opportunity for multi-temporal studies - during and between growing seasons;

- the ability to make discrete appraisals;

- economical and convenient to handle;

- the availability of infra-red sensitive films.

Agricultural requirements

The agriculturalist requires remote sensing techniques to assist in answering one or more of the following questions (Colwell, 1964):-

- What type of crop is growing in the field?

- What is the vigour of the crop and its likely yield?

- What is the agent responsible for any loss of vigour?

- What is the area of the crop?

In advanced agricultural systems this information usually contributes to the long term understanding of crop problems and leads to modifications in management in subsequent years. In under-developed and extensive agricultural systems, the synoptic view may be substituted for 'crop walking', indicating the extent of the problem and pinpointing the areas where treatment, such as spraying, are likely to be most effective (Hogg and Weste, 1975; Baker and Flowerday, 1979).

Crop identification

The identification of crops by a photo-interpreter relies on a combination of objective and subjective decisions. Identification is usually possible from carefully timed photography, using crop tone, texture, the pattern of row direction and spacing, and a knowledge of crop phenology and the husbandry operations involved in growing different crops.

Crop vigour and yield

There is no universal approach to determining crop vigour and estimating yield from aerial photography. It is often possible to make subjective assessments of comparative vigour but there are many inter-relating factors to take into account when predicting commercial yields. Even yield estimates based on objective field measurements, linked to predictive crop growth models, have proved unreliable due to climatic variability, differences in soil fertility within and between fields, insect pests and diseases and management practice. Aerial photographs are subject to the effects of lighting and haze, and also have relatively poor resolution when compared with field observations. Whilst they can help in the identification of vigour differences and in the measurement of these areas, data from aerial photographs has to be integrated with field data on crop yield to give production estimates. If we concentrate on the identification of crop stress, or reduction in vigour, using aerial photographs, the interpreter will need to appreciate the:-

- appearance of a healthy crop;
- possible damaging agents;
- manifestations of the damage;
- effect of the damage on spectral reflectance;
- likely appearance of the damage on aerial photographs.

An experienced agriculturalist will be familiar with the appearance of healthy and stressed crops on the ground but will need training to recognise them on aerial photographs. Field work is an essential part of the initial interpretation process, preferably carried out on a sample basis at the time of the photography. While the film can show a stressed condition it is necessary to check the crop to validate the diagnosis. As the number of ground checks increase and the correct identification of a particular problem is confirmed, then the greater the confidence in future photo-interpretation. However, several different damaging agents may show similar symptoms and a wide variety of other factors may change the appearance of a problem, both within and between growing seasons (Murtha, 1978). Damaging agents can be considered as any factor influencing the vigour of the crop. These include vertebrate and invertebrate pests, nematodes, disease, insufficient or excess water and nutrients, poor soil structure, mechanical and chemical damage. The manifestation of crop damage must have an effect on the above-ground foliage if it is to be recorded on aerial photographs. The effects may be categorised as changes in:-

- the tone or colour of leaves;
- leaf condition, including wilting and distortion;
- leaf area, including defoliation;
- leaf or stem orientation such as lodging.

These changes, either individually or in combination, will alter the electromagnetic reflectance characteristics of the crop. The change may be in the visible (0.4-0.7 μm) part of the spectrum and be recorded on conventional black and white or colour film, or in the solar infrared (0.7 - 0.9 μm) part of the spectrum which, although invisible to the eye, can be represented as a visible image on infra-red sensitive film.

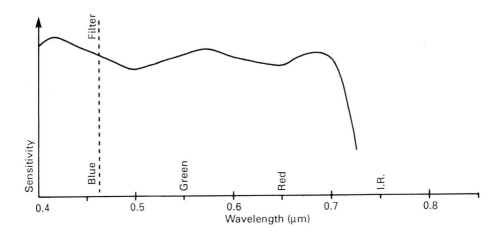

FIGURE 15.1. Spectral sensitivity of panchromatic film

Film types

Panchromatic

Panchromatic film is the film most commonly used in aerial survey cameras. It has a spectral sensitivity extending throughout the visible spectrum (*Figure 15.1*), all colours being represented as different shades of grey, producing an image familiar to the human eye. A yellow Wratten 12 filter is used with panchromatic film to absorb the blue light scattered by the water and dust in the atmosphere. This improves the contrast and resolution of the resultant image. Panchromatic aerial film emulsions have high Effective Aerial Film Speeds (EAFS), for example 320 for Kodak Double X Aerographic Film, giving them versatility over a wide range of lighting conditions (Kodak, 1971).

True colour

True colour film is again sensitive to reflectance in the visible portion of the spectrum (*Figure 15.2*), but has the advantage of representing colours in their 'true' form. As the human eye can discriminate many thousands of colour hues and only a few hundred shades of grey, colour film can be useful for plant species identification and land cover mapping (Paine, 1981). Aerial true colour films are 'slower' than panchromatic, Kodak Aerocolour 2445 having an EAFS of 100, and they are much more sensitive to haze because a yellow filter cannot be used. Producing colour prints is more expensive, and they tend to have lower resolution and a shorter archival life. However, the availability of 'fast' small format 35 mm and 70 mm colour film, and low cost printing and processing, make colour an attractive proposition for the amateur aerial photographer.

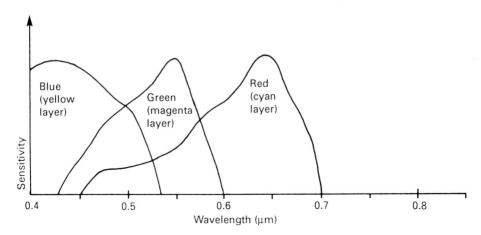

FIGURE 15.2. Spectral sensitivity of true colour film

Black and white infra-red

Infra-red black and white film has a sensitivity range from the ultra-violet (0.3 μm) extending into the near infra-red portion of the spectrum (*Figure 15.3*). A deep red filter, Wratten 89B, is usually used to absorb all but the far-red and infra-red parts of the spectrum. It is a 'fast' film, Kodak 2424 having an EAFS of 400. Although infra-red film is less affected by haze than panchromatic or colour film it does have a number of disadvantages. Shadows are recorded as black, resolution is relatively poor, and the bright tones produced from healthy agricultural crops can initially cause confusion to the interpreter. Film processing is straightforward but the film has to be handled in complete darkness. It is also necessary to recalibrate the lens for focussing, although many lenses are marked for use with infra-red films.

Colour infra-red

Colour infra-red film, sometimes known as 'false colour' film, has the same emulsion layers as colour film, but in this case they record green, red and near infra-red radiation (*Figure 15.4*). A yellow filter is essential to absorb blue wavelengths because all three emulsion layers are sensitive to blue light. On the final image green light is represented as blue, red as green, and reflected infra-red energy as red. This film penetrates haze better than true colour but is 'slow' and requires bright weather, for example Kodak 2443 has an EAFS of only 40 with a yellow filter fitted. Aerial false colour film is usually processed as a reversal film so that the film in the camera becomes the final diapositive used for interpretation. Duplication is expensive, so great care has to be taken when handling this material.

Choice of film

Stress detection is more likely when the crop canopy is well enough developed so that its reflectance predominates over that of the soil background (Lillesand and Kiefer, 1979; Curran, 1989). The choice of film to record crop stress and disease is made easier by

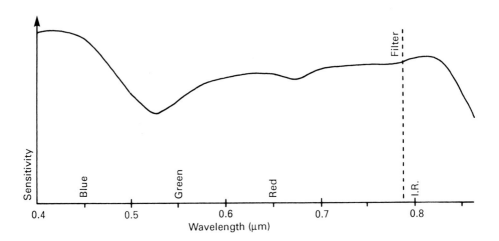

FIGURE 15.3. Spectral sensitivity of black and white infra-red film

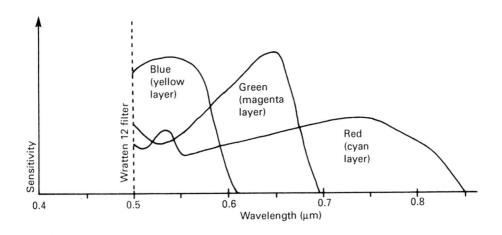

FIGURE 15.4. Spectral sensitivity of colour infra-red film

a prior understanding of the effects of the problem. The aim is to choose a film which maximises the contrast between a healthy and unhealthy crop (Hooper, 1980). In most cases comparisons have to be made between healthy and stressed crops growing in the same or adjacent fields. This enables finer differences in spectral response to be identified than would be possible without this comparison. It also helps to differentiate between damaging agents and reflectance variations due to plant variety, plant maturity, seed rate, nutrient status and soil differences.

Loss of infra-red reflectance is one of the earliest symptoms of a reduction in vigour in many plants (Colwell, 1964; Murtha, 1978). Turgid spongy and palisade mesophyll cells in the leaf of a healthy plant reflect up to 60% of incident wavelengths in the near infra-red range, 0.7 - 0.9 μm. When these cells become flaccid, as a result of wilting at times of drought or due to a restriction in water supply by a damaging agent, the infra-red reflectance is reduced. In the case of fungal infection, the leaf air spaces may be invaded by fungal hyphae, further reducing the infra-red reflectance from the leaves.

Visible changes in the appearance of leaves, such as yellowing, usually occur later when chloroplasts in the upper layers of the leaf become less reflective to green light. It is theoretically possible for crops to appear healthy in the field and to appear 'normal' on panchromatic and true colour films, but to show symptoms of stress on infra-red films. However, in practice, most aerial surveys will be at 1:5000 scale or smaller, which makes it impossible to identify individual stressed or diseased plants. It is the aggregation of affected plants that is identifiable on photographs, and at this stage there is likely to be some visible evidence on the ground. Nevertheless, because of the capability of aerial surveys to provide a synoptic view over large areas in a short time, there have been many examples in the course of our work where pest and disease problems have been found before farmers were aware of their existence.

The choice of film type will depend on the way a disease or stress affects the plant at a particular stage in its growth, taking account of the characteristics of different films. In cases where the first detectable symptom is a colour change, then panchromatic or true colour film will usually be suitable. Occasionally diseases produce a subtle characteristic change that can only be reliably detected on true colour film (Haglund and Jarmin, 1978). In cases where changes in leaf condition rather than colour predominate, infra-red film will give greater tonal contrast between healthy and unhealthy plants. *Plate 15.1* shows a field of potatoes with some areas suffering from early senescence and wilting, mainly due to differences in available soil water. Whilst an experienced interpreter would probably detect subtle textural and tonal differences on the panchromatic and colour films, the high reflectance from the healthy plants contrasts strongly with the lack of reflectance from the dying and wilting plants on the infra-red films.

Sometimes different photo-interpreters will express a preference for a different type of film even though they are dealing with similar applications (Myers, 1974). Although the features of interest may be decipherable on two or more film types, the final choice reflects the interpreters skill, experience and the fatigue element involved.

The ADAS APU has always retained the ability to use up to four different film types simultaneously. Currently two 70mm and one 35mm camera are used to provide back-up material for the higher resolution Wild 240mm metric camera, which is usually loaded with panchromatic film.

Crop disease

Although it is convenient to discuss the applications of aerial photography under specific headings, such as crop disease and stress, it is important to remember that in dealing with the dynamics of crop growth it is the interaction of many different factors which determines how a crop appears. In studying the epidemiology of disease we are seeking information on the development of disease in time and space.

Potato Blight

Brenchley (1966; 1968) identified the early stages of potato blight using black and white infra-red film. He was able to study field outbreaks of this airborne disease and monitor its spread in relation to local meteorological conditions. On infra-red film the foci of infection were seen as dark patches, contrasting with the light tone of the healthy crop. As the foci developed, secondary spread could be seen as an assymetric fanning out from the initial foci, a pattern typical of a wind borne disease (*Figure 15.5*). From the size and distribution of the foci, he was able to locate the likely source of infection. Detailed ground survey confirmed the sources as waste potato dumps and old clamps (Brenchley, 1968). The results of this work were incorporated in the advice offered to farmers recommending the destruction of dumps and earlier preventative spraying (Bell, 1974).

Sugar Beet Virus Yellows

More recently, aerial survey has been used to investigate the incidence of sugar beet virus yellows. This virus is transmitted to sugar beet by aphids feeding on the leaves. Waste beet, red and fodder beet clamps can act as reservoirs for overwintering aphids who subsequently infect the new season crop in spring and summer. *Figure 15.6* shows the distribution of virus yellows in a field of sugar beet, and illustrates how aerial photography can help pinpoint diseases carried by an airborne insect vector (MAFF, 1985).

Yellow Rust

Yellow rust, *Puccinia striiformis*, is an air-borne fungal disease of wheat and shows similar focal patterns to potato blight. However, in this case the colour changes in the leaves can be clearly seen on colour or panchromatic film. Infected fields are often identified on prints taken for other purposes and distinct differences in the level of infection can be seen between adjacent crops. The effect of variety can be seen in *Figure 15.7*, where cv Maris Kinsman has foci with 30 - 65% infection compared with cv Maris Huntsman where no foci are evident (MAFF, 1978). In 1988 yellow rust was a major problem and this was noted on survey photography. Varietal susceptibility can change as new races of the disease develop and varietal diversification is being recommended as part of future control programmes (Blake, 1989). Aerial photography could be an effective way of studying future yellow rust outbreaks and their spatial relationship with different wheat varieties in the surrounding area.

FIGURE 15.5. Development of potato blight in a field photographed on 24th June and 24th July 1964. Black and white infra-red film

Barley Yellow Dwarf Virus

Another aphid-borne disease was identified from aerial photographs taken in the summer of 1976. Distinctive patterns of dark foci were seen in winter wheat crops during the routine examination of survey photographs (*Figure 15.8*). These foci were thought to be associated with cereal aphids, but only a proportion of the aphid infected crops, identified from field survey, showed these symptoms. On photographs taken between June and late July, 134 crops showed symptoms affecting between 2% and 38% of the field. In the field the foci were easily identifiable, as the upper parts of the plants had a high incidence of *Cladosporum sp* sooty moulds and were less vigourous than the surrounding healthy crop. Detailed recordings were made of crop condition, height, growth stage, variety, sowing date, pesticide application and previous cropping. These data were collated for 127 and 69 fields with and without symptoms, respectively.

It was subsequently established that Barley Yellow Dwarf Virus (BYDV) was affecting the plants in the foci. Crops sown in September and early October had the worst symptoms and later sowings were unaffected. Even small differences in sowing date within the same field had a marked effect on the incidence of BYDV (*Figure 15.9*). Ground sampling just before harvest showed yield losses of up to 55% within the foci, resulting in field losses of up to 10%. A different type of focal pattern was associated with aphid feeding, showing no relationship with sowing date and resulting in losses of less than 2% (Greaves, Hooper and Walpole, 1983; MAFF, 1983).

The critical effect of sowing date and the severely reduced yield within the foci indicated that BYDV infection was taking place in the autumn. Further work by ADAS established that control of BYDV in winter wheat could be achieved by the application

FIGURE 15.6. Distribution of virus yellows in a field of sugar beet. Panchromatic film

of an aphicide in the mid-October to mid-November period. In the early 80's, a number of transects were flown covering a 2 km wide strip to assess the extent and distribution of BYDV in commercial crops of winter wheat and winter barley. This information was used to determine the impact of this disease on winter cereals, particularly in relation to autumn weather, drilling conditions and aphicide spraying regimes. In 1985 fields were photographed where areas had been deliberately left unsprayed. This was done to help check the validity of ADAS recommendations regarding aphid populations and autumn spraying. *Figure 15.10* clearly shows the unsprayed strip as well as several other areas where BYDV is present due to the inaccuracies of sprayer operations.

This study demonstrated the use of aerial survey to gather information rapidly over large areas where cereals were thought to be at risk. Identification of suspect fields enabled the limited ground resources to be targetted, dramatically increasing the size and representation of the survey. Using the photographic record, measurement of the affected areas in individual fields was possible, and by linking this to ground yield assessments the loss of production was calculated.

Take-all

Take-all, *Gaeumannomyces graminis*, is a root-rotting fungal disease of cereals which has become increasingly significant in the U.K. as a result of the trend towards continuous cereal cropping in the last twenty years. The disease is most serious on second and third year crops, particularly if the cereal root action is limited by other factors such as impeded drainage (Brenchley, 1974). Aerial photography has been used by ADAS to provide a visual record of well documented 'case histories', which demonstrate how soil and cultural interactions influence the development of take-all. One example will serve

FIGURE 15.7. Varietal effect on the incidence of yellow rust in winter wheat (see text for details). Panchromatic film.

FIGURE 15.8. Winter wheat fields showing dark foci infected with Barley Yellow Dwarf Virus. Panchromatic film, July

FIGURE 15.9. Effect of drilling date on the incidence of Barley Yellow Dwarf Virus in winter wheat. Panchromatic film

FIGURE 15.10. Presence of Barley Yellow Dwarf Virus within unsprayed strip. Note other spray misses resulting from inaccurate driving. Panchromatic film

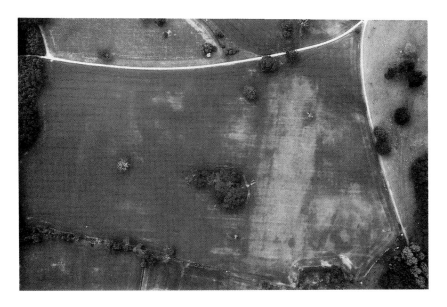

FIGURE 15.11. Take-all in winter wheat recorded as light tones on panchromatic film. June 1980

to illustrate these interactions and show how ground and aerial observations, including archival photographs, can be used in the documentation.

A severe outbreak of take-all occurred in a field growing winter wheat for the second successive year (*Figure 15.11*). It was established from ground survey that the soil varied from a well structured boulder clay at one end of the field to a sandy loam over gravel at the other. Take-all was worst at the lighter end of the field, but a straight line boundary could be seen on the aerial photograph between this and the mildly affected area, suggesting a man-made influence rather than a soil type boundary. Two years previously the mildly affected area had grown a crop of peas, whilst the worst affected area had grown second early potatoes. At the beginning of the potato harvest in July, soil moisture was high and conditions were difficult for tractors and machinery. Where the soil at the far end of the potatoes had time to dry out, harvesting conditions improved and the severity of take-all was reduced.

Archival photography, taken four years earlier, showed that the upper edge of the bad take-all patch was coincident with the boundary of a drainage system (*Figure 15.12*). This provided supporting evidence that inadequate drainage had aggravated the soil problems during the potato harvest and in the subsequent wheat crops (MAFF, 1979).

Barley Yellow Mosaic Virus

Since Barley Yellow Mosaic Virus was identified in this country in 1980, it has been recorded each year on aerial photography. The virus is associated with a common soil borne fungus *Polymyxa sp*, which infects cereal roots. Apart from the resistance shown by certain winter barley cultivars, there is no established control method. On aerial

FIGURE 15.12. Same field as *Figure 15.11* showing drainage system recently installed. Infra-red film

photographs the disease often shows 'kite' mark symptoms typical of a soil-borne organism (*Figure 15.13*). The kite often has a long axis running in the direction of primary cultivations, with a shorter axis resulting from further soil movement during seedbed preparation.

Aerial surveys have been used to identify infected fields. Follow up ground investigations have shown the influence of cropping frequency and the differences in varietal susceptibility. In 1984 a site in North Yorkshire provided clear confirmation of the importance of these factors. On this field winter barley was being grown for the third consecutive year. In the second year, the malting barley *cv* Tipper had been grown as a strip within the non-malting *cv* Sonja.

Figure 15.14 shows the situation in the third year when another malting barley, *cv* Maris Otter, was sown in a strip overlapping an area where *cv* Tipper and *cv* Sonja were grown the previous year. The remainder of the field was again sown with *cv* Sonja. Three levels of BYMV infection could be clearly seen. The highest level was in the area where two BYMV susceptible varieties followed each other, *cv* Maris Otter after *cv* Tipper. Where *cv* Maris Otter followed the non-susceptible *cv* Sonja there was a lesser degree of infection, with variations in soil type also having an influence. Where *cv* Sonja had been grown for two years no BYMV symptoms could be seen on the photograph nor were any found on the ground.

The photograph also provided evidence of the effect of past cropping and soil type. In one area, towards the end of the strip of *cv* Maris Otter, no BYMV symptoms were apparent. This area had been part of a separate field until 1968 and had been grassland until 1960. The higher organic matter in this part of the field seemed to have reduced the effect of the virus and supports previous findings that this disease is more serious on

FIGURE 15.13. Typical 'kite' pattern of Barley Yellow Mosaic Virus. Panchromatic film

FIGURE 15.14. Effect of variety and past cropping on the incidence of Barley Yellow Mosaic Virus (see text) Panchromatic film

lighter soils (MAFF, 1984).

Rhizomania

Rhizomania is a virus disease of sugar beet and, like BYMV, is associated with the fungal vector *Polymyxa sp*. Since its discovery in Italy in the 1950's it has become the major cause of lost sugar production throughout most of Europe (Richard-Mollard, 1985). In 1987 the first outbreak of rhizomania in the UK was identified in a field in Suffolk. The infected field was photographed to help establish the source of the disease. At the same time, as we had no previous experience of rhizomania symptoms in this country, we were keen to find the most suitable film type for future crop monitoring surveys. We found that changes in the spectral reflectance characteristics of the foliage were most significant in the visible part of the spectrum and that these were discernable on panchromatic and colour, but not on black and white infra-red films. This confirmed earlier research work in France (Andrieu, 1985). In the early stages of infection, rhizomania shows as pale green or translucent yellow lines and these are often elongated in the direction of soil movement. As the virus becomes more extensive, large areas of beet show this discolouration and there is more likely to be confusion with other problems such as virus yellows. Despite the obvious linear patterns of rhizomania visible on the colour photography of the infected site in Suffolk, it has proved impossible to locate the source of the disease. Aerial surveys will continue in August/September each year, particularly in East Anglia, to locate suspect fields where the symptom patterns may not be apparent from the ground.

Crop husbandry and stress

The aim in all crop growing situations is to achieve optimum plant establishment and then ensure the provision of water, nutrients and protection from pests and disease to maximise the economic return. Apart from pests and disease, the failure of crops to reach their full potential can often be related to soil characteristics and husbandry operations.

Soils and cultivations

There is little a farmer can do about underlying soil type but, as the basic growing resource, sound management is essential for long term profitability. To achieve this he needs to be aware of the soil types within and between his fields. There is an increasing awareness amongst farmers of the value of soil maps (Endacott, 1989) and aerial photographs are an important component in a soil scientist's mapping work (Carroll, Evans and Bendelow, 1977; Milfred and Kiefer, 1976). Timing of the photography is important. Tonal differences are visible in the autumn before crops become established. In the summer, soil types can be delineated as crops ripen unevenly because of differences in available soil water (*Figure 15.15*). The physical characteristics of the different soils must then be established by field sampling.

Soil structure describes the physical condition of the soil. It has an overriding influence on crop growth because it controls the environment in which the roots develop. Roots need to extend to depths of one metre or more, and farmers have to decide on the most appropriate cultivations, taking account of soil type, moisture status and the crop to be grown. The maintenance of good soil structure depends on these decisions as over-cultivation, the choice of inappropriate implements and the passage of machinery

FIGURE 15.15. Differential ripening of oil seed rape due to underlying soil differences. Soil series defined by field survey. Panchromatic film, July

can all lead to soil compaction and an increased risk of waterlogging (Davies, Eagle and Finney, 1972).

The effects of traffic and cultivations on soil structure are most apparent on photographs taken in late autumn and early spring. Areas of compaction show surface ponding after periods of rain, as well as poor crop emergence and establishment. Subsequently, weeds often invade these areas because of the lack of competition from the crop. By applying a knowledge of the field operations involved in growing a crop, patterns of poor growth can be related to the stage at which the soil damage occurred. Examples of this can be seen in a recent study of commercial sugar beet cultivation practice based on aerial photography and farmers' records (ICI Plant Protection and British Sugar, 1985).

Where drainage systems can be seen on aerial photographs this is often an indication of soil structure problems. In *Figure 15.16*, the most vigorous growth is seen directly over the drain lines, indicating a better rooting environment. This is not unusual in the first crop following the installation of drains, as these plants benefit from the soil disturbance and improved water movement through the soil profile. However, if these strong patterns persist, field examinations of soil profiles between the drains may show the need for subsoiling or moling, or there may be a compaction layer 10 - 15 cm below the surface which could be broken by a shallower working implement.

Crop husbandry operations

Mistakes in most crop husbandry operations are painfully obvious from the air. Aerial photographs can be very persuasive in convincing farmers and growers that they need to change their management practices (Blakeman, 1986; Wildman, Neja and Clark, 1976).

FIGURE 15.16. Spring barley showing vigorous growth over drains and retarded growth in areas where nitrogen fertiliser application has been missed. Panchromatic film

Photographs of cereals in late April-June will clearly show striping if fertiliser application is uneven. Misses or overlaps can be the result of poor driving accuracy, although this is less common since the widespread adoption of 'tramlines'. More often there are problems with spreader mechanisms or the machine has been incorrectly calibrated for the type of fertiliser being spread. Overlaps result in darker stripes and the extra nitrogen may cause lodging as the crop ripens. Misses show as light toned strips resulting from poorer crop vigour (*Figures 15.16* and *15.17*). Although striping may be apparent from walking crops, it is easier to diagnose the cause and take swath measurements using aerial photographs.

Herbicide overlaps are recorded as lines of scorched or retarded growth whereas misses subsequently show as areas of uncontrolled weed growth. Overlaps of fungicides and insecticides are less obvious, but misses are sometimes revealed by failure to control the offending organism (*Figure 15.10*). Striping from sprayers can also occur as a result of blocked or damaged nozzles, uneven flow to sections of the spray boom or through poor agitation. *Figure 15.17* also shows how important weather conditions can be when spraying. The chemical applied to the dark toned wheat has been carried by the wind and killed or scorched several metres of the adjacent light toned oil seed rape. Other examples of faulty husbandry operations which show up on aerial photographs include blocked drill coulters, hoeing damage in sugar beet and uneven watering by irrigation equipment.

FIGURE 15.17. Differences in crop vigour resulting from uneven application of nitrogen fertiliser. Winter wheat, panchromatic film. Note spray drift damage to adjacent oil seed rape crop

Crop trials

There are many examples where aerial photography has been used to provide researchers with a permanent record of trial plots (Harris and Haney, 1973; Wallen, Jackson, James and Smith, 1975; MAFF, 1984; 1985; 1986). The features on the photographs can be used as a cross check with field observations, and plots can be ranked on a comparative basis for factors such as crop cover, height, uniformity, vigour, disease incidence and lodging.

On occasions when there are apparent anomalies in trial results, scrutiny of the photography can help to explain these differences. In one instance a farmer had inadvertently applied split nitrogen treatments in the opposite way to that recorded for the plots. Differences in crop vigour visible on aerial photographs in June suggested a mistake had been made and this was subsequently confirmed by talking to the tractor driver (MAFF, 1984). Other benefits of using aerial photographs in crop trials work include selection of uniform soils and the placement of trials in areas known to be affected by soil-borne diseases so that control treatments can be tested.

Image analysis

Patterns of stress and disease are recorded as tonal differences on aerial photographs. While it is impossible to standardise tones on photographs taken at different times, a comparison of tones on an individual print will minimise variations due to lighting, processing and printing. Using this principle there has been some success in correlating densitometer readings with crop vigour and levels of disease infection (Wallen, Jackson, Basa, Baeriziger and Dixon, 1977; Wallen and Jackson, 1978; Wallen et al., 1975; Brenchley, 1974; Bell, 1974). More recently, computer image analysis equipment has become widely available where photographs can be digitised by 'frame grabbing' through a video camera. This enables areas to be measured using 'contrast stretching' and 'density slicing' routines. By interpreting and measuring the different grey tones on a panchromatic print and linking the results with ground observations, the commercial impact of a stress factor can be determined.

As an example, *Plate 15.2a* shows grey-tone differences in the vigour of part of a sugar beet crop due to the residual effects of a herbicide applied to the previous cereal crop. Alongside is the corresponding computed density sliced image *Plate 15.2b* where colours have been assigned to two levels of crop cover and bare soil and the total number of pixels and the number in each class calculated. It would be almost impossible to measure these discontinuous areas in the field, and the symptoms may be short-lived as backward crops often eventually establish themselves. The combination of the photographic record, field observations, photo-interpretation and an area assessment provides strong evidence in any dispute over responsibility, even when the crop has been harvested.

The role of satellite sensors

Earth observation satellites have the significant advantage of being able to provide data on land cover and vegetation at a national or even World level. However this science is at an early stage in its development and, despite the enthusiasm of its proponents, it has not made the major contributions to agriculture with which it is often credited. It is worth considering the reasons for this.

The identification of crops (crop classification) poses complex problems for skilled photo-interpreters using 1:10000 or larger scale aerial photography, with a resolution of less than one metre. Identification is usually possible, from carefully timed photography, using a combination of tone, texture, pattern, knowledge of 'crop calendars' and husbandry operations. This approach can be used to analyse satellite data visually, but effective classifiers based on computed algorithms are needed if operational systems are to contribute to crop condition and production statistics.

With satellite sensors, reflectance data are available covering discrete bands in the visible and infra-red regions of the spectrum. Crop identification based on spectral pattern recognition depends on finding reflectance signatures characteristic of specific crops. However, few crops can be identified from a single spectral signature at a single point in the growth cycle (Orsenigo, 1984). It will be necessary to devise multi-temporal indices for individual crops during their growth cycle, both to distinguish between them and to monitor variations in their development (Epema, 1987; Steven, Malthus, Demetriades-Shah, Danson and Clark, these proceedings). A number of workers have stated a requirement for three to five cloud free scenes if satellites are to realise their potential in agriculture (Odenweller and Johnson, 1984; Kleschenko, 1983). In a crop inventory study in Italy, five Landsat MSS images were needed each year. However, these were only obtained in two of the six years of the project (Menente, Azzali, Collado and Leguizamon, 1986). In the UK, lack of cloud free images has seriously hindered progress and in many areas only one or two cloud-free Landsat scenes have been obtained in any one year. Timing of acquisition cannot be predicted and even when ADAS and NRSC had priority programming in the 1987 SPOT PEPS study, the phenologically important July scene was not obtained.

The United States Department of Agriculture (USDA) has also highlighted the practical limitations of satellite data (Vogel, 1987). Entire crop seasons passed without a single cloud-free scene being obtained. Where scenes were acquired, adjacent satellite swaths had a time differential of eight days or more. Reflectance characteristics are closely linked to crop growth stage. Seasonal climatic variations, local soil differences and agricultural practices, such as choice of variety, sowing date and fertiliser regimes, all affect the phenological stage of a crop at a given date, both within and between fields in the same locality (Crist, 1984). Thus, the delay between adjacent swaths meant that the discriminant functions used to classify a crop at one date and pass could not be used on another. Equally, new functions had to be derived each year.

Delays in receiving data and the time needed for processing often mean that the results can only be used as a retrospective check on area estimates derived by conventional ground surveys. Data are required within a few days of acquisition by the satellite and, while there is no technological reason why this should not be possible, it is usually at least 1 - 2 months before delivery to the user (Ferns and Hieronimus, 1989).

It has been possible in large uniformly cropped areas, such as the Mid-West of the USA and the Steppes of the USSR, to achieve classification accuracies better than 90% (May, Holko and Andersen, 1983). In the more complex environments which exist within the EEC and other intensive agricultural regions, crop classification accuracies have been disappointing and fall below 60% (Allan, 1986; Hill and Megier, 1988;Vogel, 1987). The greater spectral and spatial sensitivity of the Landsat Thematic Mapper might be expected to increase the accuracy of identification, but the increased volume of data and the larger proportion of 'mixed' pixels present further difficulties.

The identification of crop stress and disease from satellite data is a greater challenge. Timeliness of data acquisition and repetitive coverage are even more important as symptoms are often transient and closely linked to crop growth stage (Blazquez, Elliott and Edwards, 1981). In addition, some workers consider that the wavebands currently available on satellite sensors are not only too broad to discriminate crop species but they are also unsuitable for distinguishing the early stages of chlorotic stress in the yellow-orange wavelengths (Rao, Brach and Mack, 1978). Assuming the problems of data acquisition and availability can be resolved by a combination of increased satellite numbers, radar sensors and greater attention to the logistics of getting data to users, satellite sensors are only likely to provide useful information on crop stress and disease when the symptoms are extensive (Epstein, 1975). Even then the causal agents will be difficult to identify. All too often we know of a problem and then go to the satellite imagery, see a slight spectral change, and report success. The same spectral change may well occur in other parts of the image for other reasons which we choose to ignore (Heller, 1978).

In the late 1960's it was claimed that satellite imagery enabled dying trees to be individually located and plotted accurately enough for foresters to plan chemical control and salvage operations (Pardoe, 1969). Today it is suggested that satellites such as SPOT could provide farmers with a means of continuously monitoring the health of their crops to help determine irrigation, fertiliser and disease control needs, as well as monitoring the progress of their competitors (Bescond, 1989). The reality of operational remote sensing is very different from these over-optimistic claims. Experience to date suggests that individual farmers are unlikely to have timely data available to them, even if they could afford the high cost of satellite products and their analysis. The most likely value of satellite remote sensing in agriculture will be to monitor crop production on a national and global scale. The statistics produced will be useful to policy makers and commodity traders, but even at this level there is still a lack of convincing evidence that these applications will be practical or cost effective.

Conclusion

The bird's eye view has much to offer farmers and those involved in crop studies. Aerial photography is established as an operational technique in a wide range of agricultural applications.

Aerial photography is readily available to crop scientists, farmers and growers and has been commissioned and used by them in achieving a better understanding of the factors influencing crop performance. Aerial photographs emphasize any non-uniformity, show differences in crop vigour, and allow the spatial extent and distribution of features to be determined, often an almost impossible proposition from the ground.

Multi-temporal photography continues to help in the study of disease epidemiology and provides a permanent record of crop performance where the effects of control methods and changes in management practice can be monitored.

It should be clear that aerial photography only supplements and does not replace ground data collection systems. Similarly, satellite imagery is unlikely ever to replace aerial photography in fulfilling the applications discussed in this paper and few of its proponents would claim it could do so. Satellite sensors will be used to supplement and increase the efficiency and accuracy of existing crop monitoring systems, although there are still many technical problems to overcome.

Up to two thirds of the World Population suffers from hunger and malnutrition and many national governments have to resolve the problem of feeding their people. The economic development of these countries is dependent on improvements in their agricultural production. Aerial photographs are used in some developing countries for land use planning. In others, for security reasons, they are often restricted to military use. In both cases there is usually a low level of awareness of their value in agriculture. Satellite image processing systems are unlikely to be widely adopted in these developing countries, because of a lack of resources and expertise, but an open skies policy and improved sensor resolution should serve to increase their awareness of remote sensing. The use of satellite photo-products will allow a low technology approach and perhaps result in a relaxation of the restrictions governing the use of aerial photographs. Hopefully, these developments will lead to a greater recognition of the value of combining a range of remote sensing techniques in the quest to improve food production.

References

ALLAN, J.A. (1986). Remote sensing of agriculture and forest resources from space. *Outlook on Agriculture*, **15**, 65-69

ANDRIEU, B. (1985). Utilisation de donnees photographiques et radiometriques pour la mise en evidence de la rhizomanie de la betterave. INRA, Versaille, France.

BAKER, J.J. and FLOWERDAY, A.D. (1979). Use of low altitude aerial biosensing with colour infra-red photography as a crop management service. In *Joint Proceedings of the American Society of Photogrammetry and American Congress on Survey and Mapping*, Sioux Falls. American Society of Photogrammetry, Falls Church, Virginia, USA

BELL, T.S. (1974). Remote sensing for the identification of crops and crop diseases. In *Environmental Remote Sensing: Applications and Achievements*. Ed. by Barrett, E.S and Curtis, L.F. Arnold, London

BESCOND, P. (1989). Commercialisation of remote sensing in the USA: The SPOT perspective. *International Journal of Remote Sensing*, **10**, 289-294

BLAKE, A. (1989). Yellow rust alert. *Farmers' Weekly*, Feb 24, 25

BLAKEMAN, R.H. (1986). Aerial photography - Its role in farm management. *Journal of the Centre of Management in Agriculture*, **6**, 113-117

BLAZQUEZ, C.H., ELLIOTT, R.A. and EDWARDS, G.J. (1981). Vegetable crop management with remote sensing. *Photogrammetric Engineering and Remote Sensing*, **47**, 543-547

BRENCHLEY, G.H. (1966). The aerial photography of potato blight epidemics. *Journal of The Royal Aeronautical Society*, **70**, 1082-1085

BRENCHLEY, G.H. (1968). Aerial photography for the study of plant diseases. *Annual Review of Phytopathology*, **6**, 1-22

BRENCHLEY, G.H. (1974). *Aerial photography for the study of soil conditions and crop diseases*. Soil Survey Technical Monograph No 4, Bartholomew Press, England

CARROL, D.M., EVANS, R. and BENDELOW, V.C. (1977). *Air photointerpretation for soil mapping*. Soil Survey Technical Monograph No 8, Harpenden, England

COLWELL, R.N. (1956). Determining the prevalence of certain cereal crop diseases by means of aerial photography. *Hilgardia*, **26**, 223-286

COLWELL, R.N. (1964). Aerial photography - A valuable sensor for the scientist. *American Scientist*, **52**, 16-49

CRIST, E.P. (1984). Effects of cultural and environmental factors on corn and soybean spectral development patterns. *Remote Sensing of the Environment*, **14**, 3-13

CURRAN, P.J. (1987). Remote sensing in agriculture : an introductory review. *Journal of Geography*, **86**, 147-156

CURRAN, P.J. (1989). Crop condition. In *Manual of Photographic Interpretation (second edition)*. American Society of Photogrammetry, (in press)

DAVIES, D.B., EAGLE, D.J. and FINNEY, J.B. (1972). *Soil Management*, Farming Press, Ipswich

ENDACOTT, C. (1989). Reading between the lines. *Crops*, **6**, 14-15

EPEMA, G.F. (1987). New information of second generation Landsat satellites for agricultural applications in the Netherlands. *Netherlands Journal of Agricultural Science*, **35**, 497-504

EPSTEIN, A.H. (1975). The role of remote sensing in preventing World hunger. In *Proceedings of the Fifth Biennial Workshop on Colour Photography in Plant Sciences*, American Society of Photogrammetry, Falls Church, Virginia, USA

FERNS, D.C. and HIERONIMUS, A.M. (1989). Trend analysis for the commercial future of remote sensing. *International Journal of Remote Sensing*, **10**, 335-350

GREAVES, D.A., HOOPER, A.J. and WALPOLE, B.J. (1983). Identification of barley yellow dwarf virus and cereal aphid infestations in winter wheat by aerial photography. *Plant Pathology*, **32**, 159-172

HAGLUND, W.A. and JARMIN, M.L. (1978). Aerial photography for the detection and identification of fusarium wilt of peas. *Plant Disease Reporter*, **62**, 570-572

HARRIS, J.R. and HANEY, T.G. (1973). Techniques of oblique aerial photography of agricultural field trials. Division of Soils Technical Paper No 19, Commonwealth, Scientific and Industrial Research Organisation, Australia

HELLER, R.C. (1978). Case applications of remote sensing for vegetation damage assessment. *Photogrammetric Engineering and Remote Sensing*, **44**, 1159-1166

HILL, J. and MEGIER, J. (1988). Regional land cover and agricultural area statistics and mapping in the Departement Ardeche, France, by use of Thematic Mapper data. *International Journal of Remote Sensing*, **9**, 1573-1595

HOGG, J. and WESTE, G. (1975). Detection of die-back disease in the Brisbane ranges by aerial photography. *Australian Journal of Botany*, **23**, 775-781

HOOPER, A.J. (1980). Aerial photography as an aid to pest forecasting. *Journal of the Royal Aeronautical Society*, June 1980, 136-140

ICI PLANT PROTECTION and BRITISH SUGAR (1985). *Operation Flightpath - An Aerial Survey of Sugar Beet Fields*. ICI and British Sugar Corporation joint publication. ICI, Farnham and British Sugar, Peterborough

KLESCHENKO, A.D. (1983). *Use of remote sensing for obtaining agrometeorological information*. Commission for Agricultural Meteorology of the World Meteorological Organisation, Report No.12. Geneva

KODAK (1971). *Kodak Data for Aerial Photography*, Eastman Kodak, Rochester, New York

LILLESAND, T.M. and KIEFER, R.W. (1979). *Remote Sensing and Image Interpretation*, John Wiley, New York

LO, C.P. (1986). *Applied Remote Sensing*, Longman, New York

MAFF (1978). Annual Report, ADAS Aerial Photography Unit, Cambridge

MAFF (1979). Annual Report, ADAS Aerial Photography Unit, Cambridge

MAFF (1983). Annual Report, ADAS Aerial Photography Unit, Cambridge

MAFF (1984). Annual Report, ADAS Aerial Photography Unit, Cambridge

MAFF (1985). Annual Report, ADAS Aerial Photography Unit, Cambridge

MAFF (1986). Annual Report, ADAS Aerial Photography Unit, Cambridge

MAY, G.A., HOLKO, M.L. and ANDERSEN, J.E. (1983). *Classification and area estimation of land covers in Kansas using ground gathered and Landsat digital data.* AGRISTARS technical report, NASA/USDA, DC-Y3-04441, Mississippi, USA

MENENTE, M., AZZALI, S., COLLADO, D.A. and LEGUIZAMON, S. (1986). *Multitemporal analysis of Landsat Multispectral Scanner and Thematic Mapper data to map crops in the Po valley (Italy) and in Mendoza (Argentina).* Institute for Land and Water Management Research, Technical Bulletin No.51. Institute for Land and Water Management Research, Wageningen, Netherlands

MILFRED, C.J. and KIEFER, R.W. (1976). Analysis of soil variability with repetitive aerial photography. *Journal of the Soil Science Society of America*, **40**, 533-537

MURTHA, P.A. (1978). Remote sensing and vegetation damage: A theory for detection and assessment. *Photogrammetric Engineering and Remote Sensing*, **44**, 1147-1158

MYERS, B.J. (1974). *The application of colour photography to forestry, a literature review.* Forest and Timber Bureau Leaflet 124, Australian Government Publishing Service, Canberra

NEBLETTE, C.B. (1927). Aerial photography for the study of plant diseases. *Photo Era Magazine*, **58**, 346

ODENWELLER, J.B. and JOHNSON, K.I. (1984). Crop identification using Landsat temporal spectral profiles. *Remote Sensing of the Environment*, **14**, 39-54

ORSENIGO, J.R. (1984). A history of the everglades and future applications of aerial imagery. In *Proceedings of the Ninth Biennial Workshop on Colour Aerial Photography in the Plant Sciences.* American Society of Photogrammetry, Falls Church, Virginia, USA

PAINE, D.P. (1981). *Aerial Photography and Image Interpretation For Resource Management.* John Wiley, New York

PARDOE, G.K.C. (1969). Earth resource satellites. *Science Journal*, June 1969, 58-67

RAO, V.R, BRACH, E.J. and MACK, A.R. (1978). Crop discriminability in the visible and near infra-red regions. *Photogrammetric Engineering and Remote Sensing*, **44**, 1179-1184

RICHARD-MOLLARD, M.S. (1985). Rhizomania: A world wide danger to sugar beet. *Span*, **28**, 92-94

SIMONETT, D.S. (1983). The development and principles of remote sensing. In *Manual of Remote Sensing, 2nd edition.* American Society of Photogrammetry, Falls Church, Virginia, USA

VOGEL, F.A. (1987). Applications of remote sensing at the Department of Agriculture of the USA. Statistical assessment of land use: In *The impact of remote sensing and other recent developments in methodology.* Eurostat, Luxembourg

WALLEN, V.R., JACKSON, H.R., JAMES, W.C. and SMITH, A.M. (1975). Optical density variation in aerial photographs of potato plots infected with Phytopthora infestans. *American Potato Journal*, **52**, 233-238

WALLEN, V.R., JACKSON, H.R., BASA, P.K., BAERIZIGER, H. and DIXON, R.G. (1977). An electronically scanned aerial photographic technique to measure winter injury in alfafa. *Canadian Journal of Plant Sciences*, **57**, 647-651

WALLEN, V.R. and JACKSON, H.R. (1978). Alfafa winter injury, survival and vigour determined from aerial photographs. *Agronomy Journal*, **70**, 922-924

WILDMAN, W.E., NEJA, R.A. and CLARK, J.K. (1976). Low cost aerial photography for agricultural management. *California Agriculture*, April 1976, 4-7

16

ESTIMATION OF PLANT WATER STATUS FROM CANOPY TEMPERATURE: AN ANALYSIS OF THE INVERSE PROBLEM

GAYLON S. CAMPBELL

Department of Agronomy and Soils,
Washington State University, Pullman, Washington, U.S.A.

and

JOHN M. NORMAN

Department of Soil Science,
University of Wisconsin, Madison, Wisconsin, U.S.A.

Remote sensing is often thought to apply only to radiometric measurements made from satellites or aircraft, but a number of important remote sensing applications are ground-based. One of these is the remote sensing of crop water stress using infrared radiation (IR) thermometers.

Monteith and Szeicz (1962) and Tanner (1963) were among the first to use radiation thermometry to measure canopy temperatures. Monteith and Szeicz developed a theory relating canopy temperature to canopy stomatal resistance and other variables. That analysis is the basis for the inverse method now used to determine crop water stress from canopy temperature measurements. Tanner (1963) used IR thermometer measurements of canopy temperature to investigate the possibility of detecting crop water stress.

Field application of the IR thermometer for quantification of crop water stress has increased markedly over the past 10 years, due primarily to the efforts of scientists at USDA, Phoenix, AZ, USA. Early work, reported by Idso, Jackson and Reginato (1977) and Jackson, Reginato and Idso (1977) established a method for quantification of stress based on canopy and air temperature measurements. Later analyses included other environmental variables. The history of these developments was reviewed by Jackson (1982), with additional material presented by Hatfield (1983). Jackson (1982) also developed a Crop Water Stress Index (CWSI), based on the equations of Monteith and Szeicz (1962), which provides a rational basis for relating crop water stress and canopy temperatures. His analysis also gives some indication of factors which affect the sensitivity of the technique for detecting water stress as well as sources of uncertainty in the calculation. Smith, Barrs and Steiner (1986) conducted a field study to investigate the effect of various meteorological variables on their ability to predict canopy-air temperature differences. They found that both vapour deficit and net radiation were important in determining the magnitude of the temperature difference.

Within the past two years, several commercial IR thermometers, with associated environmental measurement capabilities, have become available. The design of these units is based on Jackson's analysis, and their internal computers calculate and display a CWSI based on canopy and air temperature, vapor deficit, and some function of solar radiation. These units have found relatively wide application in research, as well as in commercial irrigation scheduling.

In spite of the extensive interest in IR thermometry for monitoring crop water stress, no systematic and complete investigation of sensitivities and errors in the method appears to have been done. It even appears that a complete model relating soil water status to canopy temperature variations is lacking. The purpose of this paper is therefore to derive a relatively complete model relating soil and atmospheric factors to canopy temperature, and then to use the model to investigate the sensitivity of canopy temperature to water stress and the influence on canopy temperature of factors which are not stress related.

Water stress measurement as an inverse problem

Menke (1984) defines inverse theory as "an organized set of mathematical techniques for reducing data to obtain useful information about the physical world on the basis of inferences drawn from observations". The "physical world" in this case is represented by a model, and the quantitative behaviour of the model is specified by model parameters. The forward modeling problem is familiar, and can be stated as:

model parameters ⟶ model ⟶ prediction of data.

The inverse problem uses the model and the data to determine model parameters:

data ⟶ model ⟶ estimates of model parameters.

Note that inverse theory does not state what the model should be (though it can be useful in determining whether one model fits data better than another model). Once a model has been chosen, however, inverse methods can be used to determine the model parameters which most closely fit a particular set of data.

The prediction of crop water stress from measurements of canopy temperature is a good example of an inverse problem. The forward problem is well documented in various forms (Campbell, 1977; Jackson, 1982; Hipps, Asrar and Kanemasu, 1985). Canopy temperature is predicted as a function of environmental variables and canopy resistance to vapour diffusion. The inverse problem takes measurements of canopy temperature and attempts to infer canopy resistance, or crop water status, from them. The questions which require investigation concern uniqueness of the results of the water status calculation, and resolution of water status from canopy temperature measurements. To answer these questions, we will first consider the models and the forward problem of predicting canopy temperature from water status. We will then use these models to investigate questions of uniqueness and resolution in the inverse problem.

Ideally, the models used to investigate inversion of canopy temperature measurements to determine crop water stress would be relatively complete and realistic, such as the model CUPID (Norman and Campbell, 1984). The complexity of such models, however, sometimes obscures the physics of the problem. The investigation we undertake here uses simpler models, in hopes that the analysis will be clear. Though approximate, these models should be sufficiently accurate to provide useful information about prediction of crop water stress from IR thermometer measurements.

An index of crop water stress

Water stress in crops is probably best measured in terms of the extent to which soil and plant water are able to meet evaporative demand for the crop. The stress index of Hiler and Clark (1971) is

$$S = 1 - E/E_p, \tag{16.1}$$

where E (g m^{-2} s^{-1}) is actual measured transpiration rate, and E_p is potential transpiration rate when water is freely available. This stress index is therefore directly related to the decrease in transpiration due to the unavailability of water. Since transpiration and dry matter production are tightly coupled (Tanner and Sinclair, 1983), this stress index is also closely related to reduction in dry matter production. It therefore appears, from both theoretical and practical grounds, to be a useful measure of water stress in crops.

Campbell (1977) shows that the ratio, E/E_p can be expressed as

$$E/E_p = (\Delta + \gamma_o^*)/(\Delta + \gamma^*), \tag{16.2}$$

where Δ (g m^{-3} C^{-1}) is the slope of the saturation vapour density function and $\gamma^* = \gamma(r_{va} + r_{vc})/r_e$; γ (g m^{-3} C^{-1}) is the psychrometer constant, r_{va} (s m^{-1}) is the eddy diffusion resistance of the turbulent boundary layer of the canopy, r_{vc} is the combined stomatal resistance of the leaves in the canopy, and r_e is the equivalent resistance for the combined transfer of sensible and radiant heat from the canopy to the atmosphere. The apparent psychrometer constant, $\gamma_o^* = \gamma(r_{va} + r_{vc}^o)/r_e$, where r_{vc}^o is the canopy resistance without water stress.

Substituting *Equation 16.2* into *Equation 16.1* (assuming $r_e \simeq r_{va}$), gives

$$S = 1 - E/E_p = \frac{(r_{vc} - r_{vc}^o)/r_{vc}}{1 + (1 + \Delta/\gamma)r_{va}/r_{vc}}, \tag{16.3}$$

which relates the stress index to the fractional change in canopy resistance. The reciprocal of the denominator is the ratio of stress index to fractional change in canopy resistance. It is a function of temperature and the ratio of boundary to canopy resistance, as shown in *Figure 16.1*.

When boundary layer and canopy resistances are equal, a change of 10% in canopy resistance results in a change of only 2 to 3% in stress index. This fact will be used later to relate changes in stress index to changes in canopy temperature.

Crop water stress response to soil water and transpiration rate

The stress index, S, is ultimately meant to reflect the water status of the soil-plant system. It is therefore necessary to relate stomatal resistance to soil and plant water status. The development given here is brief, and is based on a more extensive treatment by Campbell (1985). The analysis in Campbell (1985) shows that any combination of root distribution and water potential distribution in a soil can be expressed in terms of a single equivalent weighted mean soil water potential, $\overline{\psi}_s$ (J kg^{-1}), and a single total plant resistance to liquid flow, R_p (m^4 s^{-1} kg^{-1}). Steady state water flow from the soil

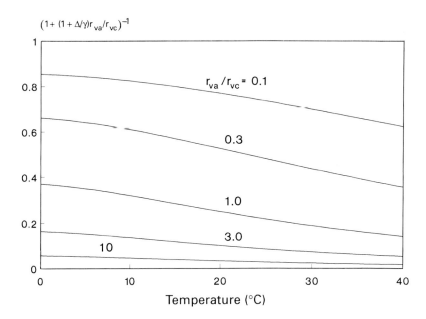

FIGURE 16.1. Reciprocal of the denominator of *Equation 16.3* as a function of temperature for several ratios of boundary to canopy resistance. This factor, multiplied by the fractional change in canopy resistance, gives a stress index for the crop

to the leaf can therefore be represented as:

$$E = (\overline{\psi}_s - \psi_L)/R_p, \qquad (16.4)$$

where ψ_L is the water potential of the leaf. Since flow is assumed to be steady, the water uptake rate is assumed to equal the transpiration rate, **E**.

The connection between *Equations 16.4* and *16.3* is through a relationship between leaf water potential and stomatal diffusion resistance. The physiological basis for the relationship has been assumed to be a direct effect of leaf turgor on stomatal opening, but recent work, reviewed by Schulze (1986), calls that view into question and suggests hormones, particularly ABA, as playing a possible role. It is still not clear, however, that hormone responses play a primary role in field-grown plants.

Whatever the physiological bases for the response, measurements on field-grown plants indicate a rapid increase in stomatal resistance of most species once leaf water potential reaches some critical value (Turner, 1973; Cline and Campbell, 1976). The canopy resistance, r_{vc}, is the reciprocal of the weighted reciprocals of leaf resistances for the canopy (Choudhury and Monteith, 1988). The increase in canopy resistance due to decreased leaf water potential should therefore be a function similar to that for the individual leaves. The function suggested in Campbell (1985) is

$$r_{vc} = r_{vc}^o[1 + (\psi_L/\psi_c)^n], \qquad (16.5)$$

where ψ_c is the critical water potential at which canopy resistance is twice its minimum value (r_{vc}^o), and **n** is an empirical constant to match field data.

If *Equation 16.4* is substituted for ψ_L in *Equation 16.5*, and *Equation 16.5* is substituted for r_{vc} in the numerator of *Equation 16.3*, we obtain

$$S = \frac{[(\frac{\psi_c}{\overline{\psi}_s - ER_p})^n + 1]^{-1}}{1 + (1 + \Delta/\psi)r_{va}/r_{vc}}. \tag{16.6}$$

Equation 16.6 allows us to investigate the relationship between the crop water stress index (**S**) and soil water potential. Values for the numerator are shown as a function of transpiration rate (**E**) in *Figure 16.2* for several soil water potentials.

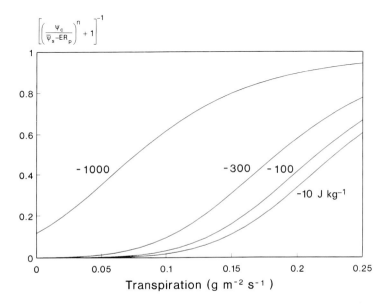

FIGURE 16.2. Numerator of *Equation 16.6* as a function of transpiration rate for several values of weighted mean soil water potential. This value, multiplied by the appropriate value from *Figure 16.1* gives the stress index for the crop based on soil moisture and transpiration

These values need to be multiplied by the appropriate values from *Figure 16.1* to give values of **S**. The highest value of **E** shown, 0.25 g m^{-2} s^{-1}, is 0.9 mm h^{-1}, and is intended to represent maximum transpiration experienced by crops at mid-day. *Figure 16.2* shows that soil water deficits do not result in increased values of **S** unless **E** is large, and that variation in **S** due to variation in **E** can be as large as variation due to changes in soil water potential.

Response of canopy temperature to canopy resistance

The basis for the assumption that canopy temperature might indicate water stress is the fact that, as water stress increases the canopy resistance for vapour transport, incident energy will be partitioned increasingly toward sensible heat. Canopy temperature must then rise in order to dissipate the additional sensible heat. Sensible heat transport between the canopy and the air above it is proportional to the temperature difference, $\delta T = T_s - T_a$. This temperature difference can be written in terms of the energy balance of the canopy using the familiar Penman transform as (Campbell, 1977)

$$\delta T = \frac{R_n - \lambda D/r_v}{\rho c_p/r_{Ha} + \lambda \Delta/r_v}. \tag{16.7}$$

Here, R_n is the net radiation absorbed by the canopy (W m^{-2}), D is the vapour deficit of the air above the canopy (g m^{-3}), ρc_p is the volumetric specific heat of air (1200 J m^{-3} K^{-1}), λ is the latent heat of vaporization for water (2450 J g^{-1}), and r_v is the sum of the canopy and boundary layer resistances ($r_{vc} + r_{va}$) for water vapour. The link between canopy temperature and water stress is, of course, through r_{vc}. The boundary layer resistances, r_{va} and r_{Ha} are assumed to be equal. They are functions of wind speed, surface roughness and atmospheric stability. Equations are given by Campbell (1985):

$$r_{Ha} = \frac{ln(\frac{z-d+z_M}{z_M} + \Psi_M) ln(\frac{z-d+z_H}{z_H} + \Psi_H)}{k^2 u}. \tag{16.8}$$

In *Equation 16.8*, z is the height above the ground at which T_a, D, and the wind speed, u (m s^{-1}), are measured; d (m) is the zero plane height, z_M and z_H (m) are roughness parameters for momentum and heat, k is von Karman's constant (0.4), and Ψ_M and Ψ_H are profile correction factors for atmospheric stability.

The profile correction factors are functions of the stability parameter, ζ, which can be calculated from (Campbell, 1985).

$$\zeta = -kgz\delta T/Tu^{*3}r_{Ha}. \tag{16.9}$$

The gravitational constant is g (9.8 m s^{-2}), T is the Kelvin temperature, and u* (m s^{-1}) is the friction velocity. For $\zeta \geq 0$, Ψ_M and Ψ_H are equal, and are given by (Yasuda, 1988):

$$\Psi_M = \Psi_H = 6ln(1+\zeta). \tag{16.10}$$

For $\zeta < 0$, Ψ_H is calculated from Businger's (1975) equation:

$$\Psi_H = -2ln\{[1 + (1 - 16\zeta)^{1/2}]/2\}, \tag{16.11}$$

and Ψ_M from (Campbell, 1985):

$$\Psi_M = 0.6\Psi_H. \tag{16.12}$$

In order to investigate the behaviour of these resistances and δT, it will be necessary to limit the number of variables considered. We will therefore assume that the zero plane displacement, d, the roughness parameters, z_H and z_M, and the height, z, at which measurements are made are all functions of canopy height, h. Campbell (1977) provides the empirical relationships: d = 0.77 h; z_m = 0.13 h; z_H = 0.2 z_M. In addition to these, we will assume z= 2 h. Based on these assumptions, *Figure 16.3* shows the relationship between boundary layer resistance and δT for several values of wind speed, u.

It is interesting that boundary layer resistance is small, whatever the wind speed, when δT is positive, but resistance increases rapidly for low wind speeds in stable flow.

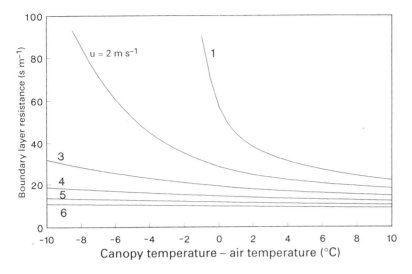

FIGURE 16.3. Boundary layer resistance as a function of canopy - air temperature difference and wind showing the effect of atmospheric stability on the resistance

This probably accounts for the fact that temperatures of dry canopies generally are not more than 5 K above air temperature (Jackson, 1982).

A short computer program was written to find δT from *Equation 16.7* and to investigate the response of δT to changes in canopy resistance and to the other main variables in *Equation 16.7* (R_n, D and u). Iterative procedures were required at two places. One was the calculation of the correct boundary layer resistance for a given value of δT, and the other was the calculation of δT from *Equation 16.7* when the boundary layer resistances are functions of δT. The substitution method given in Campbell (1985) was used to find the resistances, and a Newton-Raphson procedure with numerical derivatives was used to find δT.

The difference between canopy and air temperature is shown as a function of vapour deficit in *Figure 16.4*, for several values of canopy resistance.

These curves are similar to curves obtained by Jackson (1982) and to field measurements of Idso, Reginato, Reicosky and Hatfield (1981). *Figure 16.5a* shows δT as a function of canopy resistance, the variable of interest in predicting water stress, for different values of vapour deficit. *Figures 16.5b* and *16.5c* are similar plots showing the response of δT to radiation and wind.

In each of these graphs, default values of the variables not shown were set. The default wind speed was set to 3 m s^{-1}, the radiation at 500 W m^{-2}, and the vapour deficit at 20.1 g m^{-3}. In order to produce realistic values of Δ, the vapour deficit was calculated by setting ρ_{va} to 10 g m^{-3} and changing air temperature by 5 K increments. In each calculation, Δ was evaluated at the average of canopy and air temperatures.

The responses to radiation and vapour deficit (*Figure 16.5a* and *16.5b*) are predictable from *Equation 16.7*. The sign of δT is determined by the relative sizes of absorbed radiation and isothermal latent heat loss ($\lambda D/r_v$). Changes in either of these

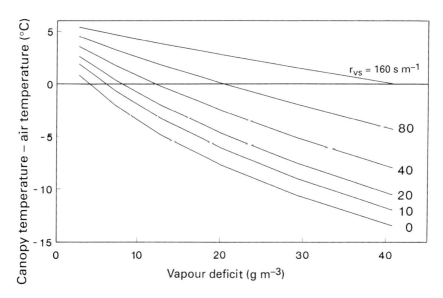

FIGURE 16.4. Difference between canopy and air temperature as a function of atmospheric vapour deficit for different canopy resistances to vapour diffusion

shift the curves up or down. The response to wind is considerably less predictable from inspection. Over a fairly wide range of wind speeds and canopy resistances, wind has little effect on δT. In general, increasing wind speed decreases canopy temperature, but there exists a small range of canopy resistances over which increasing wind increases canopy temperature. This range can be increased by changing the default values of D and R_n.

Analysis of canopy temperature response to environment

The early stress-day-degree concept (Idso et al. 1977) was based only on measurements of canopy and air temperature. Vapour deficit was later recognized as an important variable. A commercial instrument for irrigation scheduling, The Scheduler, monitors canopy and air temperatures and vapour deficit. A simple solar radiation monitor is provided which prevents the instrument from making measurements on overcast days, but no attempt is made to quantify and correct for variation in solar radiation on sunny days.

Figure 16.6 is an attempt to quantify the variation in canopy temperature which would result from uncertainty in vapour deficit, radiation or wind.

Figure 16.6a shows $d\delta T/dD$ for a range of canopy resistances and vapour deficits. Except at low canopy resistance, the sensitivity of δT to vapour deficit is almost independent of the actual value of the deficit. When canopy resistance is 50 s m^{-1}, a 1 g m^{-3} uncertainty in D would result in a 0.3 K uncertainty in δT. Since this is about the maximum uncertainty one would like to have in δT, it is obvious that D must be carefully measured since 1 g m^{-3} represents about the best accuracy that can be achieved in routine measurements.

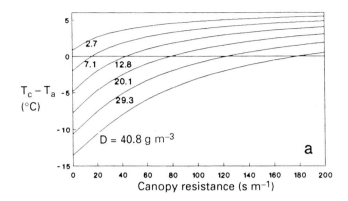
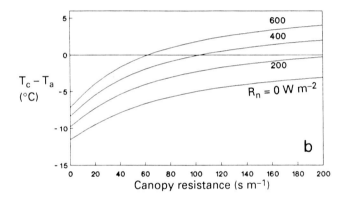
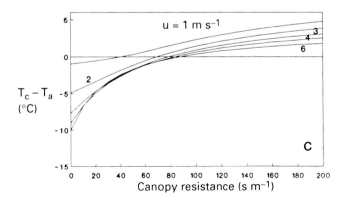

FIGURE 16.5. Difference between canopy and air temperature as a function of canopy resistance for different values of vapour deficit (**D**), radiation absorbed by the canopy (**H**) and wind speed (**u**). Default values for parameters not shown are given in the text

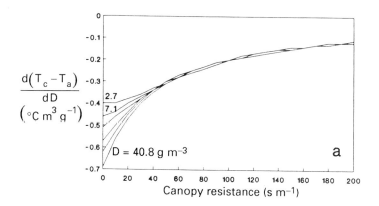
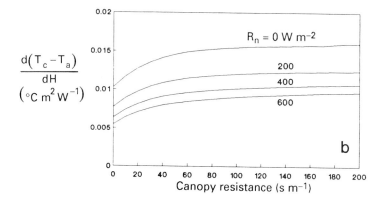
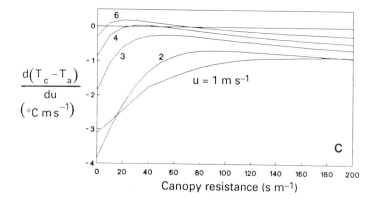

FIGURE 16.6. Changes in δT with changes in **D**, **R$_n$** and **u** as a function of canopy resistance. Default values for parameters not shown are given in the text

A similar conclusion is reached with respect to the effects of radiation. Changes in $d\delta T/dR_n$ are relatively small over the range of R_n typical of canopy temperature measurements, and are almost independent of r_{vc}. A sensitivity of around 0.01 K W^{-1} m^2 is representative of $d\delta T/dR_n$, (*Figure 16.6b*), indicating that R_n must be known with an accuracy of about 30 W m^{-2} if the uncertainty in δT from this source of variation is to be comparable to that from **D**. As with **D**, this represents about the best accuracy one is likely to achieve in net radiation measurements (especially when one considers the uncertainty associated with radiation absorption by the soil and soil heat flux). If R_n is not measured at all, then uncertainties, even on mostly sunny days, would likely cause variation in δT of 1-2 K. These results seem consistent with field measurements reported by Smith *et al.* (1986).

The situation with respect to wind is much more favourable (*Figure 16.6c*). At a wind speed of 4 m s^{-1}, and a canopy resistance of 50 s m^{-1} variations in wind speed have almost no effect on canopy temperatures. Wind speeds are, of course, extremely variable in the field, but tend to be high enough during the day to minimize effects of wind on canopy temperature measurements. As long as wind speeds are above 3 m s^{-1}, variation in wind speed appears not to contribute appreciably to variation in δT, until canopy resistances become high. It should be pointed out that these calculations have been made using a model which assumes that temperatures and other environmental variables have been averaged over at least several minutes. A quick measurement lasting only a few seconds might show substantial effects of wind speed, especially if the wind were particularly low at the time the measurement was made.

Analysis of canopy temperature response to water stress

The sensitivity of δT to changes in canopy resistance is shown in *Figure 16.7*.

The denominator is the change in canopy resistance divided by the resistance (the same as the numerator in *Equation 16.3*). To illustrate the use of this figure, assume we would like to detect changes in **S** of 0.1. If r_{va}/r_{vc} were 3, then, from *Figure 16.1*, changes in **S** would be about 15% as large as changes in $\Delta r_{vc}/r_{vc}$. The fractional change in canopy resistance corresponding to a change in **S** of 0.1 would therefore be 0.7. If **D** = 20 g m^{-3} and r_{vc} = 50 s m^{-1}, then from *Figure 16.7a*, the sensitivity of δT to canopy resistance is about 4 K. Multiplying this factor by 0.7, the fractional change in canopy resistance gives a predicted change in canopy temperature of 2.8 K.

Figures 16.7b and *16.7c* indicate that the sensitivity of δT to canopy resistance changes is insensitive to prevailing radiation, wind and canopy resistance conditions, provided wind speed is above 2-3 m s^{-1} and canopy resistance is above 30-40 s m^{-1}. High wind speeds tend to decrease the response, as do high canopy resistances. High radiation loads tend to increase the response.

The effect of vapour deficit on the sensitivity of δT to changes in canopy resistance (*Figure 16.7a*), however, is very pronounced, and has an important consequence. Locations with high vapour deficits will show a large change in δT when stomatal resistance changes, while locations with low vapour deficits are likely to show changes which are comparable in magnitude to the noise in the measurement. This decreased sensitivity of the method at low vapour deficit was pointed out by Jackson (1982), and is evident in *Figure 16.4*, where the range of δT for a given change in r_{vc} is much smaller at low **D** than at high.

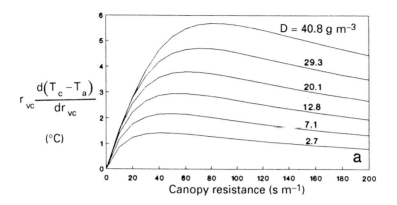
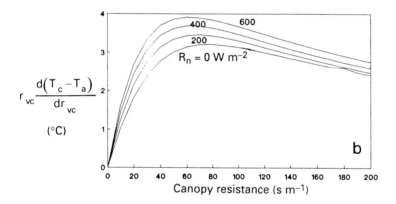
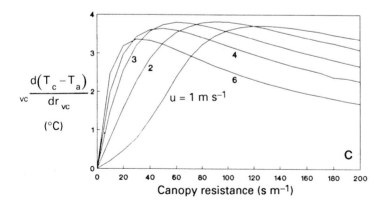

FIGURE 16.7. Sensitivity of δT to fractional changes in canopy resistance (dr_{vc}/r_{vc}) as a function of canopy resistance for different values of **D**, **R_n** and **u**. Default values for parameters not shown are given in the text

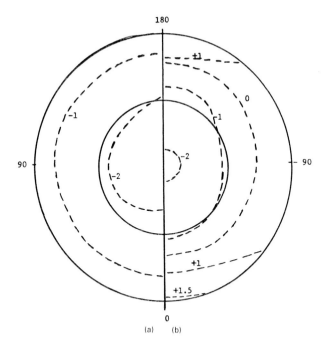

FIGURE 16.8. Polar plot of simulated directional response of δT for a corn canopy (**LAI** = 4, wet soil surface) at solar zenith angles of a) 32° and b) 76°. The 0° azimuth is with the sun behind the viewer and the centre of the polar plot represents a nadir view. Contour labels are in K. Circles represent view zenith angles of 30° and 60°

Errors in radiative surface temperature measurement

An implicit assumption in the analysis so far has been that the temperature of the exchange surface is known. In a crop canopy, the exchange surface is a complex assembly of leaf and soil surfaces. We need to consider how well a temperature measurement of the canopy with an IR thermometer approximates the exchange surface temperature.

Variation in IR thermometer measurements of canopy temperature with view azimuth and zenith angle are to be expected. A view with the sun behind the thermometer sees mostly sunlit, warm leaves, while a view facing the sun may see more shaded, cool leaves. When the thermometer has a high zenith angle, more soil and shaded leaves from deep within the canopy are in view. The contour plot in *Figure 16.8* is a simulated pattern of δT as a function of zenith and azimuth angle for two sun angles on a corn canopy.

Note that the lowest temperatures are predicted for a nadir view angle. The highest temperatures occur when the sun is behind the viewer and the thermometer zenith angle is large. These effects have been observed and reported by Monteith and Szeicz (1962), Fuchs, Kanemasu, Kerr and Tanner (1967) and Huband and Monteith (1986). Temperature variations of 1-3 K were reported for various combinations of sun and view zenith and azimuth. Huband and Monteith (1986) chose a view zenith angle of

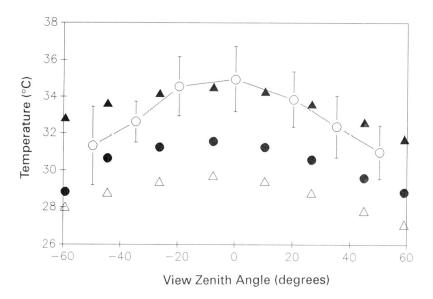

FIGURE 16.9. Measured mean and standard deviation (8 plots) of infrared canopy temperature, compared with predictions from a directional thermal radiance model for the Konza Prairie, Kansas. Model results are a) soil surface transfer coefficient typical of that for a mature corn canopy, and no humidity response of stomata ($\triangle$); b) soil surface transfer coefficient 1/3 of that in a ($\bullet$); c) transfer coefficient as in b, and with humidity response of stomata ($\blacktriangle$). **LAI** = 1.6, canopy height = 38cm, air temp. 26.5 K at 1.5 m height, relative humidity = 35% at 1.5 m height, wind speed = 2 m s^{-1} at 2.5 m height, solar radiation = 980 W m^{-2}

55° and a view azimuth at right angles to the sun to standardize reading procedures. Measurements made in this way showed radiative temperatures to be consistently 1 K cooler than the corresponding 'aerodynamic' temperatures obtained by extrapolating temperature profiles. (They assumed $z_H = z_M$ in these extrapolations. Had they used $z_H = 0.2\ z_M$, the difference would have been larger.)

Both the angular variation in radiative temperature and its sensitivity to transfer processes in the canopy are apparent in *Figure 16.9*.

The measurements are means and standard deviations of eight samples, each covering one square meter of a native prairie grass. View angles were in the principal plane defined by a vertical to the surface and the azimuth angle of the sun. The predicted results are from a detailed soil-plant-atmosphere model (Norman and Campbell, 1984). The predictions which best fit the measurements show less angular variation than the measurements, and were obtained by assuming a humidity dependent stomatal response for the plants and a heat transfer coefficient at the soil surface which is only 1/3 that which has been found suitable for corn canopies. If these adjustments are not made, the model predicts temperatures which are 3 to 5 K below measured temperatures.

Both the angular variation in apparent canopy temperature and the discrepancy between radiative and aerodynamic temperatures are problems when we try to apply IR thermometry to measurements of crop water stress. The fact that the angular variation is large compared to the precision needed for acceptable water stress measurements indicates the importance of choosing a standard view zenith angle and a standard azimuth relative to the sun for the measurements. The disagreement between radiative and aerodynamic temperatures needs further study, using both models and measurements.

The measurements of Huband and Monteith (1986) suggest limitations on precision and accuracy for surface temperature measurement of perhaps 0.4 K and 1.0 K respectively, even with very careful technique. The precision of the measurement would be the most limiting for detecting water stress, since we want to detect *changes* in δT. Using the conditions given in the previous example, uncertainty in surface temperature measurement would result in uncertainties in **S** of around 0.05, under the best conditions. Much larger errors could result from improper radiometer orientation, sparse canopy cover or radiometer calibration errors.

Other sources of error

There are a number of sources of error which have not been considered in this analysis. Substantial error is possible even in the measurement of air temperature, if care is not taken to minimize radiation absorption by the air temperature sensor. Also, variation in measurement height could have a substantial effect on the result obtained. All of the calculations presented here used a measurement height of twice the crop height. As with the radiometer orientation, whatever height is chosen, that height should be used consistently for the measurement of **D**, **T** and **u**. Errors are also possible from failure to average out fluctuations in measurements. The models used in this analysis are valid only for averages of the quantities involved over several minutes. The response times of the crop and the sensors differ, so some averaging is necessary to obtain representative results.

Errors from these sources are difficult to quantify. If all measurements were taken as accurately as is possible using good equipment and good experimental technique, these errors would still likely be a few tenths of a degree.

Summary and conclusions

We have been able to show that the difference between canopy and air temperature, by itself, is a poor measure of crop water stress. Other important factors are transpiration rate, vapour deficit of the air above the crop, and net radiation. The initial success of IR thermometry in measuring crop water stress is probably the result of its having been developed and applied in an arid environment where **D** is large, and day to day variations in R_n and **E** are small. In more humid environments, where **D**, R_n and **E** are more variable, all are expected to be important in predicting **S** accurately.

Predictions of **S** from δT are usually based on an analysis similar to *Figure 16.4* and baselines determined for completely dry and well-watered crops (Idso *et al.* 1981). The analysis here would indicate that it would be wise to abandon the use of baseline analysis and compute canopy resistance directly from the inversion of *Equation 16.7*, so that all of the environmental variables would have their proper effect. This would, of course, require accurate measurement of these variables and resolution of the discrepancy between radiometric and aerodynamic temperature measurements. This suggestion was also made by Jackson, Kustas and Choudhury (1988).

The measurement of r_v using *Equation 16.7* appears to be relatively straightforward, as does the prediction of **S** from *Equation 16.3*, using these measurements. It appears that, when **D** is large, uncertainty in **S** could be as small is 0.05. In humid conditions, when **D** is small, errors in **S** could be 2 to 3 times this large.

The most difficult question relates to the prediction of water status from **S**. As *Figure 16.2* shows, **S** increases at high transpiration rates, even when soil water is readily available. Relatively small changes in **E** can result in as large a change in **S** as occurs for quite significant changes in soil water potential. This part of the system is, at present, not well understood, and would provide a fruitful area for future research.

Acknowledgment

The data in *Figure 16.9* were collected under the supervision of Blaine Blad and Elizabeth Walter-Shea during the First ISLSCP (International Satellite Land Surface Climatology Project) Field Experiment, sponsored by NASA and supported by Grants #NAG5-894 and #NAGW-1099.

References

BUSINGER, J.A. (1975). Aerodynamics of vegetated surfaces. In *Heat and Mass Transfer in the Biosphere*. Ed. by deVRIES, D.A. and AFGAN, N.H. Wiley, New York

CAMPBELL, G.S. (1977). *An Introduction to Environmental Biophysics*. Springer Verlag, New York

CAMPBELL, G.S. (1985). *Soil Physics with BASIC: Transport Models for Soil-Plant Systems*. Elsevier, New York

CHOUDHURY, B.J. and MONTEITH, J.L. (1988). A four-layer model for the heat balance of homogeneous land surfaces. *Quarterly Journal Royal Meteorological Society*, **114**, 373-398

CLINE, R.G. and CAMPBELL, G.S. (1976). Seasonal and diurnal water relations of selected forest species. *Ecology*, **57**, 367-373

FUCHS, M., KANEMASU, E.T., KERR, J.P. and TANNER, C.B. (1967). Effect of viewing angle on canopy temperature measurements with infrared thermometers. *Agronomy Journal*, **59**, 494-496

HATFIELD, J.L. (1983). Evapotranspiration obtained from remote sensing methods. *Advances in Irrigation*, **2**, 395-416

HILER, E.A. and CLARK, R.N. (1971). Stress day index to characterize effects of water stress on crop yields. *Transactions American Society of Agricultural Engineers*, **14**, 757-761

HIPPS, L.E., ASRAR, G. and KANEMASU, E.T. (1985). A theoretically-based normalization of environmental effects on foliage temperature. *Agricultural and Forest Meteorology*, **35**, 113-122

HUBAND, N.D.S. and MONTEITH, J.L. (1986). Radiative surface temperature and energy balance of a wheat canopy I. Comparison of radiative and aerodynamic canopy temperature. *Boundary-Layer Meteorology*, **36**, 1-17

IDSO, S.B., JACKSON, R.D. and REGINATO, R.J. (1977). Remote sensing of crop yields. *Science*, **196**, 19-25

IDSO, S.B., REGINATO, R.J., REICOSKY, D.C. and HATFIELD, J.L. (1981). Determining soil-induced plant water potential depressions in alfalfa by means of infrared thermometry. *Agronomy Journal*, **73**, 826-830

JACKSON, R.D. (1982). Canopy temperature and crop water stress. *Advances in Irrigation*, **1**, 43-85

JACKSON, R.D., REGINATO, R.J., and IDSO, S.B. (1977). Wheat canopy temperature: A practical tool for evaluating water requirements. *Water Resources Research*, **13**, 651-656

JACKSON, R.D., KUSTAS, W.P. and CHOUDHURY, B.J. (1988). A re-examination of the crop water stress index. *Irrigation Science*, **9**, 309-317

MENKE, W. (1984). *Geophysical Data Analysis: Discrete Inverse Theory*. Academic Press, New York

MONTEITH, J.L. and SZEICZ, G. (1962). Radiative temperature in the heat balance of natural surfaces. *Quarterly Journal Royal Meteorological Society*, **88**, 496-507

NORMAN, J.M. and CAMPBELL, G.S. (1984). Application of a plant-environment model to problems in irrigation. *Advances in Irrigation*, **2**, 155-188

SCHULZE, E-D. (1986). Carbon dioxide and water vapor exchange in response to drought in the atmosphere and in the soil. *Annual Review Plant Physiology*, **37**, 247-274

SMITH, R.C.G., BARRS, H.D. and STEINER, J.L. (1986). Alternative models for predicting the foliage-air temperature difference of well irrigated wheat. I. Derivation of parameters. II. Accuracy of predictions. *Irrigation Science*, **7**, 225-236

TANNER, C.B. (1963). Plant temperatures. *Agronomy Journal*, **55**, 210-211

TANNER, C.B. and SINCLAIR, T.R. (1983). Efficient water use in crop production: research or re-search? In *Limitations to Efficient Water use in Crop Production*. Ed. by TAYLER, H.M., JORDAN, W.R and SINCLAIR, T.R. American Society of Agronomy, Madison, Wisconsin

TURNER, N.C. (1973). Stomatal behavior and water status of maize, sorghum and tobacco under field conditions at low soil water potential. *Plant Physiology*, **53**, 360-365

YASUDA, N. (1988). Turbulent diffusivity and diurnal variations in the atmospheric boundary layer. *Boundary-Layer Meteorology*, **43**, 209-221

17
A SIMPLIFIED ALGORITHM FOR THE EVALUATION OF FROST-AFFECTED CITRUS

M.A. GILABERT, D. SEGARRA AND J. MELIÁ

Departament de Termodinámica, Facultat de Física,
Universitat de Valencia, 46100-Burjassot, Valencia, Spain

Introduction

Citrus are probably the main cultivated crops in the Comunidad Valenciana, and therefore a frost may have adverse consequences for the Valentian economy. By using Landsat-5 Thematic Mapper images, we attempted in this work to check whether it is possible to detect and evaluate the effects of a frost over a citrus growing area. The main problem is to find a spectral parameter which can be deduced from satellite data, that can reflect the effects of cold on the vegetation canopy under study. We shall define that parameter in the next section. First we consider whether low temperatures produce spectral alterations on citrus leaves.

Effect of frost on a citrus leaf: alteration of spectral response

As a consequence of a frost, the internal structure of leaves, the water content and their pigmentation or chlorophyll level can be deeply altered (Díaz, 1983; Fuentes, 1983). Since all these factors influence leaf reflectance (Cariolis and Amodeo, 1980) it can be expected that the effect of a frost can be detected through the alteration produced in the leaf. *Figure 17.1* shows the spectral transmittance of a healthy leaf (not affected by cold) and of a leaf damaged by cold, both measured immediately after freezing at -7°C for five hours and measured a week later. The measurements were made in the laboratory using a monochromator. Since the experimental equipment did not include an integrating sphere, the measurements do not have an absolute character, that is, they are expressed in arbitrary units. Notwithstanding this, they are useful in order to show the qualitative variation that appears in the curve shapes as a consequence of cold. We measured transmittance values to obtain accurate results. However, the conclusions that can be drawn are equally valid for reflectance values because there is a close correlation between the reflectance and transmittance spectra of a leaf (Knipling, 1970).

A decrease in the near-infrared transmittance can be observed, which can be attributed to a smaller leaf vigour and to the irreversible modification in internal structure suffered due to cold damage. Changes in the visible (zone of chlorophyll absorption)

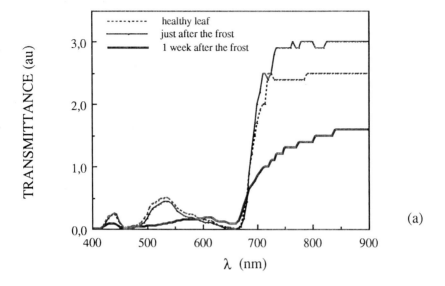

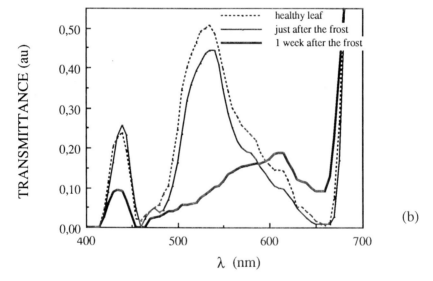

FIGURE 17.1. Spectral transmittance (in arbitrary units) of citrus leaves for a healthy leaf, a leaf damaged due to cold, measured just after freezing at -7°C for five hours, and the damaged leaf a week later; (a) shows the spectral interval betwen 400 and 900 nm, and (b) the visible band only with an expanded transmittance scale. All measurements have been made in the laboratory using a monochromator

can also be observed, which can be associated with a smaller photosynthetic activity (Schanda, 1986).

Changes in canopy reflectance and normalised difference vegetation index

The data obtained from a crop canopy by a satellite represent a reflective system which is much more complex than the reflectances of the leaves. In principle, for given conditions of incident radiation and angle of view, the reflectance of a crop can be determined through the spatial and angular distributions of the incident radiation and components of the canopy, and through the optical properties of these components and of the soil background (Colwell, 1974; Curran, 1980). These factors may also change with time. Therefore, we may conclude that the reflectance of a crop will depend on the phenological stage of the plants besides depending on the plant type and variety.

Citrus are subject to the factors mentioned above, and they will show a changeable spectral behaviour depending on the season. This was demonstrated by the study of the reflectance of citrus training areas using Landsat-5 TM images, corresponding to six different dates in two years (Segarra, Gilabert, Gandía and Meliá, 1988). Such behaviour can mask the alteration in the spectral response of the citrus as a consequence of a frost. Moreover, within an image there are several spectral signatures corresponding to the "citrus class" which are determined mainly by the degree of cover of the canopy (Gilabert and Meliá, 1990), as seen in *Figure 17.2*. This figure shows some spectral signatures corresponding to citrus parcels with different degrees of cover. As a consequence of the variability of spectral signatures, it is necessary to carry out a multi-temporal study to detect any abnormalities in the crop because, although a satellite image after a frost can show its effects, these will be masked by the other fluctuations in reflectance. Therefore, it is necessary to compare an image obtained before a frost with another after it in order to be able to determine the consequences on the study area.

The spectral parameter used for the temporal comparison of the images is the Normalised Difference Vegetation Index (**NDVI**) based on Landsat TM bands 4 and 3, given by *Equation 17.1*:

$$\text{NDVI} = \frac{\rho_{TM_4} - \rho_{TM_3}}{\rho_{TM_4} + \rho_{TM_3}} \tag{17.1}$$

where ρ is the reflectance in the corresponding TM band.

This index was chosen due to the following conditions:

1. It reduces the influence of the background on the canopy composite reflectance (Clevers, 1988) and enhances that of the trees which are the focus of our study. This vegetation index also partly compensates for illumination conditions and atmospheric effects (Tucker, Elgin and McMurtrey, 1979).

2. For citrus orchards it has been shown that the temporal changes of this index are smaller than those of other indices (Segarra *et al.*, 1988) and that it is very well correlated to the degree of crop cover (Gilabert and Meliá, 1990).

Both factors make it possible to establish the temporal evolution of this index as a function of the cover in such a way that any abnormal alteration of a parcel's spectral response can be shown. It is not possible to do this either by using other indices, because their temporal evolution shows random variations, or by using only one spectral band such as TM4, because it is not correlated with the degree of crop cover.

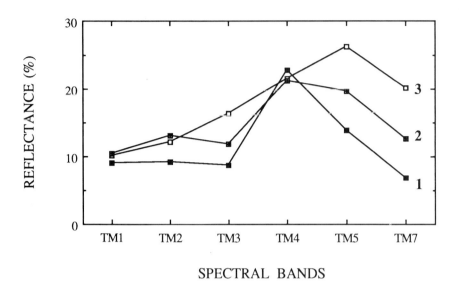

FIGURE 17.2. Reflectances in Landsat-5 TM bands for citrus parcels of different percent crop cover (PCC): 1, PCC $\geq$ 40%; 2, 20% $\leq$ PCC $\leq$ 40%; 3, PCC $\leq$ 20%

Objectives

The objectives of this work were to study the spectral alteration produced by a frost in a citrus area, with a view to developing an algorithm for the recognition, evaluation and determination of its intensity. The first aim was to find parameters capable of indicating the evolution of the chosen vegetation index for the ideal case of no frost. To carry out this study we chose a reference area of about 60 km^2 in the Plana Baixa region, to the South of the city of Castellón on the Mediterranean Coast of Spain. This area, where citrus crops predominate, was affected by intense advection frost in January 1985. The effects were very negative for the agriculture of the Comunidad Valenciana, especially for winter crops and citrus.

Experimental procedure

Images used

We have already indicated that in order to show a possible alteration in the spectral response of citrus as a consequence of a frost, it is necessary to carry out a multi-temporal study. For this reason, we have used two Landsat-5 TM images corresponding to the dates 11 January 1985 (a few days before the frost) and 16 March 1985 (two months after the frost). In these images we selected 11 training areas of citrus of different ages and degrees of cold damage. We considered the average values of TM band signals for its pixels to be the "spectral measurement" of each of the chosen parcels, and the standard deviation of the pixel values as the error of that measurement.

Atmospheric correction

In remote sensing applications involving the acquisition of data from satellites, the atmosphere between the sensor and the target and between the radiation source and the target has an effect on the data. When the atmospheric effect is constant over the frame of data, corrections for the effect may not have a significant impact on the final analysis of the frame, unless a multi-temporal study is carried out. In such a case the transformation of the digital counts (DCs) from the image into "corrected reflectance values" lets us proceed to the temporal comparison of the two scenes.

In the procedure to convert the initial TM image (in DCs) into a corrected reflectance image, we have introduced a simple algorithm that permits us to evaluate atmospheric scattering, which is an additive component. Although this correcting method is not exact, it partly corrects atmospheric effects on the images. Atmospheric absorption has a multiplicative effect and is not considered in this work: Holben (1986) showed that the effect is small for the vegetation index used in this study. One of the methods to remove the effects of scattering is the "dark-object subtraction technique", that requires only information contained in the digital image data. It supposes that there are at least a few pixels within an image which should be black (0% reflectance), corresponding to dark-shadowed pixel locations. Because of atmospheric scattering, the imaging system records a non-zero DC value at these pixels. This represents the DC value (haze value) that must be subtracted from the particular spectral band to remove the first-order scattering component. The determination of the haze value corresponding to each spectral band can be carried out by using the minimum values of the histogram for each band. By using this method, the haze value of each band is selected independently. Therefore, multi-spectral data such as that from Landsat-5 TM are spectral band dependent. For this reason we have used the method proposed by Chavez (1988), which improves this technique in such a way that allows the user to select a relative atmospheric scattering model to predict the haze values for all the spectral bands from a selected starting band haze value. The improved method normalizes the predicted haze values for the different gain and offset parameters used by the imaging system, in this case for the Thematic Mapper on board Landsat-5.

Table 17.1 shows the haze values found for both dates of study as well as the atmospheric conditions. These have been determined from the minimum value in the visible band, TM1, and coincide with those deduced from visibility values in the area of study supplied by the Meteorological Centre of Valencia. In our case, where we are just concerned with the study of the Normalised Difference Vegetation Index between spectral bands TM4 and TM3, we shall only use the haze values corresponding to those bands.

Once the haze values have been determined, and assuming that they correspond to the solar radiation backscattered by the atmosphere that reaches the satellite (Gonima, personal communication), the DC images are transformed into the corresponding vegetation index, where the vegetation index is evaluated from *Equation 17.1*. The reflectance in each band, ρ_λ is given by *Equation 17.2*:

$$\rho_\lambda = \frac{k\pi(R_\lambda - R_{0\lambda})}{E_{0\lambda}\cos\theta - \pi R_{0\lambda}}, \tag{17.2}$$

where k is the Earth-Sun distance (in astronomical units); $E_{0\lambda}$ is the spectral irradiance at the upper limit of the atmosphere; θ is the solar zenith angle; R_λ is the radiance

TABLE 17.1. Haze values found for the two dates of study in the four first bands of Thematic Mapper: (*) according to a method based on the image histogram, (**) according to the improvement proposed by Chavez (1988). Atmospheric conditions for each date are also given.

11 January 1985 (VERY CLEAR CONDITIONS)		
SPECTRAL BAND	HAZE VALUE (*)	HAZE VALUE (**)
TM1	41.0	41.0
TM2	13.0	13.5
TM3	11.0	9.9
TM4	8.0	5.2
16 MARCH 1985 (CLEAR CONDITIONS)		
SPECTRAL BAND	HAZE VALUE (*)	HAZE VALUE (**)
TM1	56.0	56.0
TM2	21.0	22.9
TM3	19.0	23.2
TM4	25.0	15.4

measured by the sensor, calculated from the digital counts using calibration constants (Price, 1987), and $R_{0\lambda}$ is the haze radiance calculated from the haze DC value. In *Equation 17.2* it has been assumed that the surface is a Lambertian reflector, that the atmosphere is horizontally stratified and that the solar zenith angle is constant in the whole image. Substitution of atmospherically corrected reflectance values for each band in *Equation 17.1* gives the equation used to transform DC images into vegetation index images.

Geometric correction

When we apply our algorithm for the identification and evaluation of frost areas in the satellite image we need to superimpose the images of different dates, and it is therefore necessary to previously apply a geometric correction to them. In order to determine the functions of co-ordinate transformation from one date to the other, shown in *Equations 17.3* and *17.4*, we have used up to 20 control points uniformly distributed on the image.

$$X_{MARCH} = 0.439 + 0.998 X_{JANUARY} \qquad (17.3)$$

$$Y_{MARCH} = -0.476 + 0.999 Y_{JANUARY} \qquad (17.4)$$

We have also tried other more complicated functions, but they did not improve the linear adjustment significantly: *Equations 17.3* and *17.4* are simple and satisfactory. The superimposition of the two images has been made by radiometrically deforming one with respect to the other by using bilinear interpolation.

Algorithm

Training-area spectral responses

We have measured the spectral response of each training area from the vegetation index images constructed, as described in the preceding paragraph. *Figure 17.3* shows the results plotted as a $NDVI_{MARCH}$ (vertical axis) / $NDVI_{JANUARY}$ (horizontal axis) scattergram. Non-frozen training areas are shown by open circles and those badly damaged as closed circles: the latter present a reduced value in March. The figure also shows the results of the measurements before and after applying the atmospheric correction algorithm.

The behaviour of frost damaged areas can be explained physically according to the processes that leaves suffer under those conditions: as the chlorophyll content decreases, absorption by that pigment also decreases and as a consequence the TM3 reflectance increases. At the same time, as the leaf dies its activity practically disappears and TM4 reflectance decreases. The net effect is therefore a vegetation index reduction.

We can also see the effect of the atmospheric correction applied: this both increases the value range of the indices and decreases the relative difference between March and January. Certainly, one should not expect a sharp increment in the March index value because, although the citrus canopy in that time has greater vigour (it is just before the flowering period which will take place in April), shadows diminish with respect to January and a reduction in shadows makes the index value decrease (Ranson, 1987; Gilabert and Meliá, 1990), .

Non-frost line: algorithm for the identification and evaluation of frost damaged areas

Figure 17.3 shows that the points representing non-frozen parcels can be adjusted by a straight line which we call the NON-FROST LINE. The physical meaning of that line in the phase space is that it indicates the temporal evolution of the citrus parcels from January to March in the case that they are not affected by frost. The equation of the non frost line is:

$$NDVI_{MARCH} = 2.1 + 1.24 NDVI_{JANUARY} \quad (17.5)$$

with an uncertainty of ± 1.3 in the intercept and ± 0.22 in the slope. According to this geometrical interpretation, the vertical distance from points representating frost affected parcels to that line is a measure of cold effects. The line permits us to design the algorithm in the following way: theoretically, if there was no frost the March image obtained would be that corresponding to the transformation of the January image using the non-frost line equation (we call this image $NDVI_{MARCH*}$). If we construct a new image which represents the difference between $NDVI_{MARCH*}$ and the true March image, $NDVI_{MARCH}$, the pixels belonging to parcels of the same spectral response (non-frost parcels) will have very small values, practically zero. The pixels affected by cold will take the highest values according to their degree of cold damage. That is, the difference between the two images (the theoretical one minus the real one) indicates the distance of the parcel pixels from the non-frost line and therefore is a measure of the degree of cold damage (**D**):

$$\mathbf{D} = NDVI_{MARCH*} - NDVI_{MARCH}. \quad (17.6)$$

The application of this algorithm requires in its final phase that the subtracted images

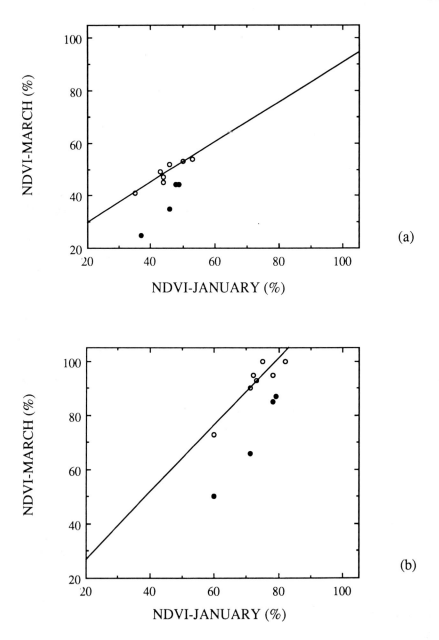

FIGURE 17.3. Comparison of values of Normalized Difference Vegetation Index (**NDVI**) for March and January, (a) without atmospheric correction (b) with atmospheric correction. Key o, non-frost parcels; •, frost parcels

be perfectly superimposable, that is, it requires the application of a geometric correction, as described earlier.

Results

The application of the algorithm described in the previous paragraph gives images which represent the degree of cold damage (**D**) of the citrus within the area of study. In order to display the degree of cold effects in citrus parcels alone we have introduced a mask (in black) to eliminate other land-use types. The mask has been constructed from the **NDVI** of January because in winter months there is the maximum contrast between citrus and other surfaces (Segarra *et al.*, 1988). Statistical analysis of the **NDVI** histogram corresponding to the smoothed image permits us to identify the pixels corresponding to non-citrus areas. We have applied a filtering technique for its usefulness in classifying agricultural areas which has been widely checked (Atkinson, Cushnie, Townshend and Wilson, 1985).

The images showing the different degrees of cold damage may also be enhanced by using different colour tones. The colour assignment has been made according to partition of the image frequency graph, as shown in *Figure 17.4*. Consideration of the uncertainties of the slope and the intercept of the non-frost line extends the limit of the **D** values representative of non-frozen pixels from 0 to 6%. The total area of frost damaged citrus that we have obtained in this work, 1508 ha, improves our previous result of 1574 ha (Universitat de València, 1987; Segarra *et al.*, 1988) that we obtained with a different algorithm also based on a multi-temporal comparison of vegetation indices. We reckon that this improvement is more qualitative than quantitative. Indeed, the introduction of a geometric correction in the images has reduced the number of pixels classified as frozen, most likely due to a better image coincidence. However, what we prefer to stand out in this work is that the present algorithm has a sounder physical basis, leading to a clearer interpretation of the degree-of-cold-damage concept through the introduction of the non-frost line.

Discussion

In this work we have tried to give an answer to the frequent problem for our region of evaluating the damage produced by frosts in citrus orchards. Due to the high spatial and temporal variability of citrus-canopy reflectance, the only way to show the alteration produced in the spectral response following frost is by means of a multi-temporal study of a vegetation index such as **NDVI**.

The definition of the non-frost line has permitted us to design an algorithm to evaluate the degree of cold damage of citrus parcels. This line, which in the phase space is essentially defined as $\mathbf{NDVI_{PRE-FROST}} / \mathbf{NDVI_{POST-FROST}}$, gives the evolution of the vegetation index in a hypothetical case where no frost has occurred. Thus, the distance from points representative of different pixels to this line represents an estimation of the intensity of the damage produced by the frost (**D**).

We have used Landsat-5 TM images, but the methodology could also suit well any other sensor with bands equivalent to TM3 and TM4 and with adequate resolution, such as SPOT. The software developed is fast and the final image is obtained quickly.

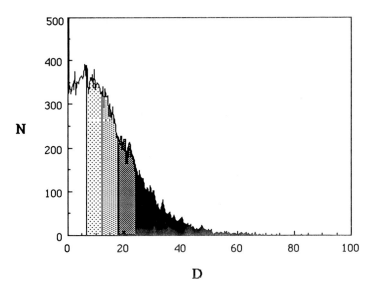

FIGURE 17.4. Frequency of class distribution for cold damage images based on **NDVI**. "N" represents the number of pixels with a given value of D (degree of cold damage). Values of D between 0 and 6% correspond to healthy citrus, and D > 6% correspond to citrus damaged by cold

However, the after-the-frost image used for comparison should be well separated in time from the frost date, for example by about a month, in order that the alteration produced by the frost can be physically noticeable. Although the damage produced by intense cold is irreversible for leaves, their death is not instantaneous and a time span is necessary to make that damage noticeable.

A common feature of remote sensing techniques is their relatively low cost when compared to direct monitoring methods, which are not always affordable and, if they are, not for complete cover. We have not been able to compare our results with official statistical surveys quantitatively because no field data were available for the frost of the 1985 winter. However, qualitative comparison of our resulting image with some data of the Consellería d'Agricultura and data obtained in our visits to the area of study, has shown the power and potential of the method.

Acknowledgements

This work has partly been supported by the Consellería d'Agricultura i Pesca de la Generalitat Valenciana and the CAICYT (Project no.A-172/85). We thank Dr. J.A. Clark for advice on revision of the manuscript. We should also acknowledge the Semiconductor Laboratory of the Department of Applied Physics (Universitat de València) for the use of the monochromator H20 Jobin-Yvon[1]. M.A. Gilabert has a research grant (PFPI 1987) from the Ministry for Science Education (Spain).

References

ATKINSON, R., CUSHNIE, J.L., TOWNSHEND, J.R.G. and WILSON, A. (1985). Improving Thematic Mapper land cover classification using filtered data. *International Journal of Remote Sensing*, 6, 955-961

CARIOLIS, C. de and AMODEO, P. (1980). Basic problems in the reflectance and emittance properties of vegetation. In *Remote Sensing Applications in Agriculture and Hydrology.* Ed. by FRAYSSE, G. A.A. Balkema, Rotterdam,

CHAVEZ, P.S. Jr (1988). An improved dark-object subtraction technique for atmospheric scattering correction of multispectral data. *Remote Sensing of Environment*, 24, 459-479

CLEVERS, J.G.P.W. (1988). The derivation of a simplified reflectance model for the estimation of leaf area index. *Remote Sensing of Environment*, 25, 53-69

COLWELL, J.E. (1974). Vegetation canopy reflectance. *Remote Sensing of Environment*, 3, 175-183

CURRAN, P.J. (1980). Multispectral remote sensing of vegetation amount. *Progress in Physical Geography*, 4, 315-341

DIAZ, F. (1983). *Práctica de la defensa contra las heladas.* Dilagio, Lérida, Spain

FUENTES, J.L. (1983). *Apuntes de Meteorología Agrícola.* Ministerio de Agricultura, Pesca y Alimentación, Madrid

GILABERT, M.A. and MELIA, J. (1990). A simple geometrical model for analysing the spectral response of a citrus canopy using satellite images. *International Journal of Remote Sensing*, (in press)

HOLBEN, B.N. (1986). Characteristics of maximum-value composite images from temporal AVHRR data. *International Journal of Remote Sensing*, 7, 1417-1434

KNIPLING, E.B. (1970). Physical and physiological basis for the reflectance of visible and near-infrared radiation from vegetation. *Remote Sensing of Environment*, 1, 155-159

PRICE, J.C. (1987). Calibration of satellite radiometers and the comparison of vegetation indices. *Remote Sensing of Environment*, 21, 15-27

RANSON, K.J. (1987). Scene shadow effects on multispectral response. *IEEE Transactions on Geoscience and Remote Sensing*, GE-25, 502-509

SCHANDA, E. (1986). *Physical Fundamentals of Remote Sensing.* Springer-Verlag, Berlin

SEGARRA, D., GILABERT, A., GANDIA, S. and MELIA, J. (1988). Signature spectrale des agrumes et son evolution - Identification d'index de vegetation de moindre variation temporelle. In *Proceedings of the 4th International Colloquium on Spectral Signatures of Objects in Remote Sensing.* Aussois, France, 18-22 January 1988. ESA SP-287, April 1988, ESTEC, Noordwijk, The Netherlands

---

[1] Trade names are for the convenience of the reader and do not imply preferential endorsement of a particular product or company over others by the University of Valencia.

TUCKER, C.J., ELGIN, J.H. and McMURTREY, R.E. (1979). Temporal spectral measurements of corn and soybean crops. *Photogrammetric Engineering and Remote Sensing*, **45**, 643-653

UNIVERSITAT DE VALENCIA (1987). Mètodes de teledetecció de distribucio de temperatures i avaluació de superfícies, Informe Final, Conveni de collaboració entre la Consellería d'Agricultura i Pesca de la Generalitat Valenciana i la Universitat de València, Facultat de Física, Valencia, Spain

VI.

NEW TECHNIQUES

18
APPLICATIONS OF CHLOROPHYLL FLUORESCENCE IN STRESS PHYSIOLOGY AND REMOTE SENSING

HARTMUT K. LICHTENTHALER

Botanisches Institut (Plant Physiology), University of Karlsruhe Kaiserstrasse 12, D-7500 Karlsruhe, F.R.G.

Introduction

In green photosynthetically active plant tissue (leaves, needles, stems), light absorbed by the photosynthetic pigments (chlorophyll and carotenoids) is primarily used in the two photosynthetic light reactions for photosynthetic quantum conversion. Under optimum conditions for photosynthesis (e.g. no water, heat, light, cold or mineral stress), only a small proportion of the absorbed light is de-excited as heat and as red to far-red chlorophyll fluorescence (Lichtenthaler, 1987; Lichtenthaler, Buschmann, Rinderle and Schmuck, 1986). Since light absorbed by the accessory pigments (chlorophyll-b, carotenoids) is transferred to chlorophyll-a, one observes *in vivo* only the fluorescence of chlorophyll-a. At room temperature this fluorescence is thought to come primarily from photosystem 2 (Papageorgiou, 1975), though photosystem 1 may also contribute to a small extent (Kyle, Baker and Arntzen, 1983).

Under various stress conditions of the plant, the rate of photosynthesis (CO_2- assimilation) is, however, considerably reduced by disturbing either the light-driven photosynthetic electron transport, the pigment apparatus, and/or the CO_2-assimilation, without affecting the process of light absorption. This then leads to an increased de-excitation of the absorbed light via chlorophyll fluorescence and heat emission. The broad inverse relationship between *in-vivo* chlorophyll fluorescence and photosynthetic carbon assimilation, first observed and described by Kautsky (Kautsky and Franck, 1943), can therefore be used to study the potential photosynthetic activity of leaves and to detect stress effects in green plants. For original literature on stress effects seen by chlorophyll fluorescence, see the recent review Lichtenthaler and Rinderle (1988a), as well as the book "Applications of Chlorophyll Fluorescence" edited by Lichtenthaler (1988a).

There are several parameters of the *in-vivo* chlorophyll fluorescence, e.g. induction kinetics (Krause and Weis, 1983), values of the fluorescence decrease ratio Rf_d-values (Lichtenthaler, Buschmann, Döll, Fietz, Bach, Kozel, Meier and Rahmsdorf, 1981, Lichtenthaler et al., 1986; Lichtenthaler and Rinderle, 1988b; Nagel, Buschmann and Lichtenthaler, 1987; Rinderle, Haitz, Lichtenthaler, Kähny, Shi and Wiesbeck, 1988; Strasser, Schwarz and Bucker, 1987), emission spectra (Lichtenthaler and Buschmann,

1988; Rinderle and Lichtenthaler, 1988), saturation pulse kinetics (Schreiber, Schliwa and Bilger, 1986; Schreiber, Neubauer and Klughammer, 1988), time-resolved chlorophyll fluorescence (Holzwarth, 1988), which change due to natural or anthropogenic stress conditions. These parameters can not only be applied for detection of stress and damage to the photosynthetic apparatus and the plant, but also for studies on the regeneration of photosynthetic function when the stressor is removed. Chlorophyll fluorescence thus opens the possibility for fast and non-destructive stress detection in plants. The different fluorescence parameters to be measured and the methods applied in field and laboratory research, and in physiological ground-truth controls associated with airborne reflection measurement, are summarised in this report, as well as possibilities of remote sensing of terrestrial vegetation via laser-induced chlorophyll fluorescence.

The chlorophyll fluorescence induction kinetics

Several fluorescence parameters, which can be used for stress detection in plants, are based upon the original registration of the light-induced chlorophyll fluorescence induction kinetics of a pre-darkened (15 to 20 min) green plant tissue (leaf, needle, stem). White, blue and red light can be applied for fluorescence induction. Green light is less favourable, since it is absorbed to a lower degree than the other light.

Upon illumination the fluorescence (measured in relative units) rises spontaneously to the ground fluorescence level ($\mathbf{F_o}$) and then increases, in general within 500ms, to a maximum (termed either $\mathbf{F_m}$ or $\mathbf{F_{max}}$) (Kautsky effect: fast component). With the onset of membrane energisation and photosynthetic oxygen evolution, the fluorescence decreases slowly and continuously to reach a steady-state level after 4 to 5 minutes (*Figure 18.1*).

The fluorescence rise from $\mathbf{F_o}$ to the maximum fluorescence $\mathbf{F_m}$ is also termed variable fluorescence, $\mathbf{F_v}$. The higher the variable fluorescence (rise above $\mathbf{F_o}$ to $\mathbf{F_m}$) and the larger the slow fluorescence decrease $\mathbf{f_d}$ (from $\mathbf{f_{max}}$ to $\mathbf{f_s}$), the higher the photosynthetic capacity of a leaf. In a darkened leaf the photosynthetic apparatus is in the non-functional state I, where the two photosynthetic photosystems are impaired. After the light-triggered fluorescence induction, the photosynthetic apparatus is in the photosynthetically active state II, with a low yield of chlorophyll fluorescence (= steady-state fluorescence). When such a leaf is re-illuminated after a dark period of only 2 or 3 seconds there is no variable fluorescence: the fluorescence then goes from $\mathbf{F_o}$ directly to the steady state $\mathbf{f_s}$.

The fluorescence decrease from $\mathbf{f_{max}}$ to $\mathbf{f_s}$ is paralleled by increasing rates of oxygen evolution and photosynthetic CO_2-fixation, as has been shown by many authors (Lichtenthaler and Rinderle, 1988a; 1988d; Walker, 1988). The relative extent of the fluorescence decrease is therefore an approximate measure of the degree of photosynthetic quantum conversion of a leaf (see below: $\mathbf{Rf_d}$-values). Several other factors may participate in this fluorescence decrease, e.g. thermal quenching, phosphorylation of the light-harvesting chlorophyll-proteins, etc. (Papageorgiou, 1975; Krause and Weis, 1983).

With respect to the application of chlorophyll fluorescence in stress physiology and ecophysiology, it is not necessary to resolve the fast fluorescence rise signal. This would require a larger experimental set up and more expensive instrumentation, which is not very suitable for field experiments. In addition, the fluorescence parameters obtained from the fast fluorescence rise ($\mathbf{F_o}$, $\mathbf{F_m}$ and $\mathbf{F_v}$ as well as the ratios $\mathbf{F_m/F_o}$ and $\mathbf{F_v/F_m}$) do

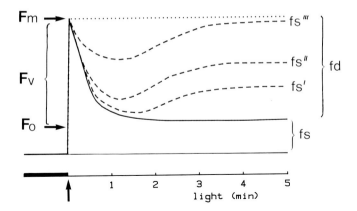

FIGURE 18.1. Light-induced chlorophyll fluorescence induction kinetic of a green pre-darkened leaf (Kautsky effect). Upon illumination the fluorescence rises very fast via F_o (ground fluorescence) to F_m (maximum fluorescence). It then declines within minutes to the steady state f_s. F_v is the variable fluorescence rise from F_o to F_m, which is not resolved in this diagram. The ratio of the fluorescence decline (f_d) from the maximum F_m to the steady state fluorescence ($Rf_d = f_d/f_s$) is a vitality index of the plant. Under continuous stress the level of f_s increases (to $f'_s \to f''_s \to f'''_s$) and correspondingly f_d and the ratio Rf_d decline

not fully reflect the actual physiological condition of the photosynthetic apparatus, since they are determined in the non-functional state I of photosynthesis. The fluorescence decrease signal f_d and the steady-state fluorescence f_s, in turn, which are sensed in state II of photosynthesis, contain more reliable physiological information (Lichtenthaler and Rinderle, 1988a). Under stress conditions both parameters change rather early and fast (increase of the steady-state fluorescence f_s and a corresponding decrease in the value of f_d), whereas the parameters F_o, F_m and F_v may yet remain unchanged. This increase in the steady-state fluorescence with increasing stress ($f_s \to f'_s \to f''_s$) after an initial decrease from f_{max} is shown in *Figure 18.1* by broken lines.

When the photosynthetic electron transport is blocked by herbicides such as diuron or bentazon, the chlorophyll fluorescence rises to f_{max} and remains on that high level (Lichtenthaler et al, 1986; Kocsanyi, Haitz and Lichtenthaler, 1988).

Rf_d-values as vitality index

The ratio of the fluorescence decrease to the steady-state fluorescence ($Rf_d = f_d/f_s$), as calculated from the slow fluorescence decline kinetics at 690nm is a very suitable indicator of vitality and stress of the plant, and has been termed a vitality index (Lichtenthaler et al., 1986; Strasser, 1986). The level of the Rf_d-values is, in fact, a measure of the potential photosynthetic activity and capacity of a leaf (Lichtenthaler, 1987; Lichtenthaler and Rinderle, 1988a; Haitz and Lichtenthaler, 1988; Nagel et al., 1987). It signals the intactness and functionality of the internal photosynthetic apparatus, even with closed stomata.

Under normal photosynthetic conditions (with open stomata), the general rule is

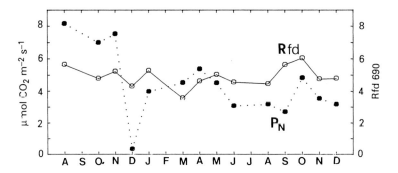

FIGURE 18.2. Seasonal changes in the Rf_d values at 690 nm (= vitality index) and in the net photosynthesis rates P_N (net CO_2 assimilation) in green needles of a healthy spruce (Northern Black Forest; Althof site; 450 m above sea level; start of measurements: August 1987; needle year 1987). The Rf_d-values are an indicator of the intactness and activity of the endogenous photosynthetic apparatus, whereas the P_N values are an indicator for the CO_2 gas uptake (measurements by U. Rinderle)

that higher Rf_d-values are correlated with higher net CO_2 fixation rates (Lichtenthaler and Rinderle, 1988a). Sun and high-light leaves exhibit higher photosynthetic rates, which are associated with higher Rf_d-values, than shade and low-light leaves (Lichtenthaler et al., 1981). The Rf_d-values measured in the 690nm region have also been applied as indicators of net CO_2 assimilation, with great success, in forest decline research (Lichtenthaler and Rinderle, 1988c; 1988d; Schmuck and Lichtenthaler, 1986; Nagel et al., 1987). When stomata are closed, due to water shortage or during frost periods in winter, the level of the Rf_d-values indicates whether the internal photosynthetic apparatus is still intact and functional under such stress conditions. In such cases the CO_2 set free by respiration in the 20 min dark period which precedes the light-induced fluorescence kinetics is used to perform photosynthesis. High Rf_d-values of 5 to 6 in young growing leaves and needles indicate that their chloroplasts are photosynthetically active, even though their stomata are not yet fully developed and functional.

Rf_d-values and net photosynthesis P_N develop independently over seasonal periods. *Figure 18.2*, for sun-exposed spruce needles of a healthy tree shows that Rf_d-values were high (4 to 6) throughout the year, and indicate that the structure and function of the internal photosynthetic apparatus was kept intact even in winter time. The net CO_2 uptake rates (P_N), however, showed a sharp decline in December 1987, due to strong frost periods with temperatures below -15°C; a revival of the CO_2 uptake was, however, seen in a very warm January of 1988. In contrast to P_N, the frost period in December 1987 decreased the Rf_d-values very little, indicating that the frost period scarcely affected the internal photosynthetic function. In spring, net CO_2 uptake P_N again exhibited higher values, which then declined during the summer (July to September), whereas the Rf_d-values remained at a high level. In October 1988 a short-term reactivation of photosynthesis, as seen in higher values for both Rf_d and P_N, took place.

Rf_d-values (690 nm) above 3 indicate a very efficient photosynthesis and very good photosynthetic rates, if the stomata are open. Stressed, declining or senescent leaves with Rf_d-values at or below 1.0 no longer exhibit a positive net CO_2 assimilation rate, even

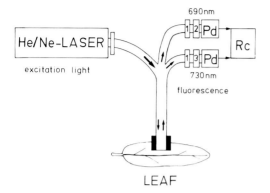

FIGURE 18.3. Diagram of laser-equipped portable two-wavelength chlorophyll fluorometer allowing simultaneous registration of the fluorescence induction kinetics (Kautsky effect: slow component) of intact or stressed leaves at two wavelengths. The red laser light is sent through fibre optics to the darkened leaf and the red chlorophyll fluorescence is sensed via fibre optics by the photodiodes (P_d) in two wavelength regions (690 and 730 nm), which correspond to the fluorescence emission maxima. The two fluorescence kinetics are registered using a two channel recorder (R_c). 1 = cutoff filter RG 665 nm; 2 and 3 = interference filters in the 690 and 730 nm region. (After Lichtenthaler and Rinderle, 1988a)

though their stomata may be open. The easy and quick determination of Rf_d-values permits a fast outdoor screening of the vitality of crop plants and trees before other more time-consuming laboratory methods (e.g. measurement of P_N via a CO_2/H_2O porometer) are applied to investigate further the type of stress and damage to the plant. In fact, the determination of Rf_d-values from the induction kinetics is today the superior method in ecophysiology to detect and describe vitality and stress (Lichtenthaler and Rinderle, 1988a).

The laser-equipped portable two-wavelength fluorometer

For fast outdoor screening of the potential photosynthetic activity of leaves we constructed a laser-equipped portable field chlorophyll fluorometer which senses the laser-induced fluorescence kinetics at two different wavelengths. It is known that the chlorophyll fluorescence emission spectra at room and outdoor temperatures exhibit two maxima near 690 and 735 nm (see below: ratio F690/F735) With the two-wavelength chlorophyll fluorometer the fluorescence kinetics are registered at both wavelength regions (Lichtenthaler and Rinderle, 1988a). Excitation is performed with a He/Ne-laser (5 to 8 mW, $\lambda = 632.8nm$) using a 3-arm glassfibre system to illuminate the pre-darkened leaf sample. The red chlorophyll fluorescence is detected by two photodiodes (SD-444-41-11-261, Silicon Detector Corp.), as shown in *Figure 18.3*.

The transmission range of the two filter systems employed is shown in *Figure 18.4*.

The intensity of the excitation light at the leaf surface should be high enough and should lie at or above 500 μE m^{-2} s^{-1}. The chlorophyll fluorescence kinetics in the 690 and 730nm regions are simultaneously registered with a two-channel recorder. One

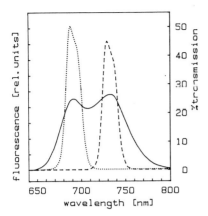

FIGURE 18.4. Chlorophyll fluorescence emission spectrum of a fully green leaf with indication of the transmission range of the filter systems used in the two-wavelength chlorophyll fluorometer

obtains in this way two Rf_d-values, the $Rf_d 690$ and the $Rf_d 730$, which contain more physiological information and in addition also permit one to estimate the chlorophyll content of the leaf sample, (see below), than data for a single wavelength.

$Rf_d 690$ and $Rf_d 730$-values and stress adaptation index

The chlorophyll fluorescence yield and the shape of chlorophyll fluorescence spectra depend on the wavelength of the excitation light. In green leaves blue excitation light results in a fluorescence spectrum of about equal intensity in the 690 nm and 735 nm region. When red laser light (632.8 nm) is applied for fluorescence excitation, the fluorescence in the 690 nm region is much lower than in the 730 nm region (Lichtenthaler and Buschmann, 1988; Lichtenthaler and Rinderle, 1988a). As a consequence, the intensity of the He/Ne laser induced fluorescence at 690 nm is only about half that in the 735 nm region, as shown in *Figure 18.5* (left side). This is also documented by higher values at 730 nm for the measured maximum fluorescence and steady-state fluorescence (f'_{max} and f'_s) as compared to the corresponding values in the 690nm region (f''_{max} and f''_s (*Table 18.1*).

The relative chlorophyll fluorescence increase and decrease during the induction kinetics are higher in the 690 nm than the 730 nm region. As a consequence, the Rf_d-values ($Rf_d = f_d/f_s$) are higher in the 690 nm (Rf''_d or $Rf_d 690$) than in the 730 nm region (Rf'_d or $Rf_d 730$). Both Rf_d-values can be taken as a vitality index. Their parallel decline, e.g. due to increasing water stress at various times after leaf abscission (*Table 18.1*), indicated the loss of photosynthetic function. In the herbaceous plant tobacco, the loss of photosynthetic function (Rf_d-values below 1.0) is seen much earlier (after 24 hours) than in fir needles (after 5 days).

TABLE 18.1. Effect of water stress (following leaf and needle abscission) on the red laser-induced (He/Ne; λ =632.8 nm) chlorophyll fluorescence signatures (f_{max}, f_s); Rf_d-values (measured in the 690 nm and 730 nm region); stress adaptation index A_p and the fluorescence ratio (f_s''/f_s'). For comparison the ratio F690/F735, obtained by blue light excitation (470 nm), is contrasted with the red-light induced ratio f_s''/f_s'. The controls were measured as attached leaves and the remaining values at listed times after leaf or needle detachment.*

* The values are the means of 6 (tobacco) and 3 (fir) determinations, respectively. The average chlorophyll contents were 46 ± 2.5 μg (tobacco) and 44 ± 2 μg (fir) per cm^2 leaf area. The fluorescence, as measured with the portable two wavelength chlorophyll fluorometer, is given in relative units

	690 nm region			730 nm region			A_p	f_s''/f_s'	$\frac{F690}{F735}$
	f''_{max}	f''_s	Rf''_d	f'_{max}	f'_s	f'_d			
Nicotiana tabacum:									
control	806	208	2.87	1639	560	1.93	0.24	0.37	0.86
after 1.5h	777	230	2.38	1536	558	1.76	0.18	0.41	0.97
after 4h	760	291	1.64	1307	588	1.24	0.15	0.49	1.14
after 24h	169	140	0.21	414	350	0.18	0.03	0.40	1.18
Abies alba:									
control	182	29	5.25	275	64	3.33	0.31	0.45	0.95
after 5h	206	37	4.66	357	85	3.22	0.25	0.44	0.93
after 24h	257	47	4.52	388	95	3.08	0.26	0.49	1.02
after 48h	202	44	3.59	311	84	2.70	0.19	0.52	1.08
after 5 d	120	74	0.64	185	129	0.46	0.11	0.57	1.55
after 5.5d	100	86	0.17	160	140	0.15	0.02	0.61	1.56

The ratio f_s''/f_s'

From the steady-state fluorescence f_s of the He/Ne laser-induction kinetics, one can calculate the ratio of the fluorescence yield in the 690 and 730 nm regions. This is shown as ratio f_s''/f_s' in Table 18.1 (= F690/F730 at f_s). The values range from 0.37 to about 0.55 in green leaves, depending on the chlorophyll content. After abscission of leaves, the values of f_s''/f_s' increase considerably due to dessication and loss of photosynthetic function (Table 18.1). The blue-light induced ratio F690/F735, as determined from the complete fluorescence spectra measured separately in a Shimadzu spectrometer, shows higher values of 0.86 to 0.95 in green leaves, and also rises with increased water stress.

The stress-adaptation index A_p

From the Rf_d-values at 690 nm and 730 nm one can determine another fluorescence parameter: the stress-adaptation index A_p, as introduced by Strasser et al. (1987). A detailed description is given in a recent review (Lichtenthaler and Rinderle, 1988a). This index is thought to be a measure of how the structure of the photosynthetic apparatus is organized for best adaptation to stress conditions. The A_p index is related to the

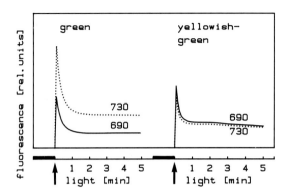

FIGURE 18.5. Induction kinetics of the *in vivo* chlorophyll fluorescence (Kautsky effect: slow component) in differently coloured spruce needles measured in the 690 and 730 nm region with the portable two-wavelength fluorometer of Lichtenthaler (Lichtenthaler and Rinderle, 1988a; Haitz and Lichtenthaler, 1988). In green needles the fluorescence emission in the 730 nm region is about twice that in the 690 nm region, whereas in the yellowish-green needles the fluorescence emission is of about equal intensity in both regions. Chlorophyll content: 41 μg (green, needle year 1988) and 18 μg cm^{-2} needle area (yellowish-green). Fluorescence excitation: He/Ne-laser, 632.8 nm. Measurements of February, 1989. The needles were darkened for 20 min prior to illumination

Rf_d-values (690 nm = Rf_d'' and 730 nm = Rf_d') by the equation:

$$A_p = \frac{Rf_d' + 1}{Rf_d'' + 1}$$

Green photosynthetically active intact leaves show A_p values of *ca.* 0.20 to 0.33. Values below 0.15, as found in senescent leaves (Lichtenthaler and Rinderle, 1988a) or during water stress (e.g. *Tables 18.1* and *18.2*), indicate very drastic damage to the photosynthetic apparatus, from which the leaf does not normally recover. The A_p-values are also higher for sun than shade leaves, and for aurea than green leaves (*Table 18.2*). From our present experience one can summarize: plants and leaves with lower A_p-values suffer from stress earlier than those with a higher A_p-value.

Chlorophyll content and fluorescence kinetics at 690 nm and 730 nm

In normal green, photosynthetically active leaf tissue the He/Ne laser-induced fluorescence kinetics of the 690 nm region are much lower than those of the 730 nm region (*Figure 18.5* left side). In light-green or yellowish-green leaves with a lower chlorophyll content, the fluorescence intensity at 690 nm is, however, only slightly lower than or equal to that in the 730 nm region (*Figure 18.5* right side). Intermediate values are found depending on the chlorophyll content. This is shown in *Figure 18.6*, where the ratio F690/F730, as determined from f_{max} of the induction kinetics, is shown for a normal green, healthy and a yellowish-green, damaged spruce.

TABLE 18.2. Values of the chlorophyll fluorescence ratio F690/F735 for excitation at 470 and 620 nm, $\mathbf{Rf_d}$-values (fluorescence vitality index measured in the 690 and 730 nm region) and stress-adaptation index $\mathbf{A_p}$ of leaves of different chlorophyll content and physiological state. Mean values of at least 3 determinations per leaf type. $\mathbf{Rf_d}$-values higher than 2.5 indicate a good to very good photosynthetic activity of the leaf (after Lichtenthaler and Rinderle, 1988a).

* The olive-green needles with some yellow spots and tips are from spruce of damage class 3/4 (ca. 85% needle loss).

** Red excitation light provided by a He/Ne-laser (5mW, 632.8 nm)

Plant	Chlorophyll a + b ($\mu g\ cm^{-2}$)	ratio F690/F735		$\mathbf{Rf_d}$-values		$\mathbf{A_p}$ index
		470 nm	620 nm	690 nm	730 nm	
Ilex aquifolium:						
dark-green leaf:	70	0.98	0.5	4.0	2.4	0.33
senescent leaf:	21	1.91	1.68	1.2	1.0	0.99
frost-injured leaf:	6	3.5-6	2-2.8	0.7	0.5	0.12
Picea abies:						
damage class 0/1 green needles						
needle year 1987:	50	1.21	0.90	6.0	3.8	0.31
needle year 1985:	90	1.09	0.72	5.1	3.5	0.26
damage class 3/4*						
needle year 1987:	52	1.10	0.86	3.9	2.8	0.22
needle year 1985:	28	1.60	1.28	1.2	1.1	0.05
Phaseolus vulgaris:						
green control leaf:	63	0.90	0.61	2.8	2.1	0.18
diuron-treated:	63	1.15	0.78	0	0	0
Zea mays:						
light green leaf:	42	1.32	0.85	2.5	1.9	0.17
diuron-treated	42	1.65	1.10	0	0	0
Nicotiana tabacum:						
green leaf (su/su)	46	1.10	0.76**	3.3	2.2	0.26
aurea leaf (su/su)	16	1.90	1.27**	4.0	2.6	0.28

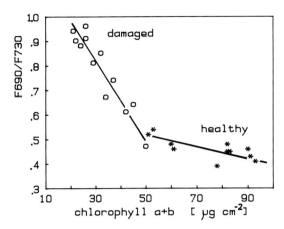

FIGURE 18.6. Decrease of the fluorescence ratio F690/F730 with increasing chlorophyll a + b content (μg cm^{-2} needle area) in a branch of a damaged spruce (*Picea abies*; yellowish-green needles) and a branch of a healthy spruce (fully green needles). The ratio F690/F730 was determined from the laser-induced fluorescence kinetics. In the branch of the damaged spruce 6 needle years (1983 to 1988) were investigated and the highest chlorophyll content was found in the 1988 needles. In the fully green healthy branch the youngest needle year, (1988), had a lower chlorophyll content (50 to 60 μg cm^{-2}) than the needle years 1985, 1986 and 1987. (Measurements performed in February, 1989)

The LICAF-Fluorometer

For laboratory use we have combined our portable two-wavelength chlorophyll fluorometer with a computer and an acousto-optic shutter. This Laser-Induced Computer Aided Fluorometer (LICAF) permits measurement of the fluorescence induction kinetics every millisecond in measuring periods of 5 μs in the 690 nm and 730 nm region (Koscanyi, Haitz and Lichtenthaler, 1988). This system also resolves the fast flourescence rise, and the values of the ratio F690/F730 are plotted together with the induction kinetics.

The ratio F690/F735 as stress indicator

The full chlorophyll fluorescence emission spectra (as excited by blue light, 450 nm) are shown in *Figure 18.7*. In the fully green leaf the two fluorescence maxima near 690 nm and 730nm are of about equal intensity, whereas in the light-green leaf the 690 nm fluorescence is much higher and the 735 nm region is only present as a shoulder. With increasing chlorophyll content of a leaf, the relative fluorescence at 690 nm becomes smaller and smaller and the shoulder at 735 nm develops to a second maximum. This is due to re-absorption of the emitted shorter-wavelength fluorescence by the *in vivo* chlorophyll, since the chlorophyll absorption and the fluorescence emission bands overlap in this region (*Figure 18.8*). As a consequence the ratio F690/F735 exhibits lower values for fully green than young developing or senescent leaves with lower chlorophyll content (Lichtenthaler, 1987).

The blue-light induced fluorescence spectra exhibit higher values for F690/F735

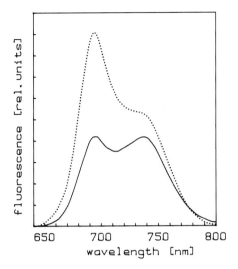

FIGURE 18.7. Chlorophyll fluorescence emission spectra of a fully green leaf (solid line) and a yellowish-green aurea leaf (dotted line) of cherry laurel (*Prunus laurocerasus L.*) showing the fluorescence emission maxima near 690 and 735 nm. Excitation by blue light (450 ± 30 nm); Chlorophyll content 52 μg (green leaf) and 6/μg cm^{-2} leaf area (aurea leaf), respectively

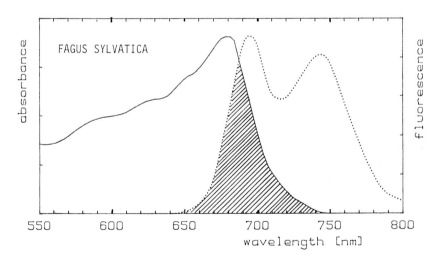

FIGURE 18.8. Absorption (solid line) and chlorophyll fluorescence emission spectra (dotted line) of a beech leaf (*Fagus sylvatica*). The overlapping of the absorption and fluorescence emision spectra, which results in a partial reabsorption of the shorter wavelength fluorescence around 690 nm, is indicated. The fluorescence spectrum was excited by blue light (wavelength 450 ± 30 nm)

(0.85 - 1.0) than the spectra induced by red-light (e.g. He/Ne laser light, λ 632.8 nm) which possess F690/F735 values of about 0.4 to 0.65 (Lichtenthaler and Rinderle, 1988b; Rinderle and Lichtenthaler, 1988). This difference in the values is due to the fact that the red excitation penetrates into deeper leaf layers than blue light before it is absorbed, and consequently the emitted shorter-wavelength fluorescence near 690 nm comes from deeper leaf layers and is re-absorbed to a higher degree that the blue light induced 690 nm fluorescence.

Continuous stress generally results in a lower accumulation of chlorophyll. In fully developed leaves stress leads to a partial breakdown of chlorophyll. Lower rates of photosynthesis are also a consequence of long-term and short-term stress. Both processes (lower chlorophyll content and/or lower photosynthetic quantum conversion) increase the values of the ratio F690/F735, no matter whether excited with blue (470 nm) or red light (620 or 632.8 nm), as shown in *Table 18.2*. In senescent leaves, in needles from damaged spruce, and in diuron-treated leaves with a block of photosynthetic electron transport, the differences in the ratio F690/F735, which can also be determined as f_s''/f_s' from the fluorescence induction kinetics with the portable two-wavelength field fluorometer, is therefore a very suitable indicator of the effects of stress on plants. High values of this ratio indicate continuous stress, which reduces the chlorophyll content. In such cases the Rf_d-values may be quite normal, indicating that the remaining chlorophyll is photosynthetically active.

PAM fluorometer and the quenching coefficients qQ and qE.

With the relatively new pulse-modulation fluorometer (PAM) of Schreiber, one obtains additional and complementary information on the photosynthetic apparatus (Schreiber et al., 1986; 1988). The PAM fluorometer permits one to determine the ground fluorescence F_o, maximum fluorescence F_m (f_{max}) as well as the photochemical (qQ) and non-photochemical (qE) quenching coefficients. The latter are calculated from the light saturation pulse kinetics shown in *Figure 18.9*. More details of this system are found in the review of Lichtenthaler and Rinderle (1988a).

The photochemical quenching coefficient qQ varies only little during stress, whereas the coefficient qE steadily increases during stress conditions. the fastest change in a fluorescence parameter is the height of the saturation fluorescence spikes (distance g-h in *Figure 18.9*. The higher these spikes, the higher the QA-reoxidation capacity of a leaf and the higher its photosynthetic activity. After leaf abscission and increasing water stress, the height of the fluorescence spikes decreases very soon, whereas the increase of the qE-values proceeds much later (*Table 18.3*).

Remote Sensing of terrestrial vegetation via chlorophyll fluorescence

Airborne sensing of the physiological state of terrestrial vegetation and forest trees is mainly based on the passive registration of reflectance signatures in the visible and near-infrared (Huss, 1984; Lichtenthaler, 1988b; Lichtenthaler, Rinderle, Kritikos and Rock, 1987; Rock, Vogelmann, J.E., Williams, Vogelmann, A.F, and Hoshizaki, 1986; Rock, Hoshizaki, Lichtenthaler and Schmuck, 1986; Schmuck, Lichtenthaler, Kritikos, Amann

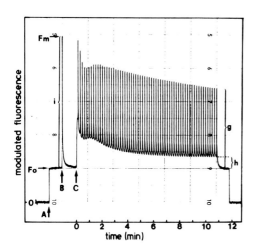

FIGURE 18.9. Chlorophyll fluorescence induction kinetics with saturation pulse spikes of a green *Platanus* leaf as measured with a pulse modulation fluorometer (PAM). A. measuring pulses of low intensity (to detect the F_o level); B. A 1 second saturation pulse (to determine maximum fluorescence F_m); C. actinic light (plus saturating light pulses every 10 seconds). (after Haitz and Lichtenthaler, 1988)

and Rock, 1987). Plants under stress, and in particular declining forest trees, usually exhibit a lower chlorophyll content and are characterized by a higher reflectivity in the visible region (500 - 750 nm) and a lower reflectance in the near-infrared region (NIR) of the reflectance spectrum (ca. 800 - 900 nm). This NIR-decrease in the reflectance spectrum is caused by a different arrangement and size of cells and aerial interspaces within the leaves of broadleaf plants or needles of conifers. These changes in the reflectance are associated with a "blue shift" of the inflection point of the red edge of the reflectance spectrum (Buschmann and Lichtenthaler, 1988; Lichtenthaler *et al.*, 1987: Rock, Vogelman *et al.*, 1986; Rock, Hoshizaki *et al.* 1986; Schmuck *et al.*, 1987). This "blue shift" near 700 nm is caused by two effects.

1. by a lower chlorophyll content of the leaves, and

2. by a higher short-wavelength chlorophyll flourescence in the 690 nm region which contributes considerably to the increase of the reflectance (Lichtenthaler and Buschmann, 1987).

An example of a changed reflectance spectrum is shown in *Figure 18.10a*, where the reflectance of 1983 needles of a damaged spruce is compared with that of 1988 needles of a healthy spruce. The "blue shift" is seen in the 2nd derivative spectrum, indicated by arrows in *Figure 18.10*.

The present problem with remote sensing of terrestrial vegetation by reflectance is that in the last three years the reflectance in the near-infrared region around 800 nm did not show up any longer in the newly formed needles of damaged trees (Lichtenthaler *et al.*, 1987): this is shown in *Figure 18.10b* for the needles of a healthy and damaged

TABLE 18.3. Effects of abscission on the water content (%H$_2$O) and on several chlorophyll-fluorescence signatures of photosynthetically functional, one year old spruce needles (needle year 1987) and two month old maple leaves. Values are given for the chlorophyll fluorescence decrease ratio (**Rf$_d$**-values at 690 and 730 nm), the stress adaptation index **A$_p$**, the blue light induced ratio F690/F735, the photochemical (**qQ**) and the non-photochemical quenching coefficients (**qE**) as well as the height of the saturation fluorescence spikes (g-h, see *Figure 18.9*). The fluorescence spikes are given in relative units. The chlorophyll contents (a+b) were 90 (needles) and 38 μg cm^{-2} (maple leaf). Mean of 3 determinations, standard deviation 5% or less (after Rinderle et al., 1988).
* Measured at the end of a 10 min saturation-pulse

Time	%H$_2$O	Rf$_d$-values		A$_p$	$\frac{F690}{F730}$	qQ*	qE*	g-h*
		690	730					
Picea (spruce)								
0h	62	4.5	3.3	0.21	1.0	0.94	0.27	98
1h	62	4.2	3.2	0.20	1.0	0.94	0.24	90
2h	60	4.2	3.0	0.22	1.0	0.95	0.24	82
6h	58	4.0	3.0	0.21	1.0	0.94	0.26	71
8h	57	3.7	2.8	0.20	1.0	0.93	0.27	60
24h	55	2.7	2.1	0.17	1.0	0.91	0.32	58
48h	50	1.2	0.8	0.19	1.1	0.46	0.39	22
Acer (maple)								
0h	76	3.5	2.4	0.24	1.0	0.92	0.48	70
1h	68	3.3	2.5	0.20	1.1	0.87	0.53	44
2h	65	3.3	2.4	0.20	1.2	0.82	0.55	22
4h	57	2.6	2.2	0.12	1.1	0.80	0.67	15
6h	45	1.2	1.0	0.12	1.1	0.79	0.68	15
8h	33	0.6	0.5	0.07	1.1	0.74	0.70	11
10h	21	0.4	0.4	0.02	1.1	0.60	0.77	6

fir, in which the "blue shift" of the inflection point at the red edge is very small or no longer detectable. These facts make the classification of agricultural crops and forest trees difficult. Reflectance data alone seem not or no longer to permit a judgement of the state of health of terrestrial vegetation, when the main damage signals (reduced reflectance near 800 nm and the "blue shift" of the inflection point) are missing. Reflectance methods should therefore be complimented by an additional remote sensing method which gives information on the physiology of the sensed green vegetation.

A remotely-sensed active laser-induced chlorophyll fluorescence could be such a complementary method. The advantage (as compared to reflectance measurements) is that the signal comes from the green plant parts. Remote sensing of the laser-induced chlorophyll fluorescence must concentrate on those fluorescence parameters which can be sensed in a very short time, e.g. from airborne systems. There are two parameters which could be measured after a short laser pulse:

- the chlorophyll fluorescence ratio F690/F735, which changes considerably due to stress, (see above) and,

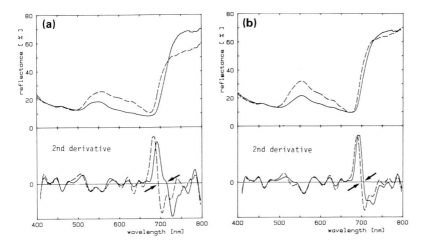

FIGURE 18.10. Reflection spectra and 2nd derivative spectra of A. spruce needles (*Picea abies*) of a damaged tree (needle year 1983, yellowish-green) and a healthy tree (needle year 1988, green) and B. of 1988 fir needles (*Abies alba*) of a damaged tree (yellowish-green) and a healthy tree (green). The arrows in the 2nd derivative spectrum indicate the "blue shift" of the inflection point at the red edge of the spectrum. The measurements were performed with the newly developed VIRAF instrumentation (see Buschmann and Lichtenthaler, 1988) in February 1989 by C. Buschmann. Solid line, green needles; broken line, yellow-green needles

- the registration of the fluorescence life-time, which increases under stress conditions (Schneckenburger and Frenz, 1986; Holzwarth, 1988).

In preliminary airborne and ground experiments the chlorophyll fluorescence was separately sensed in the 690 and 735 nm region, and differences were found between trees of different physiological states (Rosema, Cecchi, Pantani, Radicatti, Romuli, Mazzinghi, Van Kooten and Kliffen, 1988; Zimmermann and Günther, 1986). Whether the fluorescence life-time can be remotely sensed by an active system has yet to be investigated. In any case, it is clear that the development of ground and airborne laser-equipped active systems, which measure chlorophyll fluorescence, will open a new dimension in the remote sensing of terrestrial vegetation. They can complement not only remote measurements of reflectance but also radar cross-section measurements (Rinderle et al., 1988). Another aspect of remote sensing has to be mentioned here. There exists no remote sensing of agricultural crops or forests without physiological ground-truth control of selected plants and trees. Laser-induced chlorophyll fluorescence has already been applied with great success in ground-truth control (Lichtenthaler et al., 1987; Rock et al., 1986; Schmuck et al., 1987). In the future, chlorophyll fluorescence will still remain the superior method for physiological ground-truth measurements.

Conclusions

Laser-induced chlorophyll fluorescence is an extremely suitable and fast method in ecophysiology, in stress detection in crop plants and trees and offers potential for future remote sensing of the vitality of terrestrial vegetation. At present it is already the superior method in physiological ground-truth measurements of plants in connection with remote sensing of vegetation by reflectance.

Acknowledgements

Part of the work reported here was sponsored by grants from the PEF, Karlsruhe and the CEC Joint Research Centre, Ispra, which are gratefully acknowledged. I wish to thank Dr. Claus Buschmann, Ms U. Rinderle and Mr Stefan Burkart for valuable support during the preparation of this manuscript.

References

BUSCHMANN, C. and LICHTENTHALER, H.K. (1988). Correlation of reflectance and chlorophyll fluorescence signatures of healthy and damaged forest trees. In *Proceedings International Geoscience and Remote Sensing Symposium*. IGARSS-88. Edinburgh, Vol. 3, ESA Publications Division, Noordwijk

HAITZ, M. and LICHTENTHALER, H.K. (1988). The measurement of Rfd-values as plant vitality indices with the portable field fluorometer and the PAM fluorometer. In *Applications of Chlorophyll Fluorescence*. Ed. by H.K. Lichtenthaler. Kluwer Academic Publishers, Dordrecht.

HOLZWARTH, A.R. (1988). Time resolved chlorophyll fluorescence. What kind of information on photosynthetic systems does it provide? In *Applications of Chlorophyll Fluorescence*. Ed. by H.K. Lichtenthaler. Kluwer Academic Publishers, Dordrecht

HUSS, J. (1984). *Luftbildmessung und Fernerkundung in der Forstwirtschaft*. H. Wichmann Verlag, Karlsruhe

KAUTSKY, H. and FRANCK, U. (1943). Chlorophyllfluoreszenz und Kohlensäure assimilation. *Biochemische Zeitschrift*, **315**, 139-232

KOCSANYI, L., HAITZ, M. and LICHTENTHALER, H.K. (1988). Measurement of laser-induced chlorophyll fluorescence kinetics using a fast acousto-optic device. In *Applications of Chlorophyll Fluorescence*. Ed. by H.K. Lichtenthaler. pp. 99-107, Kluwer Academic Publishers, Dordrecht

KRAUSE, G.H. and WEIS, E. (1983). Chlorophyll fluorescence as a tool in plant physiology. II. Interpretation of fluorescence signals. *Photosynthesis Research*, **5**, 129-157

KYLE, D.J., BAKER, N.R. and ARNTZEN, C.J. (1983). Spectral characterisation of photosystem I fluorescence at room temperature using thylakoid protein phosphorylation. *Photobiochem. Photobiophysics*, **5**, 79-85

LICHTENTHALER, H.K. (1987). Chlorophyll fluorescence signatures of leaves during the autumnal chlorophyll breakdown. *Journal of Plant Physiology*, **131**, 101-110

LICHTENTHALER, H.K. (1988a). *Applications of Chlorophyll Fluorescence in Photosynthesis Research, Stress Physiology, Hydrobiology and Remote Sensing*. Kluwer Academic Publishers, Dordrecht

LICHTENTHALER, H.K. (1988b). In vivo chlorophyll fluorescence as a tool for stress detection in plants. In *Applications of Chlorophyll Fluorescence*. Ed. by H.K. Lichtenthaler. Kluwer Academic Publishers, Dordrecht.

LICHTENTHALER, H.K. and BUSCHMANN, C. (1987). Reflectance and chlorophyll fluorescence signatures of leaves. In *Proceedings of the International Geoscience and Remote Sensing Symposium*. IGARSS-87, Michigan, Vol II, The University of Michigan, Ann Arbor, USA

LICHTENTHALER, H.K. and BUSCHMANN, C. (1988). Changes in the chlorophyll fluorescence spectra during the Kautsky induction kinetics. In *Proceedings 4th. International Colloquium on Spectral Signatures of Objects in Remote Sensing*. Aussois. ESA Publications Division, Noordwijk

LICHTENTHALER, H.K., BUSCHMANN, C., DÖLL, M., FIETZ, H.J., BACH, T., KOZEL, U., MEIER, D. and RAHMSDORF, U. (1981). Photosynthetic activity, chloroplast ultrastructure and leaf characteristics of high-light and low-light plants and sun and shade leaves. *Photosynthesis Research*, **2**, 115-141

LICHTENTHALER, H.K., BUSCHMANN, C., RINDERLE, U. and SCHMUCK, G. (1986). Application of chlorophyll fluorescence in ecophysiology. *Radiation and Environmental Biophysics*, **25**, 297-308

LICHTENTHALER, H.K. and RINDERLE, U. (1988a). The role of chlorophyll fluorescence in the detection of stress conditions in plants. *CRC Critical Reviews in Analytic Chemistry*, **19**, Supplement 1, S29-S85

LICHTENTHALER, H.K. and RINDERLE, U. (1988b). Chlorophyll fluorescence spectra of leaves as induced by blue light and red laser light. In *Proceedings 4th. International Colloquium on Spectral Signatures of Objects in Remote Sensing*. Aussois. ESA Publications Division, Noordwijk

LICHTENTHALER, H.K. and RINDERLE, U. (1988c). Chlorophyll fluorescence signatures of healthy and damaged spruce trees. PEF-Bericht KfK-PEF **35**, 185-190, Kernforschungszentrum Karlsruhe

LICHTENTHALER, H.K. and RINDERLE, U. (1988d). Chlorophyll fluorescence as vitality indicator in forest decline research. In *Applications of Chlorophyll Fluorescence*. Ed. by H.K. Lichtenthaler. Kluwer Academic Publishers, Dordrecht.

LICHTENTHALER, H.K., RINDERLE, U., KRITIKOS, G. and ROCK, B. (1987). Classification of damaged spruce stands in the Northern Black Forest by airborne reflectance and terrestrial chlorophyll fluorescence measurements. In *Proceedings 2nd. DFVLR Statusseminar "Forest Decline"*. DFVLR, München

NAGEL, E.M., BUSCHMANN, C. and LICHTENTHALER, H.K. (1987). Photoacoustic spectra of needles as an indicator for the activity of the photosynthetic apparatus of healthy and damaged conifers. *Physiologya Plantarum*, **70**, 427-437

PAPAGEORGIOU, G. (1975). Chlorophyll fluorescence: An intrinsic probe of photosynthesis. In *Bioenergetics of Photosynthesis*. Ed. by Govindjee. Academic Press, New York, USA

RINDERLE, U., HAITZ, M., LICHTENTHALER, H.K., KÄHNY, D.H., SHI, Z. and WIESBECK, W. (1988). Correlation of radar reflectivity and chlorophyll fluorescence of forest trees. In *Proceedings International Geoscience and Remote Sensing Symposium*. IGARSS-88. Edinburgh, **Vol. 3**, ESA Publications Division, Noordwijk

RINDERLE, U. and LICHTENTHALER, H.K. (1988). The chlorophyll fluorescence ratio F690/F735 as a possible stress indicator. In *Applications of Chlorophyll Fluorescence*. Ed by H.K. Lichtenthaler. Kluwer Academic Publishers, Dordrecht.

ROCK, B.N., HOSHIZAKI, T., LICHTENTHALER, H.K. and SCHMUCK, G. (1986). Comparison of in situ spectral measurements of forest decline symptoms in Vermont, (USA) and the Schwarzwald (F.R.G.). In *Proceedings International Geoscience and Remote Sensing Symposium*. IGARSS-86, Zurich, Vol III. ESA Scientific and Technical Publications Branch, Noordwijk

ROCK, B.N., VOGELMANN, J.E., WILLIAMS, D.L., VOGELMANN, A.F. and HOSHIZAKI, T. (1986). Remote detection of forest damage. *Bioscience*. **36**, 439-445

ROSEMA, A., CECCHI, G., PANTANI, L., RADICATTI, B., ROMULI, M., MAZZINGHI, P., VAN KOOTEN, O. and KLIFFEN, C. (1988). Results of the 'LIFT' project: air polution effects on the fluorescence of douglas fir and poplar. In *Applications of Chlorophyll Fluorescence*. Ed. by H.K. Lichtenthaler. Kluwer Academic Publishers, Dordrecht

SCHMUCK, G. and LICHTENTHALER, H.K. (1986). Applications of laser-induced chlorophyll fluorescence in forest decline research. In *Proceedings International Geoscience and Remote Sensing Symposium*. IGARSS-86, Zurich, Vol III. ESA Publications Division, Noorswijk, Netherlands

SCHMUCK, G., LICHTENTHALER, H.K., KRITIKOS, G., AMANN, V. and ROCK, B. (1987). Comparison of terrestrial and airborne reflection measurements of forest trees. In *Proceedings International Geoscience and Remote Sensing Symposium*. IGARSS-87, Michigan, Vol II. The University of Michigan, Ann Arbor, USA

SCHNECKENBURGER, H. and FRENZ, M. (1986). Time resolved fluorescence of conifers exposed to environmental pollution. *Radiation and Environmental Biophysics*, **25**, 289-295

SCHREIBER, U., NEUBAUER, C. and KLUGHAMMER, C. (1988). New ways of assessing photosynthetic activity with a pulse modulation fluorometer. In *Applications of Chlorophyll Fluorescence*. Ed. by H.K. Lichtenthaler. Kluwer Academic Publishers, Dordrecht

SCHREIBER, U., SCHLIWA, U. and BILGER, W. (1986). Continuous recording of photochemical and non-photochemical chlorophyll fluorescence quenching with a new type of modulation fluorometer. *Photosynthesis Research*, **10**, 51-62

STRASSER, R.J. (1986). Laser-induced fluorescence of plants and its application in environmental research. In *Proceedings International Geoscience and Remote Sensing Symposium*. IGARSS-86, Zurich, Vol III, ESA Publications Division, Noordwijk.

STRASSER, R., SCHWARZ, B. and BUCHER, J. (1987). Simultane messung der chlorophyllfluoreszenz kinetik bei verschiedenen wellenlängen als rasches verfahren zur frü-diagnose von immissionsbelastungen an waldbäumen. *European Journal of Forest Pathology*, **17**, 149-157

WALKER, D.A. (1988). Some aspects of the relationship between chlorophyll-a fluorescence and photosynthetic carbon assimilation. In *Applications of Chlorophyll Fluorescence*. Ed. by H.K. Lichtenthaler. Kluwer Academic Publishers, Dordrecht

ZIMMERMANN, R. and GÜNTHER, K.P. (1986). Laser-induced chlorophyll-a fluorescence of terrestrial plants. In *Proceedings International Geoscience and Remote Sensing Symposium*. IGARSS-86, Vol II. ESA Publications Division, Noordwijk

19
APPLICATIONS OF RADAR IN AGRICULTURE

M.G. HOLMES

Botany School, University of Cambridge,
Downing Street, Cambridge CB2 3EA, U.K.

Introduction

Aerial photography, using the visible and the near-infrared wavebands for surveying vegetation, is well-established. The advent of satellites led to the ability to cover even larger areas than aircraft in a given time. Although early satellites suffered from relatively poor resolution by aircraft standards, system developments have led to the production of high resolution images and the computer ability to process the high data rate which this requires. The use of microwaves, or radar, for detecting objects is also long-established, but the capabilities of radar have been less exploited than those of the visible and near-infrared wavebands. The potential advantages of imaging radar have been extensively researched in the last two decades, but it is only in very recent years that commitments have been made to increase satellite radar imagery for the purpose of studying vegetation.

It is important at this stage to discriminate between active and passive forms of radar. Active microwave sensors provide their own source of energy and measure the backscattered return, whereas passive sensors measure the microwaves emitted by the target. This review restricts itself to active radar systems. These active systems can be sub-divided into two main forms. These are

- non-imaging systems (scatterometers), which transmit microwaves to the target and measure the returning energy, and

- imaging systems, which operate in a similar manner but record an image of the scattered signal returned from the target.

Imaging systems

Imaging systems fall into two broad categories. These are Real Aperture Radars (RAR), and Synthetic Aperture Radars (SAR). A common feature of the RAR and SAR is that they are both a form of side-looking airborne radar (SLAR) and this term will be used except where specific reference to real or synthetic aperture is required. SLAR functions by transmitting pulses of microwave energy to the side of the aircraft or satellite. The returning energy is then registered on photographic film using a cathode ray tube. The rate at which the film passes the tube is synchronised with the speed of the aircraft using Doppler navigational control. The result of this procedure is a continuous strip image of the ground below. The SLAR operates at an oblique angle for two main reasons. First, side transmission produces artificial shadows which highlight topographical relief. Second, side transmission ensures that the radiation which returns from the targets is primarily scattered rather than reflected. Distortion of the image caused by the sideways transmission is corrected electronically to simulate an orthogonal, or vertical view of the target area.

The growing interest in radar remote sensing is reflected in the research programmes of the last decade and in the current space proposals of a number of countries. It is evident in more recent literature that an increasing interest is being expressed in the feasibility of microwave remote sensing as an alternative to the more 'traditional' visible and infrared methods in situations where the latter prove unsuitable, such as in overcast conditions. However, this view is not universal, and it is more probable that synergistic use of the visible, infrared and microwave wavebands will provide the optimal approach to monitoring agricultural areas (e.g. see *Chapter 21*, by Paris, these proceedings).

Notable projects due to be undertaken before the end of this century include the USA's Earth Observing System (EOS). This will consist of a suite of sensing and imaging instruments, the aim of which is to understand and monitor the processes and interactions leading to changes in the Earth's environment over a period of at least 10 years, thus allowing continuity of data records. Synthetic Aperture Radar (SAR) will be one such instrument, the use of which should be most profitable in conjunction with infrared data (e.g. using the High-Resolution Imaging Spectrometer - HIRIS) for agricultural studies. EOS SAR will be the first orbital imaging radar, providing multi-frequency, multi-polarization and multi-incidence angle observations of the Earth.

The European Space Agency (ESA) proposes to launch its Earth Resources Satellite (ERS-1) by 1991, the objectives of which include ocean and ice imaging as well as land observations using, amongst others, a synthetic aperture radar imaging system.

In both arable crop and forestry studies it would seem that a more thorough understanding of the relationships between target and associated microwave backscatter is being pursued. Thus, the realization of the potential of remote sensing is only becoming truly recognised as new instruments and techniques are devised and implemented to meet current needs.

Advantages of radar

The mode of operation of the radar instrument, and the physical laws governing the propagation of microwaves in scattering media, introduce some unique characteristics for remote sensing in Agriculture. The result is that radar imagery has two substantial advantages over conventional remote sensing using the visible and near-infrared wavebands. The most important asset of radar is the use of its own energy to map the terrain. Whereas conventional methods (excluding thermal infra-red and passive microwave sensors) rely on reflected sunlight for the production of an image, SLAR's provide their own energy with the result that surveys can proceed around the clock and are not restricted to daylight hours. In addition, SLAR's are rarely hampered by the weather. The wavelengths which are used can penetrate cloud and haze, and even light rain. Problems only arise in extreme weather conditions, such as rain storms. It is therefore much easier to plan radar studies than those depending on optical sensors, and campaign timetables can usually be adhered to.

Active radar studies are by no means perfect. Even less is understood about the interaction of microwaves with soil and vegetation than is the case in the visible waveband. The objective of this chapter is to review the current state of knowledge of radar/target interactions and to emphasize those aspects which are relevant to the monitoring of agricultural areas.

Factors affecting microwave backscatter

A wide range of parameters affect the backscatter of microwaves from soils and vegetation. In a recent study undertaken for ESA, for example, it was concluded that a minimum of 4 instrument parameters, 28 vegetation parameters, 13 soil parameters, and 12 environmental parameters needed to be included in a data base which was designed for interpreting the mechanisms controlling backscatter from vegetation.

Some generalizations can be made. The important characteristics of the radiation are the frequency, the polarization and the incidence angle. The crucial features of the target in determining the proportion of radiation returning to the instrument are the biomass, the dielectric constant (which is closely linked to the water content) and the geometry of the vegetation. It therefore follows that the backscatter is dependent on the gross morphology of the crop, which in turn depends on crop species and physiological age. One of the properties of radiation in the microwave waveband is that its ability to penetrate into the target (which is wavelength-dependent) provides a potential for studying not only the surface vegetation, but also the vegetation and soil below. In the case of soil, the predominant factors affecting backscatter are the morphology (roughness) and the dielectric constant (water content).

Instrument characteristics affecting backscatter

FREQUENCY

There are reports which conclude that each of the Ku-band (Bush and Ulaby, 1978; Mehta, 1983), or the X- (Hoogeboom, 1982; Pei-yu and De-li, 1983), or the C- (Mehta, 1983; Paris, 1983), or the L-band (Mehta, 1983) are optimal for crop studies. The SIR-B mission produced data that demonstrated the sensitivity of L-HH (Horizontal transmit, Horizontal receive) imagery to surface features, e.g. forest mapping and soil moisture. There are many recommendations that multi-frequency studies are required for crop classification (e.g. Drake, Schuchman, Bryan, Larson and Liskow, 1974; Parashar, Day, Ryan, Strong, Worsfold and King, 1979; Ulaby, Dobson and Bradley, 1980) and many studies have demonstrated large improvements in classification by using multi-frequency (and multi-polarization) measurements (e.g. Brisco and Protz, 1980; Crown, 1980; King, 1980; Protz and Brisco, 1980; Bertholme, 1983; Gombeer, 1983; 1985). It is clear that an optimum wavelength for general crop studies cannot be defined because the optimum waveband depends on species.

While conclusions arrived at by many of the investigators above are application-dependent, direct comparison is not usually possible between crops because studies which eliminate all variables, including soil dielectric properties, plant age etc. are lacking. As Churchill, Horne and Kessler (1985) point out, the results of individual investigations can

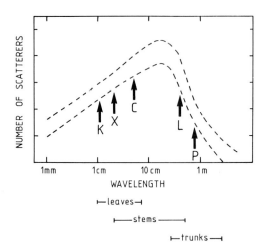

FIGURE 19.1. Radar response as a function of the size distribution of scatterers in the canopy

be seen to be unique to individual test sites due to crop stage and test area conditions at the time of measurement. Bouman's (1987) study using three agricultural crops indicated that longer wavelengths were less sensitive than shorter X-band wavelengths to changes in crop geometry of beet and potato. In both the above instances, recommendations as to which frequency is most appropriate in all circumstances are not possible. Thus, no individual wavelength can be singled out as being optimal for agricultural studies.

The long-established Canadian Centre for Remote Sensing program offers some experience in the identification of optimal frequency for discrimination of crop type. Cihlar, Brown and Guindon (1985) summarized findings to that date. They found that whereas X-band provides good discrimination, L-band was very useful for separating broad-leaved crops and fallow in some sites. However, L-band did have the disadvantage of sensitivity to row direction, which X-band did not exhibit. The ability of C-band to discriminate was intermediate between the X- and L-bands.

Field scatterometer and aircraft SAR experiments have demonstrated that radar sensitivity to vegetation type is enhanced when the wavelength used is approximately equal in size to that of the canopy components (*Figure 19.1*), this being due, in part, to a resonance effect (NASA, 1988). It follows, therefore, that by combining multi-frequency observations of a vegetation canopy, it might be possible to estimate morphological and biophysical parameters of that canopy. A combination of C- and X-band data may be more sensitive to crops, while L- and C-band data may be more useful for forest mapping. The inherent disadvantage of using only a single frequency is obvious and the potential for having multi-frequency satellite capability in the future cannot be ignored. The Earth Observing System is one such programme where this will be implemented. It is first necessary that compatible multi-frequency data should be fully exploited. This should be done by spectrally characterizing the vegetation in terms of the absorptive and scattering properties. Both field and laboratory approaches can be used to determine whether individual signatures can be obtained for different crops or for the same crop at different stages of the growing season.

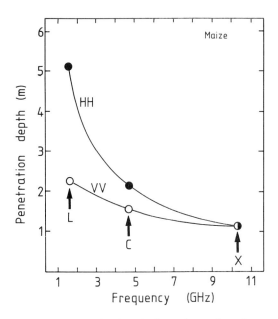

FIGURE 19.2. Wavelength dependency for microwave penetration into a corn canopy. Penetration depth is defined as the depth at which the incident power is reduced to 37% (1/e) of that incident. The data presented are for an incidence angle of 40°, LAI = 2.8, plant height = 2.7 m, leaf volumetric moisture content = 0.65, stalk volumetric moisture content = 0.47 (courtesy of F.T. Ulaby)

POLARIZATION

The penetration depth of an incident microwave source depends on its polarization and frequency, such that the optical thickness of the vegetation layer increases with increasing frequency. Whereas L-band observations are influenced by the entire crop canopy, X-band observations are generally governed by the top layers. Horizontally polarized (HH) radar couples weakly to vertical stalks, resulting in low attenuation. Vertically polarized (VV) microwaves, however, are attenuated to a greater extent causing a reduction in the penetration depth (*Figure 19.2*). Measurements using HH, therefore, give information primarilly about the underlying soil medium, while VV data are related more to canopy structure. However, this statement must be considered in the context of the wavelength used. Longer wavelengths, such as L-band, tend to penetrate deeply into vegetation, whereas shorter wavelengths, such as X-band, are scattered in the upper layers. As a result, discrimination between polarizations may be impossible at the shorter wavelengths (*Figure 19.2*).

Fewer studies of polarization optima have been made than for waveband optima, but the conclusions which can be drawn are more clearcut in agricultural crops than in forestry studies. The degree of inhomogeity of a surface or volume is strongly associated with the cross-polarization scattering coefficient of that surface or volume. The separation of crop types can be enhanced using cross-polarization data; for example, two crops having similar geometries, such as wheat and barley, may have similar like-polarization backscatter, but it is possible to separate them with observations from cross-polarization studies. Similarly, the distinction between bare soil and vegetation-covered surfaces is made easier using cross-polarization, due to the fact that the vegetation canopies depolarize the incident radiation more strongly than bare surfaces. The cross-polarization ratio (ratio of σ°_{HV} to σ°_{HH}) is the useful discriminating parameter in these studies. Earlier studies were based on higher frequency (> 8 GHz) observations and led to the conclusion that little was to be gained by using cross-polarization (Bush and Ulaby, 1977; 1978; Le Toan, 1982). Lower frequency studies in the C-band by Paris (1982) demonstrated the advantage of using cross-polarization for delineation studies of corn and soybean which could not be achieved with like-polarization. The advantages of using cross-polarization are not restricted to lower frequencies. For example, whereas Megier, Mehl and Ruppelt (1984) concluded that cross-polarization was not as good as like-polarization in the X-band, they found that multi-polarization analysis (i.e. using both cross- and like-polarized data) greatly improved the ability to discriminate between crops.

Arable crop studies require multi-polarization instrumentation if maximum information about crops is to be achieved. Because target structure influences the extent of radiation depolarization, there is a need to understand the dependence of de-polarization on the basis of individual crop structure. This implies that multi-polarization studies need to be made of different species, and at different growth stages for each species, if the causal factors which influence depolarization are to be understood.

INCIDENCE ANGLE

Almost all incidence angle studies have been restricted to correlative field observations and no substantive studies aimed directly at understanding incidence angle effects have been made. The optimal incidence angle for applied studies depends on the application. As with analysis of optimal polarization, the lack of comprehensive information for all wavebands limits interpretation. The need for more information is not only because such data will aid future monitoring programmes, but also because the data are required to aid model systems.

The influence of incidence angle depends on the polarization of the microwave source and the canopy orientation under examination. If a fully grown corn canopy is considered, using L-band radar, an increase in incidence angle from 0° to 90° has little effect on the penetration depth of HH polarized radiation due to its low attenuation, whereas it decreases with VV polarization as incidence angle increases (*Figure 19.3*). This phenomenon can be used to help choose the most suitable incidence angle and polarization, depending on the application and information required.

There is no conclusive evidence that any one angle, or narrow range of angles, is optimal for species classification purposes. Shanmugan, Ulaby, Narayanan and Dobson (1983) noted an impovement in the ability to classify crops with increasing incidence angle up to 40°, but found no improvement with higher angles. This ties in with the conclusion of Pei-yu and De-li (1983) that angles between 42° and 72° are optimal for rice, while Paris (1982) concluded that the optimal angle for delineating corn and soybean with C-band radar is 50°. Mehta's (1983) interpretation, that very low angles (10°) are best for soybean studies, underlines the inconclusive nature of studies on incidence angle. The application of radar backscatter in the monitoring of crop growth is considered by Bouman (1987), who concludes that with X-band wavelengths steep incidence angles are most suitable for this purpose.

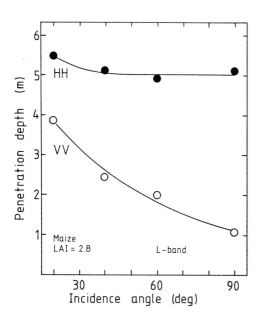

FIGURE 19.3. Polarization and incidence angle dependency for microwave penetration into a corn canopy. Penetration depth is defined as the depth at which the incident power is reduced to 37% (1/e) of that incident. LAI = 2.8, plant height = 2.7 m, leaf volumetric moisture content = 0.65, stalk volumetric moisture content 0.47 (courtesy of F.T. Ulaby)

The suggestion of Hoogeboom (1982) suitably represents the conclusion that our knowledge of the influence of incidence angle is still very poor and that future studies require an exhaustive analysis at all incidence angles to understand the subject, irrespective of the strict limitations which will be placed on air- and space-borne imagery for practical reasons.

The influence of soil on microwave backscatter

It has long been known that soil can play an important role in backscatter measurements of crops. The entire scope of the parameters influencing backscatter from soil is not understood, but soil moisture content (MacDonald and Waite, 1971; Morain and Coiner, 1972; Ulaby and Moore, 1973; Ulaby, Cihlar and Moore, 1974; Ulaby and Batlivala, 1976a; Newton, 1977; Schmugge, 1978; Ulaby, Batlivala and Dobson, 1978; Wilheit, 1978; Ulaby, Bradley and Dobson, 1979; Black and Newton, 1980; Schmugge, Jackson and McKim, 1980; Bradley and Ulaby, 1981; Dobson and Ulaby, 1981; Jackson, Chang and Schmugge, 1981; Le Toan, Flouzat, Pausander and Fluhr, 1981; Tsang and Newton, 1982; Ulaby, Aslam and Dobson, 1982; Jackson and O'Neill, 1983; Jackson, Schmugge and O'Niell, 1983; Schmugge, 1983; Ulaby, Dobson and Brunfeldt, 1983), soil roughness (Ulaby *et al.*, 1978; Choudhury, Schmugge, Newton and Chang, 1979; Ulaby *et al.* 1979; Dobson and Ulaby, 1981; Le Toan *et al.*, 1981; Bernard, Martin, Thony, Vauclin and Vidal-Madjar, 1982), and soil texture (Ulaby *et al.*, 1978; 1979; Dobson and Ulaby, 1981) are known to be three of the most important factors.

The dielectric properties of moist soils are major factors in determining the microwave scattering and absorption by a soil (e.g. Dobson and Ulaby, 1981). There have been many experimental studies of the dielectric behaviour of soil-water mixtures and several attempts have been made to model the dielectric behaviour (Wang, 1986). Although none of the existing models for soil-water mixtures is capable of fitting experimental data on the basis of the soil physical parameters which have been used, it is nevertheless clear that the dielectric constant is a complex function of the soil water content, the frequency of the radiation, and the soil type, as well as the temperature and the salinity of the soil-water mix. The water content, frequency and soil type deserve special mention here.

Soil is a heterogenous mixture of soil particles (the host material) and water. Dry soil exhibits a narrow range in real dielectric constant, or permittivity (ε'), between about 2 and 4, whereas that of water is about 81. The permittivity of soil is therefore a function of the water content of the soil and permittivity increases with increasing water content (*Figure 19.4*). In nature, the ε' range of soil is from about 3 to about 30. Also, both ε' and ε'' (the dielectric loss factor) vary with frequency, the permittivity increasing with increasing wavelength, and the loss factor decreasing with increasing wavelength. This frequency dependency is largely a function of the water, rather than the soil moiety. The backscatter increases with increasing water content of the soil.

The amount of water associated with the soil is strongly dependent on soil type. The water held by the soil consists of two main types, bound and free. Bound water is held tightly to the soil particles by matric and osmotic forces. Free water, on the other hand, is a few molecular layers away from the particles and is at a sufficient distance from them to allow relatively free movement within the soil. The amount of bound water depends on the particle surface area, which in turn depends on the particle size distribution of the soil. The dielectric properties of equal amounts of bound water and soil, and free water and soil, are very different, with the result that the soil particle size and type produce a marked effect on the soil dielectric. Referring to *Figure 19.4*, it can be seen that there is a small variation in ε' for the dry soils. It is important to note that the dielectric does not vary much with soil type *per se*, but that it varies with soil moisture content, which is a function of soil type.

Soil moisture can affect the ability to delineate between different types of vegetation (Shanmugan et al., 1983). This characteristic provides radar with a major advantage over other forms of remote sensing when information on climatic effects on soil moisture availability is required. Intensive scatterometer studies at the University of Kansas have indicated that the soil moisture content of bare soil can best be analysed with the C-band and that HH polarization and incidence angles in the range 7° to 17° gave optimum data for analysis if the confounding effects of roughness were to be minimized (Ulaby and Batlivala, 1976b). The same group found that similar instrument characteristics were optimal when the soil is covered by vegetation. They observed a strong correlation between backscatter and soil moisture content of the upper 5 cm soil layer when the soil was planted with wheat, corn, milo or soybeans (Ulaby et al., 1979; Bradley and Ulaby, 1981). This type of effect is illustrated with SIR-B data for lettuce and alfalfa in *Figure 19.5*.

Brown, Guindon, Teillet and Goodenough (1984) also observed a correspondence between wet areas (detected by aerial photography) containing no vegetation and C-

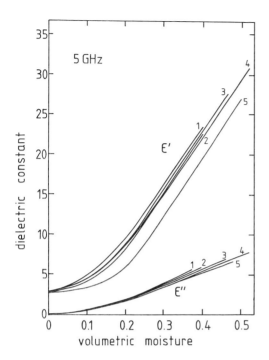

FIGURE 19.4. Dependence of dielectric constant on volumetric moisture content for a range of sandy soil types. 1 = sandy loam, 2 = loam, 3, 4 = silt loam, 5 = silty clay (redrawn from Ulaby, Moore and Fung, 1986)

band SAR backscatter at incidence angles between 30 and 45°. Similar conclusions have been reached by other workers (Schmugge, 1980; Bernard, 1981; Bernard et al., 1982; Le Toan et al., 1981; Le Toan, 1982). Anomalies do exist; for example, the airborne scatterometer studies of Paris (1982) using C-band, HH polarization and 10° incidence showed significant effects of the crop rather than the soil. Wang (1986) concluded that soil moisture could be mapped on an individual field basis, as well as within a field, using SIR-B imagery at L-band wavelengths and incidence angles of 21°. Nevertheless, it is clear that soil effects can be almost over-riding in determining the radar signal (Le Toan, 1982; Le Toan et al., 1981; Le Toan, Lopez and Huet, 1984).

The effects of soil organic matter content and the mineralogy of the soil are largely unexplored but their direct effects on the retentive capacity of the soil for water cannot be ignored. A similar argument holds for the related parameter of field capacity. All three parameters (soil organic matter content, mineralogy and field capacity) have been fairly extensively covered in ground measurements in both Europe and the United States.

Soil bulk density has been extensively recorded as a potentially important parameter for soil microwave studies. The influence of bulk density *per se* cannot be pin-pointed, it being linked in particular to the water content of the soil. Nevertheless, it is directly or indirectly an important soil property influencing the radar backscatter response (Ulaby et al., 1978; Dobson and Ulaby, 1981; Schmugge, 1980; Bernard et al., 1982).

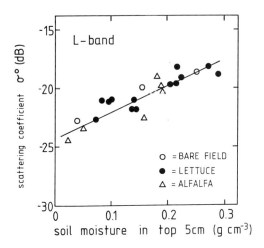

FIGURE 19.5. The influence of soil moisture content on backscatter from lettuce and alfalfa. The dependence of the SIR-B backscattering coefficient on volumetric soil moisture content in the top 5 cm of soil is shown (Wang, 1986)

SOIL ROUGHNESS

One of the main problems with the analysis of soil moisture effects on the backscatter of microwaves is the influence of variations in roughness of the vegetation and the soil. One approach to overcome this problem has been to separate the data into groupings with nearly constant roughness and vegetation parameters. By this means it is possible to compare bare versus vegetated and smooth versus rough (Ulaby, et al., 1978; Newton, 1977; Newton and Rouse, 1980; Wang, McMurtrey, Engman, Jackson, Schmugge, Gould, Fuchs and Glazar, 1982; Njokyu and O'Niell, 1982; Taube and Theiss, 1984).

The influence of soil roughness on the return signal needs to be understood and accounted for (Ulaby and Batlivala, 1976a; Ulaby et al., 1979; Schmugge, 1978; 1980; 1983; Schmugge et al., 1980; Chang, Atwater, Solomonson, Estes and Simonett, 1980). A greater understanding of the effects of soil roughness on the backscatter signal would greatly improve image analysis, especially with regard to delineation between fallow fields and grain crops. For example, Brown et al., (1984) found that confusion between fallow and grains depends strongly upon the condition of the fallow fields with C-band imagery.

Scatterometer measurements (*Figure 19.6*) have demonstrated that the effects of roughness are minimised at incidence angles of about 10°. These data were obtained over bare soil using a range of roughness commonly found in agricultural fields. Another notable feature of these observations was that the effects of different roughnesses were least with shorter wavelengths; this is because all the surfaces are relatively rough for shorter wavelengths.

The row direction resulting from soil tillage has substantial effects on the angular dependence of radar backscattering. These periodic components of soil surface roughness are related to the analysis of the soil moisture effects because they are known to have substantial influence on the angular effect of microwave backscatter (Ulaby and Bare, 1979). Thus, even when soil roughness is approximately the same, tilled and untilled

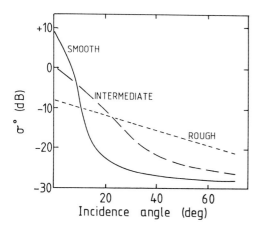

FIGURE 19.6. Typical backscatter curves against angle of incidence for smooth, intermediate and rough surfaces

fields cannot be treated as being the same. The radar signal is most sensitive to azimuthal viewing changes (Bradley and Ulaby, 1981; Ulaby and Bare, 1979; Ulaby et al., 1982; Blanchard and Jean, 1983). Elevation, frequency and polarization dependencies are also recognized (e.g. Dobson and Ulaby, 1981) and more information on these influences is still required.

Vegetation factors affecting backscatter

Penetration depth

The depth to which microwaves penetrate into a vegetation canopy depends primarily on the wavelength of the radiation and, if the vegetation exhibits a preferential orientation, on the polarization of the radiation. The dependence on wavelength was seen in *Figure 19.2* for the penetration of L-, C-, and X-band microwaves at an incidence angle of 40°. This example demonstrated clearly the general phenomenon that penetration decreases with increasing frequency. An important consequence of the wavelength-dependence of penetration depth is that longer wavelengths can provide information on the entire canopy, whereas shorter wavelengths are useful for studying the upper layers of canopies.

Polarization differences in L- band penetration depth in a corn canopy were seen in *Figure 19.3*. Vertically polarized radiation is much more strongly attenuated than horizontally polarized radiation and there is a marked increase in the attenuation of the vertically polarized radiation with increasing incidence angle. Horizontally polarized radiation, on the other hand, penetrates deeply into the canopy and shows negligible dependence on incidence angle. An important practical aspect of this type of observation is that vertically polarized data provide information which is predominantly related to the physical characteristics of the canopy, whereas the horizontally polarized radiation is largely penetrating the canopy and therefore providing information on the soil below.

CROP MORPHOLOGY

Microwave backscatter from crops is strongly dependent on the size of the scattering elements within a crop. This can be seen in *Figure 19.1*, where the size of the elements which cause maximal backscatter varies with the wavelength of the radiation. In most instances, the greatest response is shown to scatterers which are of a similar size to the wavelength. This is one of several crop characteristics which contribute to the ability of multi-frequency imagery to discriminate between targets.

In addition, crops usually exhibit preferential orientations in their geometry. In most instances these orientations can be divided into preferentially vertical or preferentially horizontal. This results in polarization-dependent differences in the penetration and return of microwaves from the canopy, which in turn permits analysis of the canopy using different instrument-polarization configurations. Useful approaches here are to use colour composites of multi-frequency images, with each frequency shown in a different colour, or to look at the phase difference between horizontally and vertically polarized imagery.

A wide range of individual plant morphological features influence the gross morphology of an arable crop, and the gross morphology of crops influences radar in a way which permits grouping into crop types. Smit (1978) defined two broad categories of crop. These were closed foliage crops (e.g. beets, potatoes) and open foliage crops (e.g. cereals, grasses). He described the closed foliage crops as exhibiting stable backscatter at various incidence angles, which led him to the inference that changing backscatter indicated changes in crop state. The open foliage crops, on the other hand, exhibited a more fluctuating return, thereby precluding reliable analysis of crop state. No information is available to indicate whether multi-temporal studies would permit a more useful exploitation of the use of broad crop characterization.

The formation of backscatter models is central to the effective use and understanding of microwave signatures, and this has led to a requirement for the development of techniques to measure and monitor plant structure and identify the dominant scattering components within a canopy. Such information could then be incorporated into these models. It is clear then, that crop and individual plant morphology affects backscatter. Bouman (1987) has shown, using short wavelengths, that radar return is sensitive to leaf size. Bouman ends his study of radar backscatter from three crops by implying that longer wavelengths are less sensitive to changes in crop geometry than shorter ones, particularly with respect to beet and potatoes.

ROW DIRECTION

Since the earliest reports on imaging radar by Porcello and Rendleman (1972), it has been clear that not just cropped land row direction (e.g. Mehta, 1983; Wooding, 1983), but also the soil row direction affects backscatter (Bradley and Ulaby, 1981; Ulaby, 1980). Although Ulaby and Bare (1979) concluded that the row direction effect was insignificant for frequencies greater than 4 GHz, other observations do not fully support this hypothesis (Bradley and Ulaby, 1981). There is circumstantial evidence regarding the effects of row direction on polarization. Both Bradley and Ulaby (using L-, C- and Ku-band) and Paris (1982; using C-band) noted the effect of row direction using like-polarization; Batlivala and Ulaby (1976) noted row effects were much stronger with like-polarized radar (HH) than with cross-polarized (HV).

It should be emphasized that row direction does not inevitably affect backscatter to a significant degree. A clear example of this is the study of Sieber, Freitag and Lawler (1982) who concluded from SAR data that row direction does not affect SAR images in a way that will cause changes of average backscatter cross-section. This study is significant because it covered three wavebands (X-, C- and L-band). Canadian studies underline the apparent confusion about effects of row direction. For example, Cihlar and Hirose (1984) found a strong effect of row direction on backscatter in grain fields at one test site (especially with airborne L-band), but no effect for equivalent fields at another test site. Also, there was no consistent conclusion to be drawn on the crop type which provided an effect of row direction on radar return signal. Similar inconsistency has been reported by Brown et al., (1984).

Effects of row spacing (range 18 to 23 cm) in barley fields and width of ridges on fallow fields are also known (e.g. Cihlar and Hirose, 1984). Canopy height effects on imagery have been difficult to detect, although limited success has been reported by Teillet, Brown, Guindon and Goodenough (1984) in relating tonal changes in C-band imagery to height.

CROP DIELECTRIC

The dielectric constant of vegetation, as with soil, is one of the most important factors in determining radar backscatter. Along with the volume fraction and the geometry of the vegetation, the permittivity and loss factors are essential for an understanding of the attenuation of microwaves by vegetation. Unfortunately, measurements of dielectric properties are not straightforward and, largely for that reason, are not common in the literature. Some measurements have been made (see El-Rayes and Ulaby, 1987, for summary). As with soils, the dielectric constant of vegetation increased with increasing water content. However, the limited information available indicates that the relationship between vegetation dielectric constant and frequency is more complex.

CROP SPECIES

Studies to date have indicated that crop species is the single most important parameter among those recorded in the field (Garron and Schubert, 1979; Remotec Applications Inc., 1979; Cihlar and Hirose, 1984; Brown et al., 1984; Cihlar et al. 1985). Crop type can result in a unique radar return (Brown et al., 1984; Cihlar and Hirose, 1984). When comparing beet, potatoes and peas, Bouman (1987) attributed the highest radar return in beet (using X-VV) to its relatively higher water content or to its general geometry and larger leaves. The reasons for many of the observed relationships between crop type and backscatter in terms of causal effects are, however, not understood. Teillet et al. (1984) ascribed differences in C-HH and C-VV imagery backscatter to broadly categorized canopy structure (preferentially vertical for grains versus lack of preferential orientation for broad leaved crops). There is general agreement that broad-leaved crops produce higher signal return than other crops for L-, C- and X-band with parallel polarization (Remotec Applications Inc., 1979; Brisco and Protz, 1982; Brown et al., 1984; Cihlar and Hirose, 1984).

The approach of considering the cross-polarization ratio is a useful method for species identification. Although two different crop species may exhibit similar backscatter in one polarization mode, there is usually a morphology dependent difference in another mode. An example is shown in *Figure 19.7*, in which discrimination between species is derived from the HV rather than the HH channel.

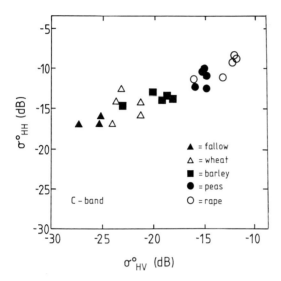

FIGURE 19.7. Classification of soil types using like- and cross-polarized backscattering measurements (courtesy of F.T. Ulaby)

Cihlar et al., (1985) point out that there is a lack of consistency between test sites in tonal range for each crop and in the ranking order of crops, and that SAR images often exhibit tonal variation within one field or crop type. There are many instances where an explanation of these variations is not possible because detailed information about the target is not available. Cihlar, Dobson, Schmugge, Hoogeboom, Janse, Baret, Guyot and Le Toan (1987) underline this need for more information by suggesting standard procedures for the description of agricultural crops and soils in microwave (and optical) studies that should be used by all investigators. An understanding of the causal relationship between crop/soil parameters and image tone is of key importance in assessing the usefulness of SAR. It allows the possibility of constructing models, whereby SAR might be used as a crop monitoring tool. It would seem that the collection of ground observations near the time of SAR data acquisition would be of importance, though limited success has been reported at present, and it has been argued that intensive within-field data collection is not practical.

Crop classification and image interpretation in SAR images using pixel-by-pixel comparisons is said to be inappropriate due to problems associated with image speckle (Cihlar et al., 1985). Durand (1987), however, suggests that crop classification is possible using this technique, so long as the image is filtered. In fact, it is suggested that filtering significantly improves visual aspects and pixel-by-pixel classification results, without losing textural information and edges.

Multi-temporal studies

The dynamic nature of a plant over short or long periods of time describes how the plant is responding to its environment, thereby providing information about that environment.

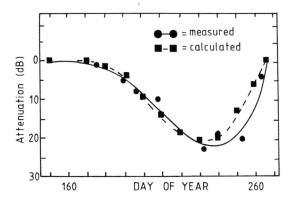

FIGURE 19.8. Seasonal variation in the measured one-way attenuation of 10.2 GHz radar (X-band) for a soybean canopy and the calculated attenuation caused by absorption by leaves and stalks (redrawn from Ulaby, Moore and Fung, 1986)

A time-series data-set of canopy scattering properties at different microwave frequencies and polarizations over diurnal and seasonal time periods could provide information about above-ground productivity in grasslands and agricultural systems and could be used to investigate temporal variability in plant phenology. Multi-temporal studies, therefore, are of great importance if agricultural systems are to be understood fully.

From the few studies which have been made, there is a clear indication that multi-temporal imagery greatly increases the classification rate of many crops, including corn, milo, soybean, wheat and alfalfa (Bush and Ulaby, 1977; 1978; Simonett, 1978; Li, Ulaby and Easton, 1980; Ulaby, Li and Shanmugan, 1981; Shanmugan et al., 1983). Bush and Ulaby (1978) acknowledge that crop classification using radar is time and site specific, and suggest that multi-date data acquisition is necessary if classification is to be more that 90% successful. The study of Brisco and Protz (1980) represents a rare exception to this general finding.

Temporal variation in one-way attenuation of X-band radiation is demonstrated in *Figure 19.8*. It can be seen that attenuation increases with time during the growth phase (days 150-220), plateaus at full growth height, then decreases from about day 240 as the maturing crop dries out. This pattern, which is in good agreement with the theoretical calculations, demonstrates the close relationship between attenuation and the water content per unit volume, which is also related to the dielectric properties of the vegetation.

It is well-known that the accuracy of crop classification is strongly dependent on acquisition date, which in turn depends on crop age. There are no clear-cut conclusions as to optimal timing of measurements for classification accuracy. For example, studies with C-VV imagery indicated that highest accuracy was obtained before three general classes reached maturity (Cihlar et al., 1985). Van Kasteren (1981) on the other hand, concluded that greatest accuracy in crop type separation was obtained when the crop had reached maturity. Contrasting studies such as these reaffirm the need for a better understanding of the parameters which influence the image. Studies in the United States (Bush and Ulaby, 1977; 1978) and in the wet tropics (Simonett, 1978) indicate revisit times at about 10 day intervals, although no direct inference for temperate regions can be drawn from this because growth rate is climate-dependent.

A potential advantage to multi-temporal studies is the ability to obtain information about the age of the crop. In comparison to forests, arable crops grow quickly, even in temperate regions. This rapid growth represents a rapid change in the size, morphology and above-ground water mass which may be expected to cause a correspondingly rapid change in backscatter signal of the appropriate wavelength/polarization combination. During the Dutch 'ROVE ' programme (de Loor, Hoogeboom and Attema, 1982), an HH-polarized X-band scatterometer at 30° incidence angle was used to study temporal changes in wheat, oats, sugar beet, and potatoes. Marked effects of changes in canopy cover were recorded throughout the growing season, along with temporal changes in soil moisture and roughness, and effects of rainfall.

It has to be recognized that results so far have indicated problems associated with the age classification of crops, and these problems can usually be traced to the reliance on correlative studies as opposed to causal studies. For example, Brown et al. (1984) reported limited success in age classification of brassicas; here backscatter decreased with maturity. Ulaby and Bush (1976) observed the reverse effect with wheat; i.e. backscatter increased with maturity. Bouman (1987) observed that backscatter from beet and potato showed a distinct pattern over a growth period; an increase up to a level of saturation occurred, about which fluctuations existed. At steep angles of incidence however, he found that saturation was reached later, leading him to conclude that such incidence angles were most suitable for monitoring crop growth.

One of the temporal changes in crops which appears to offer great commercial potential is the identification of crop damage caused by high winds, pests, disease, and drought. Few reports are available, but they do indicate that further study is worthwhile. For example, the ability to detect blight in corn using C-band radar has been demonstrated by Ulaby and Moore (1973).

Drought may be detectable using the higher frequencies because backscatter and water content of the crop are strongly correlated, as has been demonstrated long ago by Bush and Ulaby (1975; 1976) for wheat and corn. Differences were noted in the optimal incidence angles for these two studies, so any experimentation on this topic would have to include a detailed analysis of the effects of incidence angle. The detection of storm damage is unresearched, but knowledge of the effects of crop geometry leads to the suggestion that layering of a crop should be readily detectable at appropriate wavelengths.

Diurnal variation in radar backscatter from vegetation canopies has long been recognized (Ulaby and Batlivala, 1976b), but requires considerably more study. Le Toan (personal communication) has noted significant diurnal changes in backscatter from soybean. By contrast, the Dutch 'ROVE' team were unable to detect more than marginal diurnal effects (Hoogeboom, personal communication). Many plants exhibit diurnal changes in leaf angle as well as changes in petiole and flower orientation, and daily changes in moisture content. Unless the importance of plant geometry has been over-emphasized in the literature, it follows that these diurnal changes will influence backscatter. Apart from daily changes in plant geometry, it is also largely unknown to what extent dew influences the radar signal. Without such knowledge of possible changes in backscatter during the day, the often-claimed advantage of the 24-hour capability of radar may be restricted.

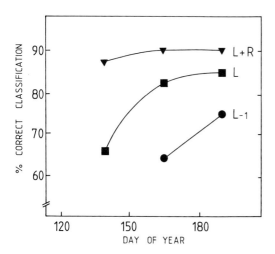

FIGURE 19.9. Simulations replacing missing Landsat data with radar data in a multi-date classification (courtesy of F.T. Ulaby)

Combining radar and optical studies

Although SAR can be used around the clock and is only restricted by the most severe weather, observations with optical sensors are limited to clear-sky conditions. Combining SAR and optical studies has the advantages of not only filling information gaps during overcast or hazy periods, but also of providing additional information which can be combined with optical data obtained during clear-sky conditions. Apart from improving our understanding of radiation/target interactions, and vegetation/atmosphere interactions, one of the most promising areas of combining radar and optical studies in an agricultural context is in the improved crop classification rates which are possible.

A simulation of replacing missing Landsat data with radar data in a multi-date classification is shown in *Figure 19.9*. The figure demonstrates the typical decrease in classification accuracy data were not obtained on the first day of a multi-temporal study. However, if radar information is substituted for the missing Landsat data, the resultant correct classification is not only recovered by the use of radar data, but is exceeded.

The improved classification rate demonstrated in the simulation in *Figure 19.9* is not unique. A comprehensive study by Li et al. (1980) showed that a combination of radar and optical data provided a better agricultural crop classification accuracy than either radar or optical data alone. The study compared classification success rates for corn (maize), wheat, and milo (*Figure 19.10*). When the land was predominantly under wheat and corn, X-band radar provided superior classification to Landsat bands 5 and 7. After the wheat had been harvested and replaced by milo, Landsat provided better crop identity separation than the radar. Overall, however, the best classification, both at the start and end of the growing season, was provided by the combined use of both radar and optical data. These, and the limited number of similar studies, indicate clearly that combined sensor data can be synergistic.

The most exciting combination of radar and optical studies in the future is the

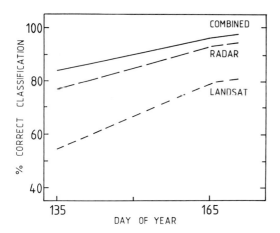

FIGURE 19.10. Classification results based on Landsat (bands 5 and 7) and radar (14.2 GHz, VV and HV) observations (after Li et al., 1980)

integrated use of HIRIS (HIgh Resolution Imaging Spectrometer) and SAR in EOS. HIRIS will provide approximately 200 spectral bands between 0.4 and 2.5 μm. Both HIRIS and SAR will provide images of similarly high resolution and therefore offer the potential for exploiting the different types of information provided by the two types of instrument. Although much more needs to be understood about the propagation of visible, IR and microwave radiation in crop canopies, the data provided by HIRIS and SAR should present an important opportunity to test compatible observations in current semi-theoretical and empirical models and to apply the conclusions to the monitoring of agricultural crops.

Apart from HIRIS and SAR, MODIS (MODerate resolution Imaging Spectrometer) should also provide useful data which can be combined with SAR imagery. MODIS has a relatively low resolution, which is only about 1/20th that of SAR and HIRIS. Nevertheless, the ability of MODIS to monitor optical, shortwave- and thermal-IR radiation, and its relatively fast global coverage of two days, presents the potential for monitoring changes in land surface temperature over large areas. For agricultural purposes, the most important information from MODIS will be the surface albedo and temperature data, which can be combined with SAR data on land surface moisture content to provide an understanding of the land/atmosphere hydrologic processes.

References

BATLIVALA, P.P. and ULABY, F.T. (1976). Radar look direction and row crops. *Photogrammetric Engineering and Remote Sensing*, **42**, 233-238

BERNARD, R. (1981). *A C-band radar calibration for determining surface moisture.* Centres de Recherches en Physique de l'Enviroment (CRPE), Rue du General Leclerc, 192131 Issy-les-Moulineaux, France

BERNARD, R., MARTIN, P., THONY, J.L., VAUCLIN, M. and VIDAL-MADJAR, D. (1982). A C-band radar calibration for determining surface soil moisture. *Remote Sensing of Environment*, **12**, 189-200

BERTHOLME, E. (1983). Optical SAR-580 data for land use studies; an evaluation for S. Belgium. European SAR-580 Investigators' Preliminary Report (ESA), Sept. (1983)

BLACK, Q.R. and NEWTON, R.W. (1980). Airborne microwave remote sensing of soil moisture. Remote Sensing Center, Texas A&M University, College Station, Technical Report RSC 108, Nov. (1980)

BLANCHARD, A.J. and JEAN, B.R. (1983). Antenna effects in depolarization measurements. *IEEE Transactions on Geoscience and Remote Sensing*, **GE-21**, 113-117

BOUMAN, B. (1987). Radar backscatter from three agricultural crops: beet, potatoes and peas. Centre for Agrobiological Research (CABO), Netherlands. Verlag Nr. 71

BRADLEY, G.A and ULABY, F.T. (1981). Aircraft radar response to soil moisture. *Remote Sensing of Environment*, **11**, 419-438

BRISCO, B. and PROTZ, R. (1980). Corn field identification accuracy using radar airborne imagery. *Canadian Journal of Remote Sensing*, **6**, 15-21

BRISCO, B. and PROTZ, R. (1982). Manual and automatic crop identification with airborne radar imagery. *Photogrammetric Engineering and Remote Sensing*, **48**, 101-109

BROWN, R.J., GUINDON, B., TEILLET, P.H. and GOODENOUGH, D.G. (1984). Crop type determination from multitemporal SAR imagery. In *Proceedings 9th Canadian Symposium on Remote Sensing*, St. John's, Newfoundland

BUSH, T.F. and ULABY, F.T. (1975). Remotely sensing wheat maturation with radar. University of Kansas Center for Research Inc., Lawrence, Kansas, USA. RSL Technical Report 177-55

BUSH, T.F and ULABY, F.T. (1976). Crop classification with radar: preliminary results. University of Kansas Center for Research Inc., Lawrence, Kansas, USA. RSL Technical Report 330-1

BUSH, T.F. and ULABY, F.T. (1977). Cropland inventories using satellite altitude imaging radar. University of Kansas Center for Research Inc., Lawrence, Kansas, USA. RSL Technical Report 330-4

BUSH, T.F. and ULABY, F.T. (1978). An evaluation of radar as a crop classifier. *Remote Sensing of Environment*, **7**, 15-36

CHOUDHURY, B.J., SCHMUGGE, T.J., NEWTON, R.W. and CHANG, A. (1979). Effect of surface roughness on the microwave emission from soils. *Journal of Geophysical Research*, **84**, 5699-5706

CHANG, A.T.C., ATWATER, S., SALOMONSON, V.V., ESTES, J.E. and SIMONETT, D.S. (1980). L-band radar sensing of soil moisture. NASA Report No. NASA-TM-80628

CHURCHILL, P.N, HORNE, A.I.D. and KESSLER, R. (1985). A review of radar analyses of woodland. In *Proceedings EARSeL Workshop 'Microwave remote sensing applied to vegetation'*, pp. 25-32. Amsterdam, 10-12 December 1984. ESA SP-227, European Space Agency, Paris

CIHLAR, J., BROWN, R.J. and GUINDON, B. (1985). Microwave remote sensing of agricultural crops in Canada. In *Proceedings EARSeL Workshop 'Microwave Remote sensing applied to vegetation'*, pp. 113-122. Amsterdam, 10-12 December 1984. ESA SP-227, European Space Agency, Paris

CIHLAR, J. and HIROSE, T. (1984). On the SAR response of agricultural targets in a northern prairie environment. In *Proceedings of the International Geoscience and Remote Sensing Symposium*, 27-30 August, Strasbourg, France

CIHLAR, J., DOBSON, M.C., SCHMUGGE, T., HOOGEBOOM, P., JANSE, A.R.P., BARET, F., GUYOT, G. and LE TOAN (1987). Procedures for the description of agricultural crops and soils in optical and microwave remote sensing studies. *International Journal of Remote Sensing*, **8**, 427-439

CROWN, P. (1980). Evaluation of radar imagery for crop identification: an interpretation of radar imagery for three spring wheat sites. Final Report of the Airborne SAR project, Intera Environmental Consultants Ltd., Report No. ASP-80-1

DE LOOR, G.P., HOOGEBOOM, P. and ATTEMA, E.P.W. (1982). The Dutch ROVE programme. *IEEE Transactions on Geoscience and Remote Sensing*, **GE-20**, 3-11

DOBSON, M.C. and ULABY, F.T. (1981). Microwave backscatter dependence on surface roughness, soil moisture, and soil texture: Part III - Soil tension. *IEEE Transactions on Geoscience Electronics*, **GE-19**, 51-61

DRAKE, B., SCHUCHMAN, R.A., BRYAN, M.L., LARSON, R.W. and LISKOW, C.L. (1974). The application of airborne imaging radars (L and X-band) to Earth resources problems. Report No. NASA-CR-139385-1; Environmental Research Institute of Michigan, 104000-1-F

DURAND, E.T. (1987). SAR data filtering for classification. *IEEE Transactions on Geoscience and Remote Sensing*, **GE-25**, 629-637

EL-RAYES, M.A. and ULABY, F.T. (1987). Microwave dielectric spectrum of vegetation—Part I: Experimental observations. *IEEE Transactions on Geoscience and Remote Sensing*, **GE-25**, 541-549

GARRON, L. and SCHUBERT, J. (1979). Preliminary analysis of microwave (radar) data of three Canadian spring wheat test sites. Report prepared by the Sibbald Group for Agriculture in Canada, contract No. ISZ 77-00209

GOMBEER, R. (1983). Crop and land use classification study on SAR data over Belgium. SAR-580 Investigators Preliminary Report. European Space Agency

GOMBEER, R. (1985). Multi-spectral (X-band and C-band) crop classification with synthetic aperture radar (SAR-580) optical data. The European SAR-580 experiment; Investigators Final Report, pp. 723-734

HOOGEBOOM, P. (1982). Classification of agricultural crops in radar images. In *Proceedings of the International Geoscience and Remote Sensing Symposium*, Vol. 2, FA-4, Munich

JACKSON, T.J., CHANG, A. and SCHMUGGE, T.J. (1981). Aircraft active microwave measurements for estimating soil moisture. *Photogrammetric Engineering and Remote Sensing*, **47**, 801-805

JACKSON, T.J. and O'NEILL, P. (1983). Aircraft scatterometer observations of soil moisture on rangeland watersheds. In *Proceedings American Photogrammetric Society Meeting*, Washington DC,

JACKSON, T.J., SCHMUGGE, T.J. and O'NEILL, P. (1983). Remote sensing of soil moisture from an aircraft platform. In *Proceedings International Association of Scientific Hydrology Symposium in Remote Sensing*, Hamburg, Germany

KING, G.I. (1980). Radar discrimination of crops. Final Report of the Airborne SAR project, Intera Environmental Consultants Ltd., Report No. ASP-80-1

LE TOAN, T. (1982). Active microwave signatures of soil and crops. Significant results of three years of experiments. IEEE publication number 82 C414723 TP-2, pp. 3.1-3.5

LE TOAN, T., FLOUZAT, G., PAUSANDER, M. and FLUHR, A. (1981). Soil backscatter experiments in the 1.5 to 9 GHz region. In *Proceedings URSI Commission F. of Signature Prob. Microwave Remote Sensing Earth*. Lawrence, Kansas

LE TOAN, T., LOPEZ, A. and HUET, M. (1984). On the relationships between radar backscattering coefficient and vegetation canopy characteristics. In *Proceedings of the International Geoscience and Remote Sensing Symposium*, Strasbourg 27-30 August 1984, pp. 155-160. ESA SP-215. European Space Agency, Paris

LI, F.K., ULABY, F.T. and EASTON, J.R. (1980). Crop classification with a Landsat-Radar combination. In *Proceedings Symposium on Machine Processing of Remotely Sensed Data*, Purdue University, West Lafayette, Indiana

MacDONALD, H.C. and WAITE, W.P. (1971). Soil moisture detection with imaging radar. *Water Resources*, **7**, 100-109

MEGIER, J., MEHL, W. and RUPPELT, R. (1984). Methodological studies of land use classification of SAR and multi-sensor imagery. The European SAR-580 experiment; Investigators Final Report, pp. 693-722

MEHTA, N. (1983). Crop identification with airborne scatterometers. In *Proceedings of the International Geoscience and Remote Sensing Symposium*, Vol 1, PS-2, San Francisco

MORAIN, S.A and COINER, J. (1972). An evaluation of fine resolution radar imagery for making agricultural determinations. University of Kansas Centre for Research Inc., Lawrence, Kansas, USA. RSL Technical Report 177-7

NASA (1988). Earth Observing System - Instrument Panel Report, Volume IIf - SAR. NASA

NEWTON, R.W. (1977). Microwave remote sensing and its application to soil moisture detection. Remote Sensing Center, Texas A&M University, College Station, Technical Report RSC 81, Jan. (1977)

NEWTON, R.W. and ROUSE, J.W. (1980). Microwave radiometer measurements of soil moisture content. *IEEE Transactions on Antennas and Propagation*, **AP-28**, 680-686

NJOKYU, E.G. and O'NEILL, P.E. (1982). Multifrequency microwave radiometer measurements of soil moisture. *IEEE Transactions on Geoscience and Remote Sensing*, **GE-20**, 468-475

PARASHAR, S., DAY, D., RYAN, J., STRONG, D., WORSFOLD, R. and KING, G. (1979). Radar discrimination of crops. In *Proceedings 13th International Symposium Remote Sensing of the Environment*, Vol. 2, pp. 813-823. University of Michigan, Ann Arbor, Michigan

PARIS, J.F. (1982). Radar remote sensing of crops. In *Proceedings of the International Geoscience and Remote Sensing Symposium*, Vol. 2. Munich, F.R.G.

PARIS, J.F. (1983). Radar backscattering properties of corn and soybeans of frequencies of 1.6, 4.75 and 13.3 GHz. *IEEE Transactions on Geoscience and Remote Sensing*, **GE-21**, 392-400

PEI-YU, H. and DE-LI, G. (1983). Radar backscattering coefficients of paddy fields. In *Proceedings of the International Geoscience and Remote Sensing Symposium*, Vol. 2, FP-5, San Francisco

PORCELLO, L.J. and RENDLEMAN, R.A. (1972). Multispectral imaging radar. In 4th Annual Earth Resources Program Review, Vol. 2, Report TM X-68397. NASA

PROTZ, R. and BRISCO, B. (1980). Analysis of soil and vegetation characteristics influencing the backscatter coefficient of a SAR system. Final Report of the Airborne SAR project, Intera Environmental Consultants Ltd., Report No. ASP-80-1

REMOTEC APPLICATIONS, Inc. (1979). Analysis of the physical characteristics of soil-plant systems which affect the backscattering coefficient of synthetic aperture radar. Final Report, Contract 075Z.01A02-8-0684

SCHMUGGE, T.J. (1978). Remote sensing of surface moisture. *Journal of Applied Meteorology*, **17**, 1550-1557

SCHMUGGE, T.J. (1980). Effect of texture on microwave emission from soils. *IEEE Transactions on Geoscience and Remote Sensing*, **GE-18**, 353-361

SCHMUGGE, T.J. (1983). Remote sensing of soil moisture: recent advances. *IEEE Transactions on Geoscience and Remote Sensing*, **GE-21**, 336-344

SCHMUGGE, T.J., JACKSON, T.J. and McKIM, H.L. (1980). Survey of methods for soil moisture determination. *Water Resources*, **16**, 961-979

SHANMUGAN, K.S., ULABY, F.T., NARAYANAN, V. and DOBSON, M.C. (1983). Identification of corn fields using multi-date radar data. *Remote Sensing of Environment*, **13**, 251-264

SIEBER, A.J., FREITAG, B. and LAWLER, K. (1982). The aspect angle dependence of SAR images. In *Proceedings of the International Geoscience and Remote Sensing Symposium*, Vol.2, FA-4, Munich, F.R.G.

SIMONETT, D.S. (1978). Potential applications of space radar for agricultural monitoring in the wet tropics. In *Proceedings 11th International Symposium on Remote Sensing of Environment*, Vol. 2. University of Michigan, Ann Arbor, Michigan

SMIT, M.K. (1978). Radar reflectometry in the Netherlands; measurement system, data holding and some results. In *Proceedings International Conference on Earth Observation from Space and Management of Planetry Resources*, Toulouse, pp. 377-388, ESA pub. ESA SP-134. European Space Agency, Paris

TAUBE, D. and THEISS, S.W. (1984). Correlation of microwave sensor returns with soil moisture. In *Proceedings of the International Geoscience and Remote Sensing Symposium '84*. Strasbourg, France

TEILLET, P.H., BROWN, R.J., GUINDON, B. and GOODENOUGH, D.G. (1984). Relating synthetic aperture radar imagery to biophysical factors. In *Proceedings 9th Canadian Symposium on Remote Sensing*, St. John's, Newfoundland

TSANG, L. and NEWTON, R.W. (1982). Microwave emission from soils with rough surfaces. *Journal of Geophysical Research*, **87**, 9017-9024

ULABY, F.T (1980). Microwave responses of vegetation. *Advanced Space Research*, **1**, 55-70

ULABY, F.T., ASLAM, A. and DOBSON, M.C. (1982). Effects of vegetation cover on the radar sensitivity to soil moisture. *IEEE Transactions on Geoscience and Remote Sensing*, **GE-20**, 476-481

ULABY, F.T. and BARE, J.E. (1979). Look direction modulation function of the radar backscattering coefficient of agricultural fields. *Photogrammetric Engineering and Remote Sensing*, **45**, 1495-1506

ULABY, F.T, and BATLIVALA, P.P. (1976a). Optimum parameters for mapping soil moisture. *IEEE Transactions on Geoscience Electronics*, **GE-14**, 81-93

ULABY, F.T, and BATLIVALA, P.P. (1976b). Diurnal variations of radar backscatter from a vegetation canopy. *IEEE Transactions on Antennas and Propagation*, **AP-24**, 11-17

ULABY, F.T., BATLIVALA, P.P. and DOBSON, M.C. (1978). Microwave backscatter dependence on surface roughness, soil moisture, and soil texture: Part I - Bare soil. *IEEE Transactions on Geoscience Electronics*, **GE-16**, 286-295

ULABY, F.T., BRADLEY, G.A. and DOBSON, M.C. (1979). Microwave backscatter dependence on surface roughness, soil moisture, and soil texture: Part II - Vegetation-covered soil. *IEEE Transactions on Geoscience Electronics*, **GE-17**, 33-40

ULABY, F.T. and BUSH, T.F. (1976). Monitoring wheat growth with radar. *Photogrammetric Engineering and Remote Sensing*, **42**, 557-568

ULABY, F.T, CIHLAR, J. and MOORE, R.K. (1974). Active microwave measurements of soil water content. *Remote Sensing of Environment*, **3**, 185-203

ULABY, F.T., DOBSON, M.C. and BRADLEY, G. (1980). Radar reflectivity of bare and vegetation covered soil. In *Proceedings 23rd Annual Conference Committee on Spatial Research.* (COSPAR). Budapest, Hungary

ULABY, F.T., DOBSON, M.C. and BRUNFELDT, D.C. (1983). Improvement of soil moisture estimation accuracy of vegetation-covered soil by combined active/passive remote sensing. *IEEE Transactions on Geoscience and Remote Sensing*, **GE-21**, 300-307

ULABY, F.T., LI, R.Y and SCHANMUGAN, K.S. (1981). Crop classification by radar. University of Kansas Center for Research Inc., Lawrence, Kansas, USA. RSL Technical Report 360-13

ULABY, F.T. and MOORE, R.K. (1973). Radar spectral measurements of vegetation. University of Kansas Center for Research Inc., Lawrence, Kansas, USA. RSL Technical Report 177-40

ULABY, F.T., MOORE, R.K. and FUNG, A.K. (1986). *Microwave Remote Sensing: Active and Passive*, Vol. III. Addison-Wesley, Reading, Massachusetts

VAN KASTEREN, H.J.W. (1981). Radar signature of crops. The effect of weather conditions and the possibility of crop discrimination with radar. In *Proceedings Colloque sur les Signatures Spectrales d'Objets en Teledetection*. Avignon, France

WANG, J.R. (1986). The SIR-B observations of microwave backscatter dependence on soil moisture, surface roughness and vegetation cover. *IEEE Transactions on Geoscience and Remote Sensing*, **GE-24**, 510-515

WANG, J.R., McMURTREY, J.E., ENGMAN, E.T., JACKSON, T.J., SCHMUGGE, T.J., GOULD, W.J., FUCHS, L.E. and GLAZAR, W.S. (1982). Radiometric measurements over bare and vegetated fields at 1.4 GHz and 5 GHz frequencies. *Remote Sensing of Environment*, **12**, 295-311

WILHEIT, T. (1978). Radiative transfer of a plane stratified dielectric. *IEEE Transactions on Geoscience Electronics*, **GE-16**, 138-143

WOODING, M.G. (1983). Preliminary results from the analysis of SAR-580 digital radar for the discrimination of crop types and crop condition, Cambridge, UK. SAR-580 Investigators Preliminary Report (European Space Agency)

20

MICROWAVE RADIOMETRY FOR MONITORING AGRICULTURAL CROPS

G. LUZI, S. PALOSCIA and P. PAMPALONI
Centre for Microwave Remote Sensing,
Viale Galileo, 32 - 50125 Firenze, Italy

Introduction

Agricultural management can benefit from remote sensing techniques to estimate both crop production and the health status of plants. Since microwave emission depends on the dielectric constant of emitting bodies, it changes according to their water content. The potential of microwave radiometry for monitoring agricultural crops and for estimating their production and water status has been investigated by our group since the beginning of the 1980's. Using a multifrequency sensor package, information about the moisture content of soil and vegetation has been collected on different agricultural test sites in Northern and Central Italy. The main results are summarized in this paper and compared with theoretical models.

Remote sensing techniques make it possible to obtain and distribute information rapidly about large areas by means of sensors operating in several spectral bands, mounted on aircraft or satellites. A satellite which orbits the Earth is able to explore the whole surface in a few days and repeat the survey of the same area at regular intervals, whilst an aircraft can give a more detailed analysis of a smaller area, if a specific need occurs (e.g. natural disasters).

The spectral bands used by these sensors cover the whole range between visible and microwaves. Microwave sensors can be passive (radiometers) and active (radars and scatterometers); the former measure the natural emission of observed bodies, whereas the latter measure the backscattered fraction of a power beam which they themselves emit. In contrast to optical frequencies, microwaves can penetrate through clouds and operate in the absence of sunlight. Moreover, microwave emission is strongly dependent on the dielectric constant of emitting bodies. Thus, since the dielectric constant of water is very different from that of dry matter, small variations in the water content of soil, for example, causes a sensible variation in microwave emissivity, mainly at low frequencies (1-5 GHz).

The main fields of application for which remote sensing techniques can be usefully employed are agriculture, meteorology, hydrology and geology. The use of remote sensing techniques in agriculture is mainly devoted to the forecasting of crop production and to the optimisation of productivity. Both objectives require a knowledge of the type and health condition of crops.

Remote sensing data, together with auxiliary data (geographic, cartographic, meteorological data) and physical or statistical models, allow the estimation of variables such as soil moisture content, plant water status, leaf area index, surface temperature, and soil roughness. From these quantities it is possible, on the one hand, to retrieve the estimate of agricultural production and, on the other hand, if data are promptly collected, to programme the irrigation schedules or fungicide treatments. In this case the main problem is to recognize the situation of plant stress, due to water shortage or parasite attacks, before the phenomenon strongly affects plant growth and, consequently, the final yield. The potential of microwave radiometry for monitoring agricultural crops and for estimating their production and water status has been investigated by our group since the beginning of the 1980's. In this paper the main results are summarised, obtained by means of a multifrequency sensor package made up of three microwave radiometers, (1.4 GHz, 10 GHz and 36 GHz) and a thermal infrared sensor (8-14 μm). Data have been collected on different agricultural test sites in Central and Northern Italy, using both ground-based and airborne platforms. The crops investigated were: alfalfa, corn, sugarbeet, sunflower, wheat and barley. The relationships between microwave emission and soil moisture, soil roughness, plant water content and leaf area index will be discussed.

Physical principles

Every natural body emits an amount of energy in the form of electromagnetic waves due to the thermal motion of its microscopic components (thermal radiation). When the emitting body is homogeneous and in thermal equilibrium, the power emitted depends on the thermodynamic temperature of the body according to Planck's law. If we restrict our analysis to the microwave range of the electromagnetic spectrum (0.1 to 30 cm) and to temperatures not much lower then 300 K, Planck's law can be well approximated by the Rayleigh-Jeans formula:

$$\mathbf{B} = 2\mathbf{k}T\epsilon/\lambda^2 \qquad (20.1)$$

where $\mathbf{k}$ is Boltzmann's constant (J K^{-1}); $\mathbf{B}$ is the brightness (W m^{-2} Hz^{-1} Sr $^{-1}$); T is the thermodynamic temperature (K); ϵ is the emissivity and λ is the wavelength. The emissivity is equal to 1 for a perfectly absorbing body (black body), whilst for a natural body it takes into account the loss of energy at the surface and depends on physical and geometric characteristics of the medium. It is convenient to define a brightness temperature, T_b related to the thermodynamic temperature by the relation:

$$T_b = \epsilon T \qquad (20.2)$$

which is rigorously valid for a homogeneous medium at a uniform temperature. As a consequence of the linear relation between the brightness and the temperature T_b, the power measured by a radiometer can be expressed in terms of T_b. From *Equation 20.2* we can write:

$$\epsilon = T_b/T \qquad (20.3)$$

The emissivity of a natural medium can be expressed by means of the normalized temperature T_n

$$T_n = T_b/T' \qquad (20.4)$$

where T' is the surface temperature of the medium measured by an infrared radiometer. Considering a plane wave impinging on a smooth surface of a medium in thermal equilibrium, the energy conservation law, under the condition that the power trasmitted in the medium is negligible, states that:

$$\alpha(\theta, p) = 1 - \rho(\theta, p) = \epsilon(\theta, p) \tag{20.5}$$

where $\alpha(\theta, p)$ is the absorption coefficient, $\rho(\theta, p)$ is the reflectivity, θ the observation angle and p refers to horizontal or vertical polarization. The behaviour of ρ and hence of ϵ, is related to the physical properties of the medium by means of the Dielectric Constant (D.C.). The emissivity, (ϵ), and the D.C. (ε) are related to each other by the following expression:

$$\epsilon(\theta, H) = 1 - \frac{\cos\theta - \sqrt{(\varepsilon - \sin^2\theta)^2}}{\cos\theta + \sqrt{(\varepsilon - \sin^2\theta)^2}} \tag{20.6}$$

$$\epsilon(\theta, V) = 1 - \frac{\varepsilon\cos\theta - \sqrt{(\varepsilon - \sin^2\theta)^2}}{\varepsilon\cos\theta + \sqrt{(\varepsilon - \sin^2\theta)^2}} \tag{20.6}$$

The D.C. can be defined as the ratio between the electric displacement and the electric field and it is related to the ability of the molecules (or of the microscopic components) to align their own or induced electric dipole moments to the electric field. In the case of an isotropic homogeneous lossy medium the D.C. is represented by a complex number, $\varepsilon = \varepsilon' - j\varepsilon''$. Since natural bodies are heterogeneous media, the usual approach for estimating the D.C. of media like soil, vegetation, etc., consists in modelling the medium as a mixture of its components (one of which is water) and assigning a global value to the D.C. by averaging the dielectric constants of the components (De Loor, 1983) . The D.C. of pure water depends strongly on the frequency of the e.m. radiation: in *Figure 20.1* the behaviour of the D.C. of pure water versus the frequency (in the microwave region) at 20°C is shown; its value is very different from that of the other typical components of natural media.

Thus the presence of liquid water very strongly affects the intensity of the microwave radiation upwelling from media like soil, vegetation canopies, snow, etc. The D.C. of water is also affected by temperature and by other parameters (e.g. salinity). The geometric and surface characteristics of the emitting medium, e.g. the roughness of the soil surface or the shape and dimensions of the leaves of a vegetation cover, also affect the brightness temperature. As an example of the strong influence of the water content on the behaviour of the soil D.C., *Figure 20.2* shows the D.C. measured at 1.4 GHz versus the volumetric water content of a soil sample (Wang and Schmugge, 1980).

In order to assign a value of D.C. to a heterogeneous medium, it is very useful to model the medium as a mixture, where the substance with the highest volume fraction is considered the host material and the others are considered inclusions. To calculate the electric field inside the inclusions further hypotheses are made, one of which is to assume that the dimensions of the inclusions are smaller than the wavelength in the host medium. Soil and vegetation represent examples of two different mixtures: for soil the host material is the dry matter (sand, clay, loam etc.) and the inclusions water molecules, with a volume fraction which can reach up to 50%; instead, for vegetation the host is the air and the inclusions are leaves, fruits and other components whose shape and dimensions can be very variable but whose volume fraction generally does not exceed

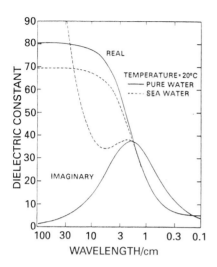

FIGURE 20.1. The dielectric constant of pure water at 20°C as a function of the wavelength in the microwave range of the electromagnetic spectrum

1%. One of the most used expressions for the D.C. of the mixture is the well known "refractive formula":

$$\sqrt{\varepsilon_m} = \sqrt{\varepsilon_h} + V_i(\sqrt{\varepsilon_i} - \sqrt{\varepsilon_h}), \qquad (20.7)$$

where ε_h, ε_i and ε_m are the D.C. of the host material, of the inclusions and of the mixture, respectively and V_i is the volume fraction of the inclusions.

In the case of dry soil the real part of the D.C. of the host material has a value of a few units and the imaginary part is slightly lower. Until the density of water molecules reaches a high volume fraction (10%) these can surround the soil particles and remain tightly attached to them (bound water). In this case the behaviour of the D.C. of the mixture is like that of dry matter, because the molecules are not free to rotate under the influence of the external electric field. When further molecules are included they can occupy the free space among the soil particles, so they behave as free water molecules and very strongly affect the D.C. of the mixture. A good estimation of the soil D.C. is obtained by means of the Debye equation (Debye, 1929), for water where the values of the parameters included are estimated empirically. The depth of soil from which the greatest part of the radiation emitted upwells, which is called the penetration or sampling depth, is an important parameter for investigating soil or vegetation: this parameter obviously depends on the wavelength. Different results have been obtained by many authors (Wilheit, 1978; Shutko, 1982), but it is a common result that the greater the water content the smaller is the penetration depth of soil, because the wavelength of the radiation decreases as the D.C. increases. Two parameters have been used as indicators of soil water content: the gravimetric and the volumetric soil moisture contents. However, more recent investigations suggest the "volumetric percentage" as the most suitable indicator for correlating microwave emission to soil moisture independently of soil texture.

Although water represents up to 98% of the inclusions in a vegetation canopy layer,

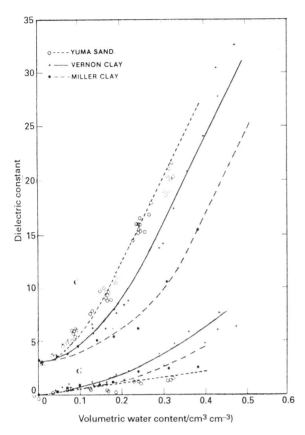

FIGURE 20.2. Measured values of the dielectric constant at 5GHz for three different soils as a function of the volumetric moisture content. Symbols represent experimental data whilst dashed and solid lines refer to the predictions of the semi-empirical model of Wang and Schmugge, (1980)

the global D.C. of the air-vegetation mixture is much lower than the soil's, because of the very low volume fraction of the inclusions. Besides, there is not an abrupt change in the dielectric profile passing from the air to the vegetation, so the interface between these two media is characterized by a negligible value of reflectivity. Despite the low global value of the D.C., extinction phenomena (absorption and scattering) strongly affect the microwave propagation. At higher frequencies of the microwave band, the upwelling radiation is slightly affected by soil characteristics and depends on biomass parameters, like leaf area index (LAI) and plant water content (**Q**) (Pampaloni and Paloscia, 1986).

Equipment and experimental results

The experimental campaigns carried out during the last few years, have utilized a sensor package which has been improved gradually. For ground-based measurements it was mounted on a hydraulic boom, whilst for air surveys small aircraft and helicopters were used. The equipment included three microwave radiometers, at Ka (36GHz), X (10GHz) and L (1.4GHz) bands, and an infrared (IR) one, 8-14 μm bandwidth, whose characteristics are summarized in *Table 20.1*. The IR radiometer was a standard commercial type. Both X and Ka radiometers were able to measure H and V polarization components of the brightness temperatures simultaneously. All the microwave radiometers were Dicke type; the temperature of the reference load was measured by means of a Pt100 thermometer with an accuracy of 0.2 K and recorded simultaneously to the RF data (Pampaloni, 1981). The calibration of the sensors was carried out by means of internal and external calibrators.

TABLE 20.1. Characteristics of radiometers employed in this study, giving Radio Frequency Bandwidth (RFBW), Integration Time, Accuracy, Angular Beamwidth (HPBWS), and Polarization

Band/Freq (GHz)	RFBW (MHz)	Int.Time (s)	Acc. (K)	HPBWS (degree)	Pol.
L / (1.4)	30	10/1/.1	2	18/21	H or V
X / (9.6)	1000	10/1/.1	1	15	H and V
Ka / (36.6)	1000	10/1/.1	1	12	H and V
IR	(8-14μm)	.1	.5	2	-

The achievable accuracy was ±0.5 K in ground based measurements, ±1 K in aircraft surveys. A colour TV camera with a video recorder was used in aircraft measurements in order to relate radiometric measurements to the actual scene correctly. A clock signal, synchronized with the computer, was simultaneously recorded on video tapes and on the data file. In this way, measured brightness temperatures could be related to time and then to the geographical location of the scene. All sensors were battery operated with a minimum working period of three hours. In total nine voltage outputs were available from the sensor package: six correspond to brightness temperatures from Ka and X band, H and V polarization, L band H or V polarization and IR respectively; three other channels correspond to the Dicke reference temperatures. The outputs from the sensors were recorded on a floppy disk by means of a Data Acquisition System controlled by a portable personal computer.

During several years of investigations a great quantity of data have been collected over different natural surfaces. Experiments were carried out on agricultural crops: alfalfa, corn, wheat, barley, sunflower and sugarbeet. Remote sensing data have been compared with the biophysical parameters of soil and plants in order to establish some relationships between them and test these relations through physical models. The following parameters of the soil and plants have been measured during the experiments: soil moisture content (**SMC**), by weight and by volume (using the gravimetric method), soil density and soil roughness, by means of a profilometer. For each crop the plant height and density, the number of leaves and the leaf dimensions, the Green Leaf Area Index (**LAI**) measured by means of a planimeter, the fresh and the dry weights and, consequently, the plant water content (**Q**) have been measured. Most of these ground-truth data were collected simultaneously to the remote sensing data.

A preliminary form of analysis was to investigate the different behaviour of bright-

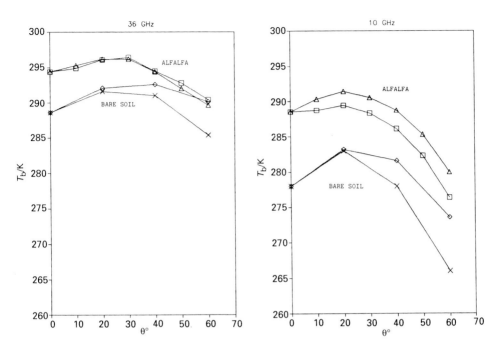

FIGURE 20.3. Brightness temperature T_b in (K) versus observation angle θ (a) at 36GHz and (b) 10GHz, for bare soil (x = horizontal polarisation; ◇ = vertical polarisation) and vegetated soil (□ = horizontal polarisation; △ = vertical polarisation) (Crop alfalfa)

ness temperature (T_b) as a function of the incidence angle (θ) on different types of targets, e.g. bare soil and vegetation. Measurement of T_b as a function of θ, at 36 GHz (a) and at 10 GHz (b) in V and H polarization and on bare and vegetated soil are shown in *Figure 20.3*. These data were collected by means of ground-based equipment at Fagna, a test-site near Florence, where three plots of corn, alfalfa and bare soil were kept under observation continuously for more than one month during the summers of 1983 and 1984.

In these diagrams vegetation is represented by fully developed alfalfa crops (when the height was 80 cm and **LAI** = 4). Bare soil had a soil moisture content equal to 7% and a moderate to high roughness, especially at these frequencies. As preliminary observations, it can be noticed that both at 36 GHz and 10 GHz the signals for the two polarizations are very close on fully developed vegetation - with a slight dominance of H polarization at 10 GHz - whereas on bare soil they are clearly separated and V polarization is higher. Moreover, at 10 GHz, the difference between the brightness temperatures on bare and vegetated soil is greater than at 36 GHz. For a comparison, a similar diagram at 1.4 GHz for soybean before and after cut time is also shown in *Figure 20.4* (Mo, Choudhury, Schmugge, Wang and Jackson, 1982). At this frequency vegetation is almost transparent and its presence does not affect very much either the difference between the two polarizations or the absolute value of T_b.

In order to investigate a wider sample of surfaces (and to validate the results obtained

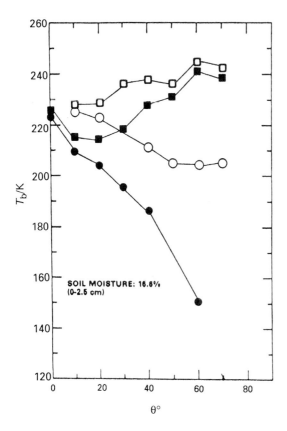

FIGURE 20.4. Brightness temperature T_b (in K) versus observation angle θ at 1.4 GHz before (o = horizontal polarisation; □ = vertical polarisation) and after cutting. (• = horizontal polarisation; ■ = vertical polarisation) (Crop soybean). (After Mo et al., 1982)

on the same crops) the sensor package was mounted on an aircraft which in 1986 and 1987 was flown on several occasions over a test site in Northern Italy (Oltrepò Pavese) and in 1988 over an area located about 20 km south of Florence along the Pesa river. This test site, on which in 1988 the AGRISCATT campaign also took place, is a flat area (2 km x 0.5 km) on both sides of the river. Fields show a square shape with a median size of about 4-6 ha. Main crops were sunflower, wheat, barley, alfalfa and corn. As an example, the T_b profiles at four frequencies (36 GHz, 10 GHz, 1.5 GHz and thermal infrared) recorded along a flight line at $\theta = 10°$, are shown in *Figure 20.5*. During this flight, carried out on July 13th 1988, the conditions of plants were the following:

Wheat : height = 80 - 90 cm, ripening stage, **LAI** and **Q** ≈ 0;
Corn : height = 220 cm ; **LAI** = 5.4 ; **Q** = 6.2 kg m^{-2};
Sunflower: height = 210 cm ; **LAI** = 5.5 ; **Q** = 8 kg m^{-2}.

We observe that the same crop tends to the same brightness temperature both at 10 GHz and 36 GHz, even in different fields. Sunflower usually shows the lowest value of T_b, ripe wheat the highest, whilst corn shows an intermediate value.

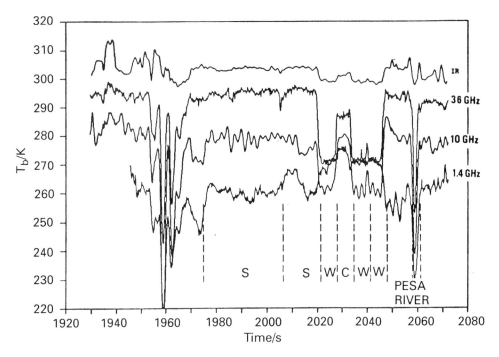

FIGURE 20.5. An example of profiles of brightness temperature T_b (K) versus time (s) at three microwave frequencies: 1.4GHz, 10GHz and 36Ghz and at IR (8–14μm) all measured along a flight line at $\theta = 40°$ over different crops: S, sunflower; C, corn; W, wheat

In general, we noticed that plants characterized by broad and horizontal leaves (e.g. sunflower, sugarbeet) show low values of T_b, whereas plants with small and randomly oriented leaves (e.g. alfalfa and wheat) show higher values of T_b. Corn, which has large leaves, especially on the upper layer, at an angle to the horizontal, usually shows intermediate values of brightness. Moreover, the boundaries of fields are clearly distinguishable, especially at 36 GHz (Bonsignori, Chiarantini, Paloscia and Pampaloni, 1987b; Bonsignori, Chiarantini, Paloscia, Pampaloni, Ferrazzoli, Mongiardo and Solimini, 1987a). At 1.4 GHz it is fairly clear that the main parameters that influence emission are soil moisture and roughness, whilst vegetation type does not influence it very much. From these diagrams it is clear that the brightness temperature depends on the crop type, but it is also true that it is influenced by soil and vegetation physical conditions, particularly water content and surface roughness. Moreover, the variations due to the crop type can be of the same order of magnitude as the variations due to different water conditions. Therefore, some combinations between the polarizations and the frequencies have been considered in order to better distinguish several effects one from the other and to quantify the sensitivity of microwave emission to different plant parameters. In particular, we considered the normalization of T_b to the infrared temperature (i.e. the normalized temperature T_n), the difference between vertical and horizontal polarizations as measured by the polarization index **PI**, $(T_{bv} - T_{bh})/\frac{1}{2}(T_{bv} + T_{bh})$ and the difference between 36 GHz and 10 GHz brightness temperatures ($\boldsymbol{\delta T_b}$) at horizontal polarization.

The ability of T_n to discriminate crops better than the brightness temperature itself is confirmed by the diagram of *Figure 20.6* where T_n values at 10 GHz are plotted against T_n values at 36 GHz. Three clusters are distinguishable, which represent three different types of surface: corn, alfalfa and bare soil (Pampaloni and Paloscia, 1985a).

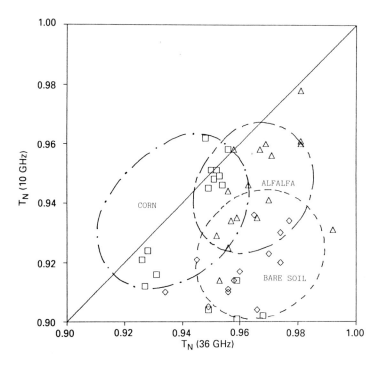

FIGURE 20.6. Normalised temperature T_n at 10 GHz versus T_n at 36 Ghz: the clusters corresponding to bare soil, corn and alfalfa. (After Paloscia and Pampaloni, 1988)

In the diagram the cluster of bare soil is below the 1:1 line (that is T_n at 36 GHz > T_n at 10 GHz), whereas the vegetation cluster is on the line (T_n at 36 GHz $\approx T_n$ at 10 GHz) or sometimes above the line (T_n at 36 GHz < T_n at 10 GHz). Therefore, we can notice that the passage from bare to vegetated soil is mainly characterised by an increase in T_n at 10 GHz. Moreover, inside each cluster, the shift of points towards higher values of T_n seems to be due to a progressive drying of the soil or to an increasing lack of water in plants. Since agricultural soils are often irrigated before sowing, we expected that the growth of plants could be detected by a rapid increase of T_n at 10 GHz. As a matter of fact, as we can see in *Figure 20.7*, T_n (10 GHz) increases from 0.88 to 0.96 when the leaf area index of alfalfa changes from 0 to 4. This behaviour has been confirmed by several sets of data collected on different fields of alfalfa, although it has been clearly found only in this crop (Bonsignori et al. 1987b).

Emission at frequencies higher than 5 GHz seems to be mostly influenced by vegetation features and by the water content of the most superficial layers of bare soil. In contrast, at lower frequencies, vegetation is almost transparent; therefore emission comes

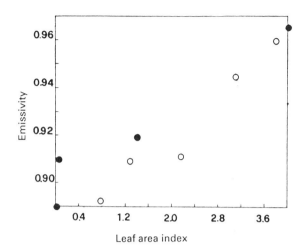

FIGURE 20.7. Alfalfa emissivity at 10 GHz as a function of Leaf Area Index (**LAI**). (o) data collected by means of truck mounted radiometers on a single field at Fagna (Tuscany); (•) data collected by aircraft campaign on Oltrepó Pavese. (After Bonsignori et al. 1987a)

from soil and can be directly correlated to soil moisture content. In *Figure 20.8* the normalized temperature at 10 GHz is represented as a function of volumetric soil moisture (SMC) of the layer 0-5 cm for bare and vegetated soil. Vegetation is represented by a wheat crop in the ripening stage.

From this diagram the sensitivity to soil moisture content, that is the slope of the regression line, turns out to be about 1.4 K per 1.0% volumetric SMC for bare soil. On the other hand, on vegetated soil the sensitivity to soil moisture becomes very low because it is masked by the presence of vegetation (Pampaloni, Paloscia and Chiarantini, 1986). If normalized temperature at 1.4 GHz is considered, as shown in *Figure 20.9*, the sensitivity to soil moisture is instead fairly high in the case of vegetated soil as well (≈ 1.5 K/% SMC). In this diagram symbols represent different types of vegetation: sunflower, wheat, corn and alfalfa. Although the sensitivity is lower than that found on bare soil (2-3 K/% SMC), this result is rather good, since all the soils were covered with tall vegetation.

Sunflower is in this case the most absorbent type of vegetation, since its biomass is equivalent to about 8 kg of water per square metre.

Let us consider now other parameters obtained by the combination of different frequencies and polarizations. From the diagrams of *Figure 20.3* we can see that the brightness temperatures at 10 GHz and 36 GHz are almost the same on fully grown vegetation and that T_b (36 GHz) is higher than T_b (10 GHz) on bare soil (smooth). Thus, the parameter δT_b (i.e. the difference between the two brightness temperatures: $T_b 36 - T_b 10$) could be useful in separating bare from vegetated soil and in identifying different stages of plant growth. In *Figure 20.10*, δT_b (H pol., $\theta = 40°$) is plotted as a function of the leaf area index (**LAI**) of different crops.

The decrease of δT_b is rather gradual as **LAI** increases and, if bare soil is smooth

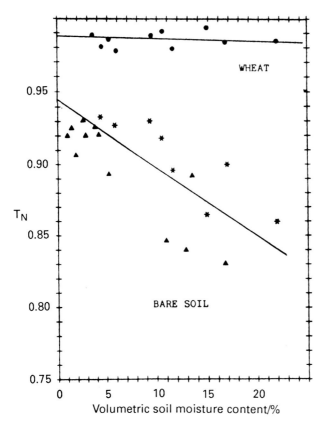

FIGURE 20.8. Normalised temperature data (T_n) at 10 GHz, $\theta = 10°$, as a function of volumetric soil moisture content (SMC) of the 0-5 cm layer, collected on bare soil (lower curve)(▲ = smooth surface, * = rough surface) and on a ripe wheat crop (upper curve)

enough, the progressive growth of vegetation can be estimated (Pampaloni and Paloscia, 1985a; Chiarantini, Paloscia and Pampaloni, 1987). Still referring to the initial diagram, it can be observed that the difference between V and H polarizations is relatively high on bare soil and nearly zero on vegetation, at both frequencies. The detection of this behaviour can be enhanced by employing the Polarization Index (**PI**). As an example, in *Figure 20.11* the **PI** computed at 10 GHz and and 36 GHz is shown as a function of incidence angle, θ, for an alfalfa crop in two contrasting situations: well developed, (**LAI** = 3.8) and just after cutting, (**LAI** = 0.2).

From this diagram we can see that **PI** at 10 GHz has a much wider dynamic range than **PI** at 36 GHz. Moreover, periodic observations carried out during the growth cycle of some crops of corn and alfalfa showed that, at 10 GHz, **PI** systematically decreases as vegetation grows until the leaf area index reaches its maximum value (Pampaloni and Paloscia, 1985b).

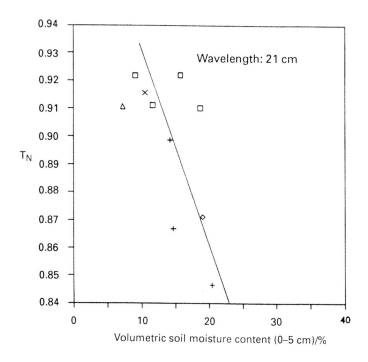

FIGURE 20.9. Normalized temperature T_n at 1.4 GHz, $\theta = 10°$, as a function of the average volumetric SMC of the 0-5 cm layer. Different symbols characterise different types of types of vegetation: □ = sunflower; + = bare soil; ◇ = wheat; x = alfalfa; △ = corn. (July 13, 1988), Val Di Pesa, (Italy)

Modelling soil and vegetation emission

The simplest approach to model soil emission is the one representing soil as a homogeneous (uniform moisture) and isothermal medium separated from the air by a plane interface. In this case, the emissivity can be calculated from the corresponding specular reflectivity using the energy conservation law. When the surface of the soil is rough, i.e. when some random irregularities are present which are not small with respect to the wavelength, the emissivity increases, since the surface area involved in the emission process becomes larger.

Semi-empirical approaches, obtained by introducing correction factors to specular reflectivity (Choudhury, Schmugge, Newton and Chang, 1979) are in good agreement with experimental data. A general theory requires knowledge of the surface parameters and of the surface scattering physical processes; using geometrical optics and Kirchhoff's approximation, Mo, Schmugge and Wang (1987), obtained a model which is in excellent agreement with experimental data collected at L band by Wang, Jackson, Engman, Gould, Fuchs, Glezer, O'Neil, Schmugge and McMurtey (1984), except for vertical polarization. The basic approach for evaluating emission from soils with non-uniform moisture profiles is to divide the soil into thin layers and then add the contributions of each layer.

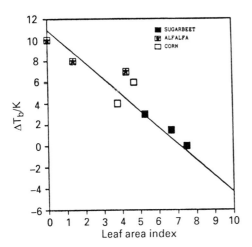

FIGURE 20.10. The difference δT_b between the brightness temperature at 36 GHz and 10 GHz as a function of Leaf Area Index (**LAI**) for different crops. (After Chiarantini et al., 1987)

In a model developed by Burke, Schmugge and Paris (1970), the soil is treated as a layered dielectric and the Radiative Transfer Theory (RTT) equation for each layer, which is assumed to be homogeneous, is solved. The transfer of energy between layers is then computed by the interface Fresnel reflectivity. In an alternative (coherent) approach by Wilheit (1978) soil is again treated as an ensemble of dielectric layers. However, the dielectric field in each layer is computed by using Maxwell equations and the boundary conditions at the interfaces. Energy fluxes are computed from the dielectric fields to obtain the fractional absorption (f_j) in each layer at temperature T_j and the resulting brightness temperature is given by $T_n = \Sigma f_j T_j$. A comparison between these two models carried out by Schmugge and Choudhury (1981) has shown that for most practical cases the two models give similar results and that the difference becomes more significant at low frequencies.

The difficulty in establishing a universal model for vegetation is due to the great variability of plant biophysical parameters, even within the same crop. In a simplified approach vegetation can be modelled in RTT as a uniform lossy dielectric layer above the soil surface at constant temperature, and its absorbing properties can be correlated to biomass characteristics. Regarding the scattering properties, different approximations have been considered: Kirdiashev, Chukhlantev and Shutko (1979) ignore this effect, Ulaby, Ratzani and Dobson (1983) consider vegetation as an ensemble of isotropic scatterers whilst Mo et al. (1982), assume that scattering is mainly forward. In the real case the vegetation is inhomogeneous and the temperature of the leaves changes both in time and in space. On the other hand, some crops are so different in shape and dimension that their scattering functions can differ a lot. If vegetation is represented as a plane parallel absorbing medium at a constant temperature T with randomly distributed scatterers above a homogeneous media (soil), the one dimensional RTT equation for the brightness

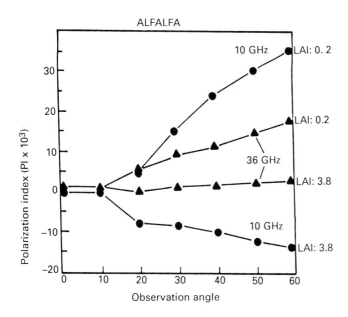

FIGURE 20.11. Polarisation Index **PI** computed at 10 GHz and 36 GHz, as a function of observation angle θ at different values of Leaf Area Index (**LAI**) of an alfalfa crop. (After Pampaloni and Paloscia, 1986)

temperature T_{bp} becomes;

$$\frac{dT_{bp}(\tau,\mu)}{d\tau} = -T_{bp}(\tau,\mu) + (1-\omega)T + \frac{\omega}{2}\int_{-1}^{1} P(\mu,\mu')T_{bp}(\tau,\mu')d\mu' \qquad (20.8)$$

where **P** is an index of polarization, $\mu = \cos\theta$, θ is the observation angle, and τ is the optical depth:

$$\tau = \int_{z'}^{z_o} K\epsilon dz \qquad (20.9)$$

ω is the albedo $\omega = K_s/(K_a + K_s)$ and K_a, K_s are the absorption and scattering coefficients, which take into account the decrease of power respectively due to absorption and scattering phenomena, and $P(\mu,\mu')$ is the scattering function which describes the behaviour of the scattered power along each direction. The solution of this equation requires many approximations; if we, like Mo et al. (1982), consider that the scattered power is mainly forward, we can write:

$$\mathbf{P}(\mu,\mu') = 2\mathbf{f}\delta(\mu - \mu') \qquad (20.10)$$

where **f** is a parameter related to the percentage of the whole scattered power which is forward directed; and $\delta(\mu - \mu')$ is the Dirac function. Transforming τ and ω to τ' and ω' by means of relations (introduced by Joseph, Wiscome and Weinman, 1976):

$$\tau' = \tau(1 - \mathbf{f}\omega) \qquad (20.11)$$

$$\omega' = \omega(1-f)/(1-f\omega). \qquad (20.11)$$

The integration of *Equation 20.8* leads to this expression for T_{bpd}

$$T_{bpd} = (1-\omega')T(1-e^{-\tau'/\mu}) \qquad (20.12)$$

where:
$$T_{bpi} = (1-R_p)T_s + R_p T_{bpd}; \qquad (20.13)$$

$R_p i$ is the soil reflectivity; and T_s is soil temperature.

In *Equation 20.12* we can identify the first term on the right as the upward vegetation emission (T_{bpu}) and the second as the upward soil emission (($1-R_p)T_s$) plus the downward vegetation emission T_{bpd} reflected by soil surface, shown in *Figure 20.12*.

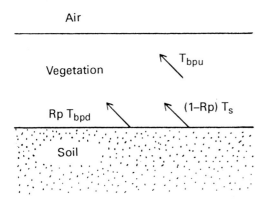

FIGURE 20.12. Different contributions to global soil-vegetation system emission described in *Equation 20.12*

Using the further conditions that vegetation and soil are in thermal equilibrium ($T \approx T_s$) and the soil reflectivity is so low that $T_{bpd} \approx 0$, *Equation 20.12* becomes:

$$T_{bp} = (1-\omega')(1-e^{\tau'/\mu}) + (1-R_p)Te^{\tau'/\mu} \qquad (20.14)$$

An evaluation of the two unknowns τ' and ω' was carried out at X and Ka bands by means of experimental data and an iterative procedure developed by Pampaloni and Paloscia (1986). To characterize more rigorously the electromagnetic behaviour of vegetation, Eom and Fung (1984) represent it as an ensemble of thin elementary layers in which the D.C. and the geometric characteristics of leaves are input data. This model is based on the matrix doubling algorithm (MDA), which represents the medium as the ensemble of many identical layers, each one thin enough so that multiple scattering within it can be neglected. The plant components are schematized by dielectric entities of simple geometrical form such as disks or cylinders, to make simpler the form of the scattering function; it is also supposed that they do not interact with each other. In the case of crops with large leaves, these can be approximated by an ensemble of several small disks. The MDA operates by doubling the thickness of the layers a number of times until a depth equal to the one of the whole layer of vegetation is reached. Multiple scattering effects are considered, compounding the scattering matrices of each layer; the bistatic scattering coefficient is thus calculated and, consequently, the emissivity is obtained by means of the reciprocity principle. This approach is very interesting, because the input parameters of the model are closely related to the biophysical properties of plants, and it seems to be suitable in the evaluation of the influence of soil moisture and roughness as well (Ferrazzoli, Pampaloni, Paloscia and Solimini, 1988a).

Comparison of model with data

The model described by *Equation 20.12* has been used to compute the equivalent optical depth (τ') and the equivalent single scattering albedo (ω') (Pampaloni and Paloscia, 1986). The data were collected during clear days, so that the sky brightness temperature reflected by soil had a negligible effect on the brightness temperature measured by the radiometer at the highest frequency. In *Equation 20.12* T_{bp}, T and T_s are measured, $\mathbf{R_p}$ is obtained from measurements on bare soil, with the same characteristics as vegetated soil, and τ' and ω' are the unknowns. The equation has been solved by means of an iterative fitting method. Computations showed that the equivalent single scattering albedo both at 36 GHz and 10 GHz is $\ll 1$, whereas the equivalent optical depth is often higher than 1 and can even reach values of 3 or 4 at 36 GHz. In addition, the following relation between τ' and the plant water content $\mathbf{Q}$ was established:

$$\tau' = (\mathbf{k}/\sqrt{\lambda})ln(1+\mathbf{Q}) \qquad (20.15)$$

where $\mathbf{k}$ ($m^{-0.5}$) is independent of wavelength and depends only on the crop type: it is 0.25 for alfalfa and 0.16 for corn; λ is the wavelength (m) and $\mathbf{Q}$ is the plant water content in kg m^{-2}. In *Figure 20.13*, τ' multiplied by the square root of the wavelength is represented as a function of plant water content.

The continuous lines represent *Equation 20.15* and the points are the experimental values collected on alfalfa and corn. In this diagram two points obtained by Mo et al., (1982), at 5 GHz and 1.5 GHz for corn are also shown. Although the extension of this equation to lower frequencies can be arbitrary, the measured points are in reasonable agreement with the model. In this model the radiation component due to vegetation is assumed to be unpolarized, whereas the radiation emitted by the whole canopy-soil system is polarized. The polarization of the radiation can be characterized by the polarization index (**PI**). Using the same model and relationship between the optical depth and the plant water content, *Equation 20.15*, and assuming that vegetation is in thermal equilibrium with soil, the **PI** can be defined by the following equation:

$$\mathrm{PI}(\mathbf{Q},\mu) = \mathrm{PI}(\mathbf{O},\mu)(1+\mathbf{Q})^{-\mathbf{k}/\mu\sqrt{\lambda}} \qquad (20.16)$$

where $\mathrm{PI}(\mathbf{O},\mu)$ is the polarization index of bare soil.

We observe that, whereas the plant water content $\mathbf{Q}$ is a good indicator of the plant physical state, the leaf area index is a parameter which better characterizes plants for agricultural applications, since it is representative of photosynthetic capacity. Considering data collected on several fields, we found that the relation between **LAI** and $\mathbf{Q}$ can be expressed by a logarithmic law (Paloscia and Pampaloni, 1986). The relationship

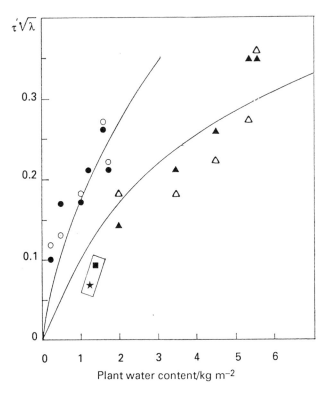

FIGURE 20.13. The quantity $\tau'\sqrt{\lambda}$ as a function of plant water content Q (corn: o = Ka, • = X; alfalfa: △ = Ka, ▲ = X). The computed values of $\tau\sqrt{\lambda}$ are approximated by the equation $\tau'\sqrt{\lambda} = k \ln(1 + Q)$ where $k = 0.16$ (m$^{0.5}$) for corn, and $k = 0.25$ (m$^{0.5}$) for alfalfa. (After Pampaloni and Paloscia, 1986). ■ and ∗ indicate data collected at L and C bands, respectively, by Mo et al. (1982)

between **PI** and **LAI** can be written as follows:

$$\mathbf{LAI} = ((v\mu\sqrt{\lambda}/k)ln(\mathbf{PI}(0,\mu)/\mathbf{PI_{(LAI)}})) \tag{20.17}$$

where **v** is the coefficient which best fits data of **LAI** and **Q** for alfalfa and corn. Therefore, **LAI** can be estimated from the crop polarization index once vegetation type and the polarization index on bare soil (**PI(O,μ)**) are known. The relationship between **PI** at 10 GHz ($\theta = 50°$) and **LAI** is shown in *Figure 20.14*, where the model is represented by the continuous lines and compared with the experimental values (points) collected on corn and alfalfa.

The measured **LAI** has then been compared with the computed LAI by means of *Equation 20.17* at 10 GHz, to test the ability of the model to predict **LAI**. The result is shown in *Figure 20.15*, where we can see that the model is quite able to predict the **LAI** of corn and alfalfa, although with some underestimation, at least during the first phase of the growth cycle when **LAI** is lower than 5 for corn and lower than 2 for alfalfa.

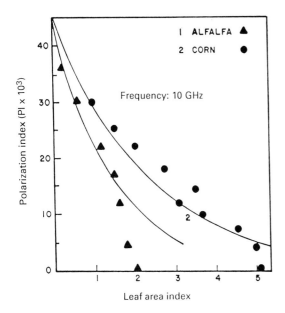

FIGURE 20.14. Comparison of calculated (solid lines) and observed (points) polarisation index as a function of leaf area index (**LAI**) (10GHz, $\theta = 50°$); ▲ = alfalfa, • = corn. (After Paloscia and Pampaloni, 1986)

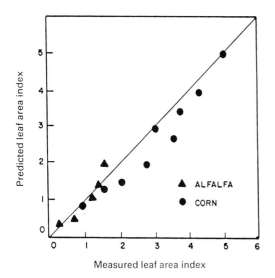

FIGURE 20.15. Predicted leaf area index (**LAI**) extracted from radiometric measurements at 10 GHz of polarisation index as a function of measured **LAI**: ▲ = alfalfa, • = corn. (After Paloscia and Pampaloni, 1986)

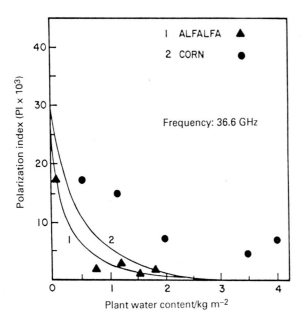

FIGURE 20.16. Comparison of calculated (solid line) and observed (points) polarisation index **PI** as a function of plant water content **Q** per unit area (kg m^{-2}) at 36 GHz, $\theta = 50°$ (▲ = alfalfa, • = corn.) (After Paloscia and Pampaloni, 1986)

At 36 GHz the dynamic range of **PI** is rather small, because the depolarization effect of soil roughness and the attenuation of soil emission induced by vegetation are higher than at 10 GHz. In *Figure 20.16* the polarization index at 36 GHz is represented as a function of the plant water content per square meter. In this case, the model predicts a sharp decrease of **PI** as vegetation grows; therefore, we can distinguish bare from vegetated soil, but we are unable to detect different growth stages.

If the model based on the matrix doubling algorithm is taken into consideration, emissivity (ϵ) can be computed after summation by the following equation (Ferrazzoli, Schiavon, Solimini, Luzi, Pampaloni and Paloscia, 1988b):

$$\epsilon(\theta, \Phi) = 1 - \frac{1}{4\pi \cos\theta} \iint \sigma°(\theta, \Phi; \theta_s, \Phi_s) \sin\theta_s d\theta_s d\Phi_s, \qquad (20.18)$$

where θ and Φ represent the incidence direction, θ_s, Φ_s the scattering direction and $\sigma°$ is the bistatic scattering coefficient.

Computations carried out by Ferrazzoli *et al.* (1988b), using this approach are shown in *Figure 20.17* where the emissivity computed by means of *Equation 20.18*, is represented as a function of leaf area index of alfalfa and compared with experimental data of T_n collected at 10 GHz, H polarization, $\theta = 40°$. The variation of T_n as vegetation grows shows the same trend as the model considered; the range of measured T_n is smaller than the computed one. However we observe that, at low values of **LAI**, the emission from the field surface is strongly affected by soil parameters and by the presence of short alfalfa

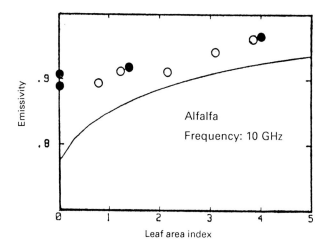

FIGURE 20.17. Comparison between emissivity obtained by the Matrix Doubling Algorithm (solid line) and the experimental data of *Figure 20.7* (After Ferrazzoli *et al.*, 1988b)

stems, whereas the model considers only soil and leaves. In particular, soil roughness was found to influence the theoretical curve heavily.

Conclusions

Theoretical and experimental investigations have shown some ability of microwave radiometry in detecting soil and plant parameters. In particular, we observed that multifrequency multipolarization observations are able to distinguish different crops, detect the water status of soil and the biomass of plants. The more the observing wavelength decreases, the more the emission comes from the upper layer of vegetation and changes according to its water content. Monitoring the vegetation water content can be very important because the relationship between the real and potential evapotranspiration does not depend entirely on soil moisture content but also on plant and weather conditions. A comparison of experimental results with models based on the radiative transfer theory showed that, at least for some crops, these models are able to explain fairly well several relations between radiometric and agricultural parameters.

References

BONSIGNORI, R., CHIARANTINI, L., PALOSCIA, S., PAMPALONI, P., FERRAZZOLI, P., MONGIARDO, M. and SOLIMINI, D. (1987a). Aircraft microwave radiometry of land. In *Proceedings IGARSS 1987* Ann Arbor, Michigan. **Vol. II**, THA-7, IEEE, Piscataway, (NJ), USA

BONSIGNORI, R., CHIARANTINI, L., PALOSCIA, S. and PAMPALONI, P. (1987b), The use of microwave radiometry for remote sensing of soil and vegetation. In *Proceedings*

International Microwave Symposium, Rio de Janeiro. **Vol. II**, 795-800. IEEE, Piscataway, (NJ), USA

BURKE, W.J., SCHMUGGE, T.J. and PARIS, J.F. (1979). Comparison of 2.8 and 21 cm microwave radiometer observations over soil with emission model calculations. *Journal Geophysical Research*, **84**, 287-294

CHIARANTINI, L., PALOSCIA, S. and PAMPALONI, P. (1987). The use of microwave radiometry in watershed hydrology, In *Proceedings ERIM*. **Vol.II**, 875. Environmental Research Institute of Michigan, Ann Arbor, Michigan, USA

CHOUDHURY, B.J., SCHMUGGE, T.J., NEWTON, R.W. and CHANG, A. (1979). Effects of surface roughness of the microwave emission from soil. *Journal Geophysical Research*, **81**, 5699-5706

DEBYE, P. (1929). *Polar molecules*. Chemical Catalog Co., New York

De LOOR, G.P. (1983), The dielectric properties of wet materials. *IEEE Transactions Geoscience and Remote Sensing*, **GE-21**, 364

EOM, H.J. and FUNG, A.K. (1984), A scatter model for vegetation up to Ku band. *Remote Sensing of Environment*, **15**, 185-200

FERRAZZOLI, P., PAMPALONI, P., PALOSCIA, S. and SOLIMINI, D. (1988a) Modelling microwave emission from vegetation. In *Proceedings Symposium on Microwave Radiometry and Remote Sensing Applications*. Ed. by P. Pampaloni. International Science Publisher VSP, The Netherlands

FERRAZZOLI, P., SCHIAVON, G., SOLIMINI, D., LUZI, G., PAMPALONI, P. and PALOSCIA, S. (1988b). Comparison between microwave emissivity and backscattering coefficient of agricultural fields. In *Procedings IGARSS '88 Symposium*. Edinburgh, Scotland, ESA SP-284, 667 - . ESA, ESTEC, Noordwijk, Netherlands

JOSEPH, H.J., WISCOME, W.J. and WEINMAN, J.A. (1976) The Delta-Eddington approximation for radiative flux transfer. *Journal Atmospheric Science*, **33**, 2452-2459

KIRDIASHEV, K.P., CHUKHLANTEV, A.A. and SHUTKO, A.M. (1979). Microwave radiation of the Earth's surface in the presence of vegetation cover. *Radio Engineering Electronics Physics (English Translation)*, **24**, 256-264

MO, T., CHOUDHURY, B.J., SCHMUGGE, T.J., WANG, J.R. and JACKSON, T.J. (1982). A model for microwave emission from vegetation covered fields. *Journal Geophysical Research*, **87**, 287-294

MO, T., SCHMUGGE, T.J. and WANG, J.R. (1987). Calculations of the microwave brightness temperature of rough soil surface: Bare soil. *IEEE Transactions on Geoscience and Remote Sensing*, **GE-25**, 47-54

PALOSCIA, S. and PAMPALONI, P. (1988). Microwave polarization index for monitoring vegetation growth. *IEEE Transactions on Geoscience and Remote Sensing*, **GE-26**, 617-621

PAMPALONI, P. (1981). *Microwave Radiometry for Remote Sensing in Agriculture*. Proc. International Conference Remote Sensing Society, London 1981

PAMPALONI, P. and PALOSCIA, S. (1985a). Microwave emission from vegetation: general aspects and experimental results. In *Proceedings of EARSeL Workshop*. ESA SP-227, 53-60. ESA. Amsterdam, Netherlands

PAMPALONI, P. and PALOSCIA, S. (1985b). Experimental relationships between microwave emission and vegetation features. *International Journal of Remote Sensing*, **6**, 315-323

PAMPALONI, P. and PALOSCIA, S. (1986). Microwave emmission and plant water content: a comparison between field measurements and theory. *IEEE Transactions on Geoscience and Remote Sensing*, **24**, 900-904

PAMPALONI, P., PALOSCIA, S. and CHIARANTINI, L. (1986). Contribution of passive microwave remote sensing in soil moisture and evapotranspiration measurements. *Proceedings International Conference on Parametrization of land surface characteristics*. ESA SP-248, 327-331 ESA, ESTEC, Noordwijk, Netherlands

SCHMUGGE, T.J. and CHOUDHURY, B.J. (1981). A comparison of radiative transfer models for predicting the microwave emission from soils. *Radio Science*, **16**, 927-938

SHUTKO, A.M. (1982). Microwave radiometry of land under natural and artificial moistening. *IEEE Transactions on Geoscience and Remote Sensing*, **GE-20**, 18-26

ULABY, F.T., RATZANI, M. and DOBSON, M.C. (1983). Effects of vegetation cover on the microwave radiometer sensitivity to soil moisture. *IEEE Transactions on Geoscience and Remote Sensing*, **GE-21**, 51-61

WANG, J.R. and SCHMUGGE, T.J. (1980). An empirical method for the complex dielectric permittivity of soils as a function of water content. *IEEE Transactions on Geoscience and Remote Sensing*, **GE-18**, 288-295

WANG, J.R., JACKSON, T.J., ENGMAN, E., GOULD, W., FUCHS, J., GLEZER, W., O'NEIL, P., SCHMUGGE, T.J. and McMURTEY, J. (1984). Microwave radiometer experiments of soil moisture during summer 1981. NASA Tech. 86056. NASA, Greenbelt, (MD), USA

WILHEIT, T.T. (1978). Radiative transfer in a plane stratified dielectric. *IEEE Transactions on Geoscience and Electronics*, **GE-16**, 138-143

21

ON THE USES OF COMBINED OPTICAL AND ACTIVE-MICROWAVE IMAGE DATA FOR AGRICULTURAL APPLICATIONS

J.F. PARIS.

Department of Geography, California State University, Fresno, CA., 93740-0069, U.S.A.

Introduction

Quantitative measurement of key vegetation properties, the identification of vegetation types, and the mapping of their areal extents through the aid of remotely-sensed data from Earth-orbiting spacecraft have been the grand goals of Remote Sensing for several decades. Many investigators have pursued these goals by studying only the information content of common broad-band optical sensors such as the Landsat Thematic Mapper (TM) and the French SPOT High Resolution Visible (HRV) imaging systems. Papers abound in the published literature on this topic (see Colwell, 1983, for a summary). Other investigators (see Ulaby, Moore, and Fung, 1986, for a summary) have pursued these grand goals by studying the information content of the signals from certain active-microwave sensors, mostly those that operate at wavelengths near 24 cm (the L-band). Few researchers have treated the combination of these sensors in a multi-sensor approach.

My purpose here is to attempt to provide a general framework of knowledge and understanding for approaching the task of agricultural-information extraction. This framework is mostly based on recent multi-sensor experiments involving broad-band optical and active-microwave sensors in agricultural settings. Many of the conclusions are based on my personal research in large team efforts; however, I have attempted to integrate the results of other completely-independent investigations into the overall framework. Through this process of synthesis, I hope to provide a set of basic principles which investigators at large can use when using combined optical and active-microwave image data for agricultural applications. In this paper, I do not address the use of low spatial-resolution imagers, such as the National Oceanographic and Atmosphere Agency (NOAA) satellite's Advanced Very-High Resolution Radiometer (AVHRR). The 1.1 km spatial resolution of the AVHRR greatly limits the application of these data to most agricultural problems. Also, I do not address the potential uses of future high-spatial and high-spectral resolution sensors, such as those on the several polar platforms of the Earth Observing System (EOS). Nor, do I treat the notion of using multi-angle optical data for vegetation studies: proposed EOS sensors may include this approach. Rather, I confine my attention to existing broad-band optical imaging systems and to the class of active-microwave sensors that several nations will place in Earth orbit in the 1990's.

Lessons from broad-band optical studies

Landsat MSS

During the 1970's and early 1980's, the National Aeronautics and Space Administration (NASA) sponsored massive investigations of the uses of broad-band optical data for agricultural applications. After an initial set of small and exploratory studies of the then new Landsat Multispectral Scanner System (MSS), NASA began the Large-Area Crop Inventory Experiment (LACIE) that existed from 1975 to 1980. In LACIE, MSS data over several countries were used to identify wheat-producing fields in a number of cropland regions, to measure their areal extent, and to predict their yield through the combination of the remotely-sensed crop area and the yield per unit area. The latter was estimated by using commonly available surface weather data (e.g. temperature and precipitation) in yield models.

Perpendicular measures of vegetation amount based on MSS

While many contributions were made during this era to the task of extracting quantitative information about vegetation condition from MSS data, especially the condition of agricultural crops, the most notable, in my opinion, was that of Kauth and Thomas (1976). With their Tasseled Cap (TC) transformation, they defined a set of coefficients that could be applied to raw MSS data to produce measures of:

1. soil brightness;

2. greenness (or green vegetativeness) and

3. yellowness (or a measure of senescence).

The TC transformation is a member of several "perpendicular" transformations which produce measures of (green) vegetation amount by rotating the coordinate axes in n-dimensional feature space. Other members of this class include the Perpendicular Vegetation Index (PVI) and Transformed Vegetation Index (TVI).

The basic idea behind these is as follows: In the four-dimensional feature space of the MSS (on a specific date), various soil spectra appear to exist close to a line in that space that extends from the origin (locus of zero reflectance factors) of the 4-space toward domains in that 4-space which represent sets of ever increasing bare-soil brightnesses (similar to albedo). The fact that nature seems to have provided a convenient reference line (Line of Soils) in that space for all bare soils (especially agricultural soils) leads to a convenient way of measuring the amount of vegetation, the Kauth-Thomas Greenness, which is the perpendicular distance (in 4-space) from the Line of Soils to regions in 4-space occupied by green vegetated areas. Due to the effects of absorption by chlorophyll in vegetation of visible-wavelength photons, in the wavelength bands 500 - 600 nm and 600 - 700 nm, corresponding to MSS bands 4 and 5, both of these band brightnesses decrease as areal green cover (AGC, %) increases and as green leaf-area index (GLAI, $m^2\ m^{-2}$) increases. When AGC reaches 100% and as GLAI continues to increase, usually the near-infrared brightnesses in the bands MSS6 (700-800 nm) and MSS7 (800-1100 nm)

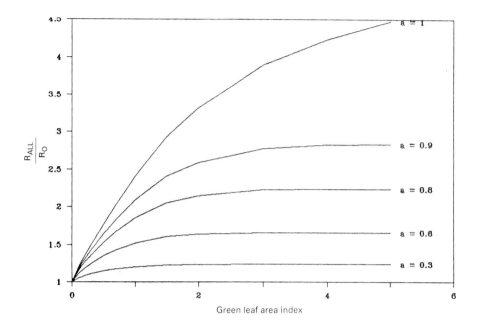

FIGURE 21.1. Variation of Relative Reflectance factor R_{ALL}/R_0 with Green Leaf Area Index, where the reflectance factor due to single scattering only is R_0, and that due to all orders of multiple scattering is R_{ALL}. Based on analysis by author of data presented by van de Hulst (1980). ρ is the single scattering albedo

continue to increase to values as high as 60% of that of a perfect Lambertian reference. An important exception to this rule is the case of woody vegetation such as orchards and forests. Trees having the same GLAI as herbaceous vegetation will have markedly lower values of reflectance in the near infrared. I propose that this is due to the effect of stems being present and also due to the tight clustering of leaves in trees. Both inhibit multiple scattering, the primary cause of elevated near-infrared reflectances of crops (see below).

The high brightness in MSS6 and MSS7 is caused by the fact that absorption is weak (about 10%) at these wavelengths in green leaves and that the corresponding leaf reflectance (backward scattering) and leaf transmittance (forward scattering) are large. Low absorption in a volume scattering medium allows brightness to rise due to multiple scattering interactions (van de Hulst, 1980). Based on an analysis of the extensive data given by van de Hulst, I developed a simple multiscattering model.

One result is shown in *Figure 21.1*. In the figure, the Reflectance Factor R_{ALL} (%) produced by all orders of scattering is compared to that produced by single-interaction scattering only, R_o (%). The single scattering albedo, ρ, is a measure of the probability that a given photon will survive an interaction with a leaf without being absorbed (i.e., $\rho = 1$ is the case of no-absorption, the so-called conservative case; $\rho = 0$ is the case of no-scattering). In this example model, the optical depth of the scattering medium has been scaled directly to the GLAI. In bands MSS6 and MSS7, ρ is approximately 0.9. Thus, multiple scattering can elevate the reflectance factor to values about 2.8 times greater than that of a single leaf. In the chlorophyll-absorption bands (MSS4 and MSS5), ρ is closer to 0.1. *Figure 21.1* would predict that multiple scattering does not elevate the reflectance factors in these bands above that of a single leaf. Thus, the observed reflectance factor of a field will decrease with increasing AGC, since the intrinsic reflectance factor of most soil in the 500-700 nm wavelength range is greater than that of green leaves. These simple and differing responses in the visible bands and in the near infrared bands produce the familiar result that as AGC and GLAI increase, MSS4 and, especially, MSS5 brightnesses decrease while MSS6 and MSS7 brightness remain the same or increase.

Some investigators (e.g. Huete, 1986) have proposed techniques to estimate the proportions of soil, shadow, and vegetation in regions of low plant densities (e.g. arid lands). The curves in *Figure 21.1* suggest that such simple, linear mixing model approaches will encounter difficulties when AGC approaches 100% and GLAI is high.

Simple ratio and normalized difference measures of vegetation amount based on MSS

The aforementioned increase with canopy cover in near infrared reflectances (**IR**), accompanied by decreasing visible-region reflectances **R** led many (e.g. Tucker, 1979) to the definition of a Simple-Ratio Vegetation Index, **SRVI**,

$$\text{SRVI} = \text{IR}/\text{R}. \quad (21.1)$$

Due to the path radiance of the solar illuminated (and irradiated) atmosphere above a scene, the **SRVI** runs into problems when applied to raw Landsat MSS or similar data, e.g., from Landsat Thematic Mapper (TM) and SPOT sensors. Another ratio-like measure, the Normalised Difference Vegetation Index, **NDVI**, defined by Tucker (1979), is:

$$\text{NDVI} = (\text{IR} - \text{R})/(\text{IR} + \text{R}). \quad (21.2)$$

This appears to remedy the effects of path radiance when applied to raw Landsat MSS data. However, **NDVI** and **SRVI** are related (albeit, in a non-linear fashion) and are therefore not independent measures.

One should note that while **SRVI** and **NDVI** as well as some of the perpendicular measures of vegetation amount (e.g. the PVI) operate only in the near infrared-red plane of the 4-space of MSS, they still capture most of the variance of the 4-space. In other words, the dimensionality of MSS data for soils and growing vegetation is only two (i.e. MSS vegetation data exist mostly in a hyperplane in 4-space). Near the end of the season, the changes in the visible spectral channels become uncorrelated; this leads to the so-called Yellowness measure of Kauth and Thomas (1976), a measure that ventures into MSS 3-space for vegetation rather than 4-space.

Measures of vegetation amount based on SPOT

Since the SPOT HRV imager uses bands similar to those of the MSS, one can use the same principle for developing measures of vegetation amount as for the MSS. The improved spatial resolution of SPOT lessens the problem of trying to estimate field areas due to mixtures of spectral signatures that occur at field boundaries. This is an important improvement for agricultural sensing since, in many cases, half of the total MSS pixels covering a given field can be contaminated by boundary objects (e.g. roads or objects in the adjacent fields).

Measures of vegetation amount and condition based on the Landsat TM

Coinciding almost with the launch of Landsat 4 (and continuing with Landsat 5), another large research project began – the Agricultural Resources and Inventory Surveys Through Aerospace Remote Sensing (AgRISTARS) Project from 1980 to 1985. On Landsat 4 (and on Landsat 5) is the Thematic Mapper (TM). The TM extended the spectral range beyond the near infrared into the middle-infrared bands. Also, an additional blue-light sensitive band was added (TM1: 450-520 nm). TM2 (520-600 nm) is similar to MSS4; TM3 (630-690 nm) is similar to MSS5; and TM4 (760-900) overlaps somewhat with MSS6 and, especially, MSS7, but avoids some of the less-transparent "window" region where water vapour effects are moderately strong. TM5 (1550-1750 nm) and TM7 (2090-2380) broke new spectral ground in the middle-infrared bands, and TM6 provided high-spatial resolution thermal infrared for the first time.

In a development similar to the Kauth-Thomas Tasseled Cap transformation, Crist and Cicone (1984) calculated a new set of transformation coefficients to apply to TM data to yield measures of vegetation condition. These included the familiar measures of Soil Brightness, Greenness, and Yellowness, but went further to Wetness. This extension was based on the fact that increases in leaf moisture content lead to increases in leaf absorbance (and thus to decreases in leaf reflectance) in the middle-infrared (TM5 and TM6).

The old ratio measures survived in the TM's 6-space with the ever present **SRVI** and **NDVI**. Jackson (1983) suggested several other indices involving ratios of various bands of TM including a moisture-stress index (**MSI**) calculated as the ratio of the middle-infra red reflectance, (IR_M) to IR:

$$MSI = IR_M/IR. \qquad (21.3)$$

While the idea of a moisture-stress index has been used in several studies, including studies of acid rain effects, the **MSI** can be a misnomer. The inverse of the **MSI** is a ratio that behaves much like the **SRVI**; only the red-light reflectance has been replaced by the middle-infrared reflectance. Since leaves contain both chlorophyll (red-light absorber) and water (middle-infrared absorber), these two spectral bands respond in a similar fashion to changes in green foliage amount. Thus, the inverse of the **MSI** is a simple-ratio vegetation index and may contain no independent information about changes in leaf moisture. In fact, it is difficult to find vegetation that retains chlorophyll while losing significant amounts of water from its leaves.

While TM may appear to be a 6-space sensor, the highly correlated responses of TM1, TM2, TM3, TM5, and TM7 to changes in green foliage reduce this to an effective 2-space case, similar to that of the MSS. The fact that path radiances are much lower in the TM5 and TM7 bands than in the visible bands would suggest that the reflectance in bands 5 or 7 would serve well as the denominator of *Equation 21.1*) or in place of R in *Equation 21.2*). Senescence and changes in leaf moisture condition not accompanied by changes in leaf chlorophyll condition can move spectral responses into 3rd and 4th dimensions, as happens with MSS data.

Spectral vegetation identification: A hope unfulfilled

Early in remote sensing and persisting to this day, investigators tried to establish a characteristic pattern of reflectances (a spectral "signature") to associate with specific types (e.g. species or varieties) of vegetation. In some cases, the results were excellent; in others, very poor. In fact, this approach was not robust – it depended too much on luck and fortuitous circumstance. As a general class of objects, green vegetation itself can be identified as a spectrally unique object: it possesses rather unique visible-band absorbers (chlorophylls and other pigments) and a not-so-unique middle-infrared absorber (water). Nevertheless, the whole scheme of converting MSS, and especially TM data, to a single vegetation measure (index) rests on the fact that vegetation has a unique composition compared to objects such as soils and rocks. However, specific types of vegetation, e.g. corn and soybeans, can (and do) have similar optical broad-band spectral properties at a given location and given time in the season. In a particular region, one might count on a particular vegetation type having a particularly high value of a primary measure, such as alfalfa having a large GLAI with no heavy stalks to interfere with multiple scattering (thus, alfalfa has a very high near-infrared reflectance compared to many types of crops). Nevertheless, the occasionally successful spectral identification of a crop like alfalfa depends on a set of circumstances that may not hold in the next region or at a different stage of alfalfa growth.

What are the alternatives to the identification of crops based on single-date spectral measurements (no matter how well calibrated they may be)? The answer seems to be the element of time (seasonal change). In the AgRISTARS Project, researchers found that one could relate the seasonal patterns of emergence of a green crop, its growth, and its senescence or harvest, as seen by Landsat MSS or TM (or any other sensor that responds to changes in standing biomass, especially foliage biomass), to crop type. An excellent example was the identification of corn and soybeans. Using the Greenness measure from the Kauth-Thomas or Crist and Cicone transformations (MSS and TM, respectively), it is possible to fit a simple two-parameter model to the season variations of changes in Greenness in a particular field. Then, using a model incorporating a non-linear distribution in Greenness having characteristics of emergence date, peak Greenness, and length of season, corn could be distinguished from soybeans in every crop growing region without changing the decision rule or its parameters! This is a robust method – one that works everywhere.

Another approach, that I used successfully, was to consider information in another part of the electromagnetic spectrum – the microwave region.

Lessons from active-microwave studies

Space-based L-band synthetic aperture radar (L-SAR)

Starting in 1978 with the Seasat Synthetic Aperture Radar (SAR), NASA began a programme of exploring the potential uses of a class of active-microwave imagers that promised to add new information (or provide similar information under cloudy sky conditions or in the dark) about vegetation (and other objects) relative to information present in optical data. In 1981, NASA experimented with another SAR on a Space Shuttle flight [the Shuttle Imaging Radar (SIR-A)]. Based on these two flights, investigators noted that forests (trees and other dense woody plants) were consistently brighter than pastures and general cropland (herbaceous plants). Waite, MacDonald, Kaupp and Demarche (1981) noted that vegetated wetlands were much brighter in the Seasat SAR images than surrounding vegetated lands (vegetated drylands). Stone and Woodwell (1988) noted the obvious usefulness of SIR-A data in detecting and mapping disturbed tropical forests. These results were for L-band (24-cm wavelength) SARs that used the horizontal-transmit, horizontal-receive mode (HH). The relatively long wavelength of the Seasat SAR and of the SIR-A enabled both:

- good penetration of moderately dense vegetation (to interact with the surface, e.g. a flooded surface under trees) and

- size discrimination (due to efficient backscattering by stems having diameters comparable to the wavelength as opposed to inefficient backscattering by leaves).

Interestingly, in Stone and Woodwell's images, felled trees were brighter than standing trees. An explanation for this, I believe, is that in standing forest, in addition to the effects of the drastic change in stem orientation, the felled trees lacked a cover of foliage, that appeared to absorb microwaves otherwise being scattered by the supporting stems.

In 1984, NASA completed the SIR-B mission, another L-band SAR having HH polarization. With the SIR-B mission, however, two differences existed compared to previous missions. First, the SIR-B Project funded 12 investigators (for the first time) out of the 43, specifically for vegetation application exploration. Second, the angle of irradiation (incidence) could be varied between 15 and 60 degrees on successive passes (days); this allowed one to explore and quantify the effects of incidence angle on the performance of the L-band HH SAR. Selection of the angle of incidence allowed one to choose good penetration with strong potential interactions with the substrate (small angles) or to reduce surface contributions in favour of overlying vegetation (large angles).

The Fresno County experiment: A multi-sensor experiment over an irrigated cropland

To explore the combined uses of TM and SIR-B (L-band HH SAR) data (i.e. broadband optical and single-band active-microwave), I chose a test site in Fresno County, California, for my SIR-B study (Paris and Kwong, 1988). We later used the same site for several airborne imaging radar investigations (AIR-1 and AIR-2). AIR-1 was a quad-polarization L-band SAR that crashed in 1985! The replacement to AIR-1, the AIR-2, is a multi-frequency, quad-polarization (full amplitude and phase) SAR that operates at C-band (6-cm), L-band (24-cm), and UHF or P-band (67-cm). Thus, we have a great deal of experience in the appearance of this site in these various combinations of SAR.

The site contains many herbaceous and woody plants. The primary herbaceous plants present during the 1984 SIR-C overpass on October 10 were alfalfa and cotton. The primary 'woody' plants were corn (maize), almond orchards, walnut orchards, peach orchards, plum orchards, and vineyards. I include corn in this class due to the large size of the corn stalk. On October 5, 1984, the Landsat TM acquired a data set under clear skies. Thus, we have a multi-sensor data set (with a caution concerning the five-day interval of time between). The October 10th SIR-B pass was for an incidence angle of 22 degrees. An October 9th SIR-B pass at 40-degree incidence existed also; however, the signal-to-noise was poor (0 dB!), so we did not use these data.

During the Fresno experiment, my assistants and I collected information on the ground from some 160 fields on crop type and quality, crop row direction, irrigation pattern, general soil wetness, and general crop appearance. For the orchards, we measured main-stem (trunk) diameters, row and tree spacing, and tree height. For particular crops we also obtained data on crop condition, e.g. alfalfa height and quality and cotton defoliation.

To prepare the data for analysis, we extracted the raw numbers, N, for each field and sensor band and computed the means and standard deviations. We identified field boundaries in the TM data and extracted both TM and SIR-B data from the registered and combined multisensor data set. Using a spreadsheet program, we combined the field-averaged data from the SIR-B (L-band, HH, 22 degrees, October 10, 1984), from the TM data (6-bands, October 5, 1984), and from the ground survey. I used the vicarious method to convert the mean intensity values to mean reflectance factors. This consisted of finding fields (areas) of known reflectance spectra (bare fields and water bodies). With the known reflectance factors and the measured values of N, one can easily define coefficients to convert N to **R**.

To put the SIR-B data on the same definitional footing as the TM, I converted the SIR-B data (expressed as backscattering coefficient, ρ°, m² m⁻²), to an equivalent reflectance factor, **R**, by:

$$\mathbf{R} = \rho^{\circ}/(4\pi \cos^2 \theta), \tag{21.4}$$

where θ is the angle of incidence.

Now the data were ready for analysis. I examined the data from the fields by presenting a scatter plot of the reflectance factors in TM4 (i.e. $\mathbf{R}_4$) versus $\mathbf{R}_3$ (see *Figure 21.2* and the list of labels in *Table 21.1*).

Note the expected distribution of the herbaceous crops (and bare fields) in this figure. Spectral variations range from values near the Line of Soils for bare and short herbaceous vegetation to fully developed herbaceous vegetation having very high near-infrared reflectances and very low red reflectances. An unexpected result was the spectral location of trees and orchards (A, K, L, N, and P) on this diagram. Many orchards are near the 'Line of Soils'; thus, by any measure of greenness one might reason that these had little foliage—in fact, they were fully foliated!

To analyse the data further, we developed the Progressive Transformation Method (PT) (Paris and Kwong, 1988). In the spirit of the Tasseled Cap Transformations for MSS and TM, the PT Method allows one to develop a translational (not in the Tasseled Cap) and rotational transform with rescaling to produce one or more measures of chosen biophysical condition up to the number of bands involved.

The calculation of the coefficients is based on selected spectra (see *Table 21.1*) designed to emphasize certain biophysical differences and to generate measures that relate to the differences while maintaining orthogonality in the transformation. Furthermore, the PT Method allows one to calculate the transformation coefficients one measure at a time without having the complete set. With the PT Method, one can tailor a transformation to a very specific need without being controlled by the statistics of the whole selected image, as is the case with popular transformation methods such as principal components.

Using the distribution of alfalfa fields in the 7-space (the 6-space of TM plus the 7th feature-space dimension of the SIR-B data "reflectances"), I defined the first two scalar

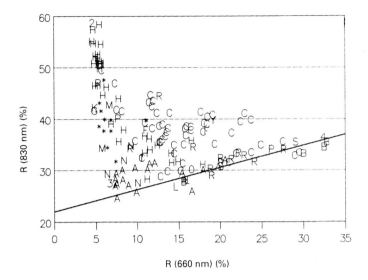

FIGURE 21.2. Distribution of field means of reflectance factors in the near infrared (R(830nm)) and red (R(660nm)) bands of the Thematic Mapper (TM) for an agricultural test site in Fresno County, California, on October 5, 1984 (Paris and Kwong, 1988). See *Table 21.1* for key

measures with the PT Method to provide measures of areal green cover (AGC, %) and green leaf area index (GLAI, $m^2\ m^{-2}$).

These measures are called $\mathbf{F_1}$ and $\mathbf{F_2}$ (see *Figure 21.3*) and were calculated from calibrated TM data, expressed as **R** values, by means of a set of transformation coefficients listed in *Table 21.2*.

$\mathbf{F_1}$ and $\mathbf{F_2}$ were rescaled (linear reassignment) to provide measures of AGC and GLAI, respectively (see *Figure 21.4*).

$\mathbf{F_1}$ depends mostly on the absorptive bands of TM (TM1, TM2, TM3, TM5, and TM7); $\mathbf{F_2}$ depends almost entirely on the scattering band, TM4 (near-infrared). The microwave (L-band HH 22 degree) brightness did not enter significantly into the determination of $\mathbf{F_1}$ and $\mathbf{F_2}$. Note that most orchards have negative $\mathbf{F_2}$ (GLAI) measures! This is due to the fact that the enhanced reflectance of the near-infrared band TM4, which dominates $\mathbf{F_2}$, decreases in orchards due to the existence of stems and shadows. Thus, no useful estimate of GLAI could be made for orchards from the TM data alone, since SIR-B data did not affect $\mathbf{F_1}$ and $\mathbf{F_2}$. The estimates of AGC for orchards are good (from $\mathbf{F_1}$).

To take the PT analysis one step further, I used a mature almond orchard as the basis for a measure of woody vegetation amount, $\mathbf{F_3}$. This measure depends mostly on the SIR-B SAR band (see *Table 21.2*).

When the distribution of the fields is viewed in the $\mathbf{F_1}$ - $\mathbf{F_3}$ plane (*Figure 21.5*), one sees a clear pattern: the mature orchards have uniquely large values of $\mathbf{F_3}$. This is apparent in *Figure 21.6*, which shows the orchard data in *Figure 21.5* on the same scale. Using a simple allometric equation, I used the basal area of the orchard trees (total trunk cross-sectional areas per unit ground area) to estimate the standing woody

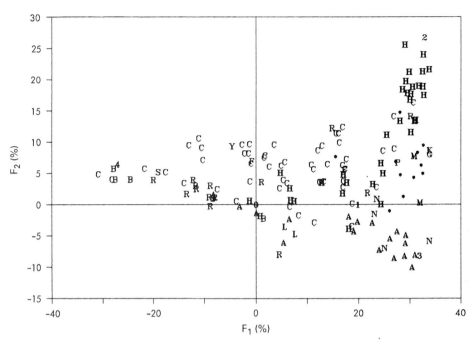

FIGURE 21.3. The same data as in *Figure 21.2* after applying the Progressive Transformation (PT) Method, (Paris and Kwong, 1988) F_1 was designed to respond to changes in areal green cover (AGC); F_2 to changes in green leaf area index (GLAI). See *Table 21.1* for key

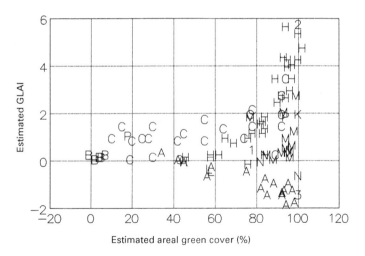

FIGURE 21.4. Subset of data in *Figure 21.3* after F_1 and F_2 have been rescaled to provide estimates of AGC and GLAI. See *Table 21.1* for key

TABLE 21.1. Codes used in *Figures 21.2* to *21.9* to indicate ground cover type.

Code	Meaning
0	Base reference spectrum (cut alfalfa) for Progressive Transformation Method
1	1st reference spectrum (mid-cycle alfalfa) for PT Method
2	2nd reference spectrum (end-of-cycle alfalfa) for PT Method
3	3rd reference spectrum (mature almond orchard) for PT Method
4	4th reference spectrum (bare field) for PT Method
A	Almond orchards
B	Bare fields
C	Cotton flields
F	Fallow fields
G	(Small) grain fields (mixed wheat, barley, and oats for dairy feed)
H	(Hay) alfalfa fields
K	Oak groves
L	Plum orchards
M	Corn (maize) fields (present on both October 5 and October 10, 1984)
N	Walnut orchards
P	Peach orchards
R	Pastures
S	Stubble fields (Spring small grains)
Y	Blackeye fields
*	Changed field: Corn on October 5, 1984, and bare on Ocober 10, 1984

biomass per unit area (SWB, kg m²). A plot of F_3 versus SWB (*Figure 21.7*) shows that an approximately linear relationship existed between the two. Thus, I could use the rescaled features from the PT Method, F_1, F_2 and F_3, to estimate three biophysical characteristics, AGC, GLAI, and SWB, respectively (see *Figures 21.8* and *21.9* and *Figure 21.4*).

In conclusion, I found in the Fresno County SIR-B/TM Experiment that having the L-band HH 22-degree data in addition to the TM data allowed me to quantify three important biophysical variables: AGC, GLAI and SWB. Without the TM data, I could only estimate SWB. With the SIR-B data, I could only estimate AGC and GLAI for herbaceous crops and only AGC for woody crops. In fact, I could not separate spectrally those fields with woody crops from those carrying herbaceous crops; this separation

TABLE 21.2. Transformation coefficients from progresive transformation analysis of combined Thematic Mapper (TM) bands and Shuttle Imaging Radar (SIR-B) data for Fresno County, California, experiment, October 1984.

Parameter	TM1	TM2	TM3	TM4	TM5	TM7	SIR-B
Initial Offset	11.3%	14.4%	16.6%	30.4%	33.7%	29.0%	1.82%
F1	-0.19	-0.20	-0.30	+0.12	-0.53	-0.73	+0.08
F2	-0.03	-0.05	-0.08	+0.97	+0.14	+0.10	-0.11
F3	-0.03	-0.09	-0.05	+0.08	-0.02	+0.19	+0.97

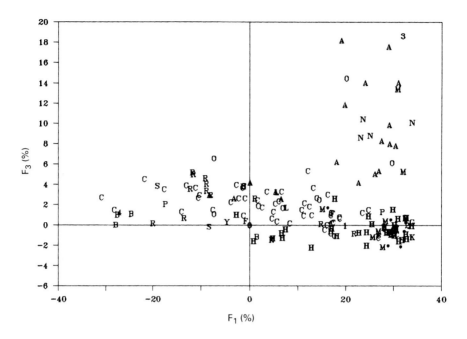

FIGURE 21.5. Distribution of indices F_3 versus F_1 for fields in an agricultural area of Fresno County, California, on October 5 and 10, 1984. F_3 was designed to respond to changes in standing woody biomass (SWB); it depends mostly on the brightness of the Shuttle Imaging Radar (SIR-B) L-band (24cm), HH, 22 degree incidence image data. See *Table 21.1* for key

required the L-band SAR data. *Figure 21.10*, for example, illustrates a specific case where the TM spectra of a short alfalfa field and a mature almond orchard were almost identical, but where the active-microwave brightnesses were quite different.

Field studies of vegetation with C-band sensors

At an agricultural research field-laboratory in Fresno County, California, I have been using a truck-based C-band (6.3 cm) multi-polarization field scatterometer [Microwave Scatterometer, C-band (MS-C)] to explore the scattering properties of plum and walnut orchards. Most of these measurements have been made at an incidence angle of 55 degrees due to the limitations of truck-bucket height and the requirement for avoidance of near-range effects. In addition, I have made measurements in arid lands (shrubs) and on a pigmy forest and artificial model forest (Westman and Paris, 1987).

Before these studies, little was known about the C-band backscattering properties of trees, especially of deciduous trees that have leaf development and leaf fall changes in condition over a season. One interesting result is shown in *Figure 21.11*.

My research assistants and I made a series of MS-C measurements on three orchards, two plum orchards having two levels of saline water application (in the spring of 1986) and a walnut orchard (spring of 1987). The seasonal changes in backscatter for VV polarization are shown. Note that in each of the three cases, backscatter de-

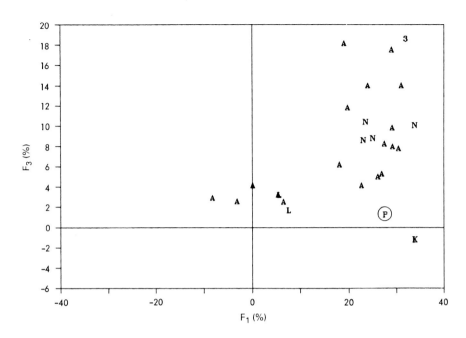

FIGURE 21.6. Subset of data in *Figure 21.5* for orchards only. The oak field (K) is a very sparse field; the circled plum field is also very sparse with a cover of herbaceous vegetation. See *Table 21.1* for key

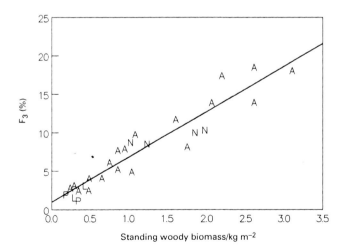

FIGURE 21.7. Relationship between F_3 and Standing Woody Biomass (SWB) for selected orchards in the Fresno County test site. Estimates of SWB were based on allometric equations involving trunk diameters and tree spacings. See *Table 21.1* for key

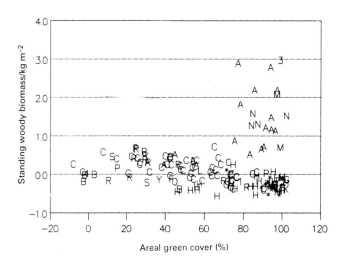

FIGURE 21.8. Plot of Standing Woody Biomass (SWB) versus Areal Green Cover as derived from combined set of Thematic Mapper (TM) and Shuttle Imaging Radar (SIR-B) acquired over Fresno County test site, California, on October 5 and 10, 1984, respectively. See *Table 21.1* for key

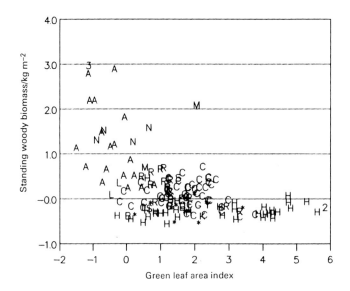

FIGURE 21.9. Plot of Standing Woody Biomass (SWB) versus Green Leaf Area Index (GLAI), as derived from combined set of Thematic Mapper (TM) and Shuttle Imaging Radar (SIR-B) data acquired over Fresno County, California, on October 5 and 10, 1984 respectively. See *Table 21.1* for key

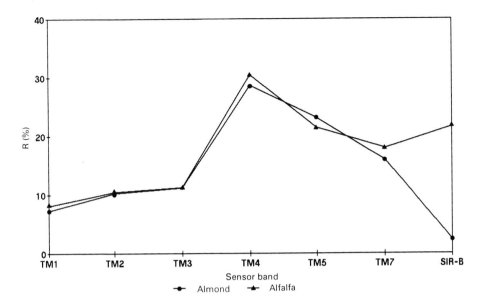

FIGURE 21.10. Examples of combined Thematic Mapper (TM) and Shuttle Imaging Radar (SIR-B) spectra of two quite different surfaces in the Fresno County test site. These were selected for their similarity in the TM bands and vast differences in the SIR-B band. They show quite clearly the need for both data sets for studies of the classification of vegetation

creased markedly when leaves emerged (after bud break). This decrease was followed by an increase in backscatter (recovery) when leaves expanded in walnuts (walnut leaves are comparable in size to the 6-cm wavelength of C-band) and when stems extended in plums (plum leaves are small compared to the C-band wavelength). The changes in the water-stressed plums (higher levels of irrigation water salinity) were smaller than for the healthy-plum case. Other evidence clearly shows that scattering at C-band is dominated by higher-order stems, and the evidence appears to show that small leaves (compared to wavelength) cause less intrinsic scattering (greater absorption) than do stems at C-band. This behavior is clearly different from that at L-band. Many studies at X-band have shown that foliage dominates the backscattering by plants (both woody and purely herbaceous) for the relatively short wavelengths involved (near 3-cm). Thus, the results of these C-band experiments and of other L-band and X-band experiments show that the size discrimination ability of multispectral SAR will be an important aspect of SARs on the EOS. Much work is needed to extend these results to natural vegetation and to other types of cultivated vegetation. Kasischke and Larson (1986) showed that having multi-wavelength SAR data over forests of different ages (and standing biomasses and sizes of plant parts) will enable vegetation identification and characterizations not possible with one wavelength alone, or even with one band of SAR combined with TM bands.

Multipolarization SAR can produce useful additional information. Consider *Figure 21.12* which shows a collection of measurements made with the MS-C in several different experiments (see *Table 21.3* for the key). Note that having both VV and HH

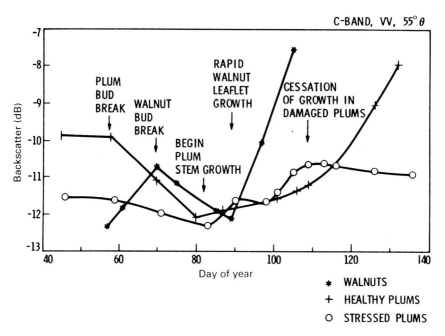

FIGURE 21.11. Time course of the seasonal changes in backscattering coefficients of three orchards in the springs of 1986 (plum) and 1987 (walnut), as measured by the Microwave Scatterometer, C-band (MS-C). Orchards were at the University of California, Kearney Agricultural Center, Fresno County, California

allows one to obtain a measure of stem orientation, with more vertical-stem trees having a higher VV backscattering (compared to HH) and vice versa. In fact, one can use a simple measure such as the Normalised-Difference Stem-Angle Index (**NDSAI**),

$$\textbf{NDSAI} = (\sigma_{vv}^o - \sigma_{hh}^o)/(\sigma_{vv}^o + \sigma_{hh}^o) \qquad (21.5)$$

for backscatter intensities σ measured at 55 degrees incidence to indicate the stem-angle structure (see lines of constant **NDSAI** in percent in *Figure 21.12*). Conversely, an additive or combined magnitude derived parameter using VV and HH data can indicate standing woody biomass per unit area (distance along radial lines from the origin). For example, one measure of SWB might be

$$\textbf{SWB} = ((\sigma_{vv}^o)^2 + (\sigma_{hh}^o)^2)^{\frac{1}{2}} \qquad (21.6)$$

This diagram also shows that having only HH or VV can lead to false indications of standing woody biomass due to the unknown effect of stem angle in some cases. Many stands of trees will fall along the line where **NDSAI** is zero; however, exceptions will occur (as for the plum trees). If the substrate is rough, the additive combination of VV and HH can falsely indicate that standing woody biomass is high (see the contrast between BS and BR in *Figure 21.12*). This confusion can be reduced if one also has cross polarization data (see *Figure 21.13*) where the effect of surface roughness and changes in standing woody biomass are decoupled.

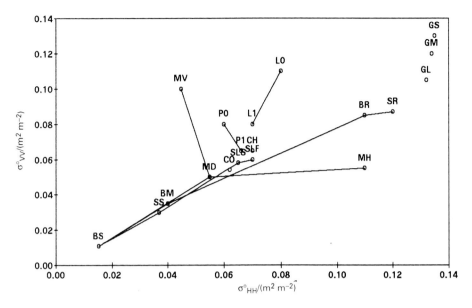

FIGURE 21.12. Comparison of backscattering coefficients of selected woody vegetation measurements in the C-band (6.3 cm). Measured by the C-band Microwave Scatterometer (MS-C) for vertical polarization, vertical reception (VV) and horizontal polarization, horizontal reception (HH). (See *Table 21.3* for key)

TABLE 21.3. Symbols used in *Figures 21.12* and *21.13*

Symbol	Meaning
BS	Bare, smooth surface (aridlands)
BM	Bare, medium-rough (aridlands)
BR	Bare, rough surface (aridlands)
CH	Chaparral (aridlands)
CO	Blackbush (aridlands)
GS	Smallest pigmy forest
GM	Mid-sized pigmy forest
GL	Largest pigmy forest
L0	Plum trees without leaves
L1	Plum trees with full leaves
MD	Model tree, dry stems
MH	Model tree, wet stems, 60-deg. from vertical
MV	Model tree, wet stems, 30-deg. from vertical
P0	Peach tree with no leaves
P1	Peach tree with full leaves
SS	Short sagebrush (aridlands)
SLS	Large sagebrush in spring with leaves
SLF	Large sagebrush in fall without leaves on smooth surface
SR	Large sagebrush in fall without leaves on rough surface

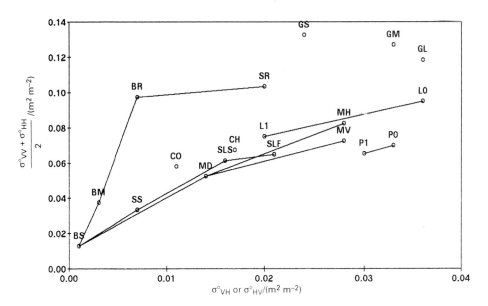

FIGURE 21.13. Measurements of the backscattering coefficients from woody vegetation, as in *Figure 21.12*. Like (average of VV and HH) polarization data and cross polarization data are shown. VH and HV indicate cross polarization signals, vertical polarization, horizontal reception, and *vice versa*. See *Table 21.3* for key

Diurnal C-band measurements of a walnut orchard

In August, 1987, a large team of scientists and I made multi-sensor measurements of a walnut orchard over several diurnal cycles (over two weeks). *Figure 21.14* shows the results at C-band.

Note the regular and quite large variations in backscattering coefficient for all channels, with a range of 3 dB. (VV, HH, and cross). These data are still under study by the team (EOS Synergism Study). However, they clearly show that time-of-day is an important variable in the active-microwave measurements. Whether or not the range of diurnal variations or timing of the pattern of change can be used to derive useful information remains to be seen.

Summary

Based on several studies with common broad-band optical sensors [e.g., the Landsat Thematic Mapper (TM)] and with multifrequency, multipolarization, active-microwave sensors, I conclude that the primary measures of vegetation condition that can be extracted on a specific date from the combination of these are as follows: areal green cover (AGC, %), green leaf-area index (GLAI, $m^2\ m^{-2}$), and standing woody biomass per unit area (SWB, kg m^{-2}). Other vegetation parameters (e.g. stem angular orientation) and substrate parameters (e.g. soil type and condition, surface soil moisture condition) affect the optical and active-microwave properties; however, these are usually param-

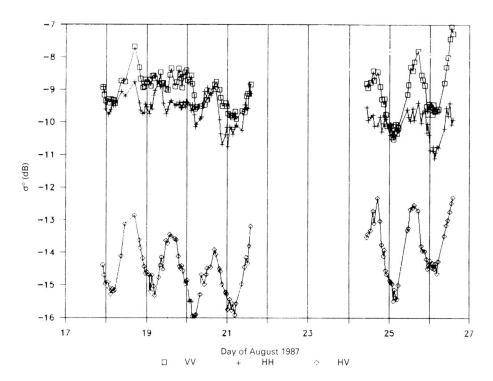

FIGURE 21.14. Diurnal variations in backscattering coefficients (VV, HH, and cross polarization combinations) for a walnut orchard in Fresno County, California, for two weeks in August, 1987. Measured by the Microwave Scatterometer, C-band (MS-C)

eters that one should suppress in extracting important vegetation characteristics from the remotely-sensed data. An important exception is the detection of the existence of very wet soil, snow or standing water beneath vegetation (i.e. the detection of vegetated wetlands). Multi-date sets of these primary measures of vegetation condition, acquired on appropriate dates, can be used to identify the type of vegetation, at least in broad categories, by interpreting the temporal patterns and spatial patterns in these data over a region.

References

COLWELL, R.N. (ed.), (1983). *Manual of Remote Sensing*, 2nd Ed., American Society of Photogrammetry. Washington D.C.

CRIST, E.P. and CICONE, C.R. (1984). A physically-based transformation of Thematic Mapper data - the TM Tasseled Cap. *IEEE Transactions on Geoscience and Remote Sensing*, **GE-22**, 256-267

HEUTE, A.R. (1986). Separation of soil-plant spectral mixtures by factor analysis. *Remote Sensing of Environment*, **19**, 237-251

JACKSON, R.D. (1983). Spectral indices in n-space. *Remote Sensing of Enivronment*, **13**, 409-421

KASISCHKE, E.S. and LARSON, R.W. (1986). Calibrated X- and L- band scattering coefficients from a southern U.S. forest. In *Proceedings International Geoscience and Remote Sensing Symposium*

KAUTH, R.J. and THOMAS, G.S. (1976). The Tasseled Cap a graphic description of the spectral-temporal development of agricultural crops as seen by Landsat. In *Proceedings Symposium on Machine Processing of Remotely-Sensed Data*, Laboratory for Applications of Remote Sensing, Purdue Univ., West Lafayette, Indiana

PARIS, J.F. and KWONG, H.H. (1988). Characterization of vegetation with combined Thematic Mapper (TM) and Shuttle Imaging Radar (SIR-B) image data. *Photogrammetric Engineering and Remote Sensing*, **54**, 1187-1193

STONE, T.A. and WOODWELL, G.M. (1988). Shuttle Imaging Radar: an analysis of land use in Amazonia. *International Journal of Remote Sensing*, **9**, 95-105

TUCKER, C.J. (1979). Red and photographic infrared linear combinations for monitoring vegetation. *Remote Sensing of Environment*, **8**, 127-150

ULABY, F.T., MOORE, R.K. and FUNG, A.K. (1986). *Microwave Remote Sensing: Active and Passive*. Vol. III: From Theory to Applications. Artech House, Inc., Dedham, MA, 1065-2162

VAN DE HULST, H.C. (1980). *Multiple Light Scattering: Tables, Formulas, and Applications*. Vols. 1 and 2, Academic Press, New York.

WAITE, W.P., MacDONALD, H.C., KAUPP, V.H. and DEMARCHE, J.S. (1981). Wetland mapping with imaging radar. In *International Geoscience Remote Sensing Symposium Digest*, **2**, 794-799

WESTMAN, W.E. and PARIS, J.F. (1987). Detecting forest structure and biomass with C-band multipolarization radar: physical model and field tests. *Remote Sensing of Environment*, **22**, 249-269

VII.

OPPORTUNITIES, PROGRESS AND PROSPECTS

22
REMOTE SENSING IN AGRICULTURE: FROM RESEARCH TO APPLICATIONS

C. KING
Remote Sensing Department,
and
J. MEYER-ROUX
Project Manager, Institute for Remote Sensing Applications,
CEC Joint Research Centre, 21020 Ispra, Italy

Introduction

For each type of application of remote sensing, crop inventories, yield prediction related to disease and stresses, climate or soils, there are constraints which often prevent the transition from research to operational projects. When pilot projects of applications are set, they in turn induce new kinds of research, for new methodologies or instruments. In this chapter, both application and research studies will be illustrated by actual work within the Institute of the Application of Remote Sensing, particularly the Agriculture Project.

Going from research to applications seems to be difficult for remote sensing techniques using space data. For agricultural applications, in particular for inventories and monitoring of crops, major efforts have been made and programmes set up, especially in the United States of America. However, we are still rather far from a profitable service-oriented sector based on earth observation. What are the state of applications, and the trend in fundamental research and applied programmes? A complete overview could be difficult. If we limit ourselves to specific subjects like:

- applications in zoning, stratification, localisation;

- inventories of the main crops;

- crop monitoring;

over temperate countries, we can try to establish some perspectives.

Earth observation from space requires satellites and sensors. Because of the complex technology involved, the decision of joining the exclusive club of high-tech nations can be taken for reasons other than responding to the needs for specific applications. When the decision to launch a space system is taken, this strongly influences both research and application projects.

The United States with the Landsat series, Europe, and France in particular, with SPOT, have realized heavy investments for earth observation programmes. In both cases, monitoring agriculture was one of the main objectives. It is important to assess the state of applications and the trends in research.

Location, zoning and stratification

The synoptic and instantaneous vision over large regions given by satellite images such as the Multi-Spectral Scanner (MSS) (Beaubien, 1983), has given us immediately a new perception of our planet. Besides this new perception, important in itself, satellite images appeared rapidly as a new tool, an intermediary between a map and a mosaic of infrared aerial photographs which could be very useful for all kinds of applications, in particular, land cover ones. We can differentiate three kinds of applications, location, thematic zones and stratification.

Location

Aerial photographs have been used for a long time to locate specific features. In forest inventories they are used to discriminate forest types but also to locate sites where investigations will be carried out. In the French survey "TERUTI" they are used to locate points where, each year, enumerators go and check the land cover and land use information. In some countries or regions where cadaster maps don't exist they have been used as a substitute to locate fields. For such a purpose, Landsat MSS had to be used in combination with maps or aerial photographs, but it could not replace them. Second generation satellites and sensors such as SPOT HRV or Landsat TM have allowed more applications. Landsat TM, and especially SPOT with its 10 m resolution of the panchromatic band, can even be a substitute for ground observations in some cases. Those applications which need increased resolution seem to be the leading force in the technology trend. MSS, with its 80 m resolution, will not be kept with Landsat 6, which will have a 5 m band. SPOT XS and P, with their 20 m and 10 m resolution have allowed ground enumerations which were very difficult with TM and its 30 m resolution. Besides the increased resolution, new satellite images are a better mapping product. The strong internal deformations of Landsat MSS are much reduced with SPOT and TM. Instead of complicated transformations and resampling, first order polynomial transformations are now generally sufficient for geometric corrections.

Thematic maps or zoning

Besides the mapping quality, a satellite image can be used to discriminate objects, vegetation in particular, since the main bands for earth observation systems are chosen in the visible and near-infrared according to the spectral behaviour of vegetation. Such a zoning approach is used for many disciplines,— agronomy, forestry, ecology, soil maps, geology — since it is generally the first step before any in-depth investigation.

Applications

Dividing satellite images into homogenous zones has been considered a rapid approach to construct thematic maps of land cover. Visual discrimination of satellite images has not really improved in 15 years, but such methods are well defined and have proved to be efficient. For example, in Canada maps of forest types have been made by the

Remote Sensing Center of Quebec in cooperation with the Ministry of Forestry. Ground observations have been limited to a description of forest stands on homogenous zones (Beaubien, 1983)

The land cover CORINE project (CEC, 1988) is another typical example of this work. The objective of the project is to construct a map of ecozones at least 50 ha in size according to a common classification throughout the European Community. The choice of image type depends a lot upon economic considerations and objectives, such as the minimum size of homogenous zones which should be related to the heterogeneity of landscape and the scale of map products.

STUDIES TO IMPROVE ZONING METHODS

Studies linked with this application are of two types: the first is oriented towards image analysis, contrasts, stretching, geometric corrections. the second is oriented towards the control or validation of zoning work. This second exercise is difficult, since the objectives or ultimate applications of such maps are not well defined. They are often considered a multi-purpose tool.

Two examples of thematic validations can be shown:

The SCEES, (under contract with the JRC) checked the quality of such partitioning of space as compared to the official one of agricultural regions (SCEES, 1983). The study showed that there was a strong link between farm-related criteria and land use which appeared on Landsat images. Moreover, for every province where the two partitions differed, the new one was considered much better. Another example of validation can be shown by the study of the Canadian Center of Remote Sensing (CCRS) (Bain, Mack and Prevost, 1986). Uniform productivity areas were defined with Landsat images using photo-interpreting techniques. These units were checked with other data (pedology, farming methods and meteorology) and show that partitioning based on images corresponds to that based on these other criteria.

Satellite images are therefore a very powerful tool to partition space in homogeneous zones according to land cover and soil types. Consequently, most large companies or governmental agencies dealing with soil inventories, agronomic inventories or management have created a remote sensing unit for internal purposes to support investigations.

Stratification and area sampling frames

DEFINITION

Stratification is a statistical term which means a partition of the population leading to a sampling strategy. Satellite images can be used to partition statistical units spatially. Good stratification will lead to strata which will be as different as possible from one another and as homogeneous as possible within the characteristics of interest. Stratification is an intermediary tool to draw a sample which will represent the whole population. It depends a lot on the size and type of the units. Remote sensing techniques can be useful in the case of spatial units. Stratification is then called an area sampling frame.

APPLICATIONS OF STRATIFICATION

In many countries, e.g. the USA, Canada, Morocco, Tunisia and France, the statistical system used to obtain crop inventories is based on a sample of spatial units. Most of these countries use a sample of portions of land called segments. Those statistical units of 1 square km or mile in size are enumerated each year and give series of crop areas and productions. Since approximately 6-7 years ago, these methods have been introduced in Europe and are used by the SCEES (1983), the Ministry of Agriculture in Italy (Consortium I.T.A., 1989) or the Agriculture Project of the EEC (Meyer-Roux, 1987).

If enough documents or thematic maps exist, satellite images are not strictly necessary. However, they have proved to be an important tool to differentiate strata. In the U.S.A., Landsat TM images are used operationally by the National Agricultural Statistical Service (NASS) to update the area sampling frame in two states per year, (Hanuschak and Morrissey, 1987). Such use of remote sensing information is even more important for developing countries where basic maps are often old (Fecso, Gardner, Hale, Johnson and Pavlasek, 1982).

Building area sampling frames is an important operational application of satellite images. Improved resolution has led to the expansion of this technique, which could replace a 10 year agricultural census in many countries, especially in developing ones. In developed countries such surveys are more interesting, since they can be linked with global land-cover studies and environmental statistics.

Studies for area sampling frames have been of two types. The first is related to image analysis. Image segmentation could derive the population of statistical units. Until now, this has not been successful. Automated stratification should be an easier task, especially if we use square segments. Second, most of the studies are more practical, leading to adapted methods of stratification sampling strategy, for example, in terms of adjusting the size and shape of segments, and choosing equipment to compute the surfaces easily within the segments.

Conclusions

Satellite images are very useful in all applications which require the drawing of homogeneous zones, stratifying, etc. If MSS was sufficient for certain applications, second generation satellites have increased the number of applications. They tend to replace high altitude aerial photographs not only for the quality of the product but also for the simplicity of ordering products and the rapidity of delivery. Methods of location, zoning and stratification generally require a low technological level in remote sensing but good knowledge in the thematic field. For statistical systems, second generation satellites are very important since they facilitate the building of an area sampling frame: this becomes an interesting alternative for crop statistics and general environmental surveys in a number of cases, especially in developing countries.

Inventories of crops

The potential of discrimination from satellite images allows the differentiation of main categories of land cover and sometimes main crops. The size of fields being well matched to Landsat MSS resolution in North America, studies of inventories have been rapidly successful in this part of the world (Hanuschak, Allen and Wigton, 1982). In Europe more systematic studies and operational programmes were launched only with the second generation satellite sensor systems, Landsat TM and SPOT.

Studies and research

DISCRIMINATION

The variety of spectral responses of plants can be used to discriminate crops, and for the last 15 years all congresses and conferences concerned with research in this area have presented examples of such inventories of crops. It has, however, been difficult to get clear answers from these studies, such as land class categories, since the discrimination depends a lot upon the combination of crops, farming methods, phenological stages, etc. In order to understand spectral signatures better, ground radiometric campaigns were conducted in the USA within the LACIE (1978) and AgRISTARS (1986) programs and in Europe (Saint, Podaire, Fournier, Meyer-Roux and Cordier, 1983). Other campaigns were more oriented towards the selection of the best possible bands to discriminate crops. These more fundamental studies led to the addition of middle infrared bands for TM, also planned with Spot 4 (Ferns, Zara and Barber, 1984; Guyot and Baret, 1988)

SATELLITE DATA CLASSIFICATION

A number of studies have been presented about supervised or unsupervised satellite data classification, with *a priori* probabilities or without (Hill and Megier, 1987a; 1987b). Their main results show that:

- the best results are obtained with prior knowledge, homogenous zones, large fields, flat terrain, limited size studies, multi-temporal images and specific windows;

- some classification algorithms, either per field or contextual, work better than others but external data are often necessary, like field limits;

- the results of classifications in terms of surfaces of crops depend a lot upon the skill of the analyst and methods used;

- it is difficult with completely automated methods to get more than 70% of well-classified pixels for 4 or 5 crop categories with one image and 85% with two coverages over a large region.

ANALYSIS OF RESULTS

Remote sensing inventories are being influenced strongly by cartographers, in the sense that the most frequently used criterion of the quality of results is a cartographic one: the percentage of correctly classified pixels, which is one of the results of the confusion matrix. For large areas, statisticians from the USDA along with scientists from Purdue University introduced the regression estimator (Sigman, Hanuschak, Craig, Cook and Cardenas, 1978), which depends upon ground observations obtained from an independent sample. The direct expansion estimate given by this conventional survey is modified by satellite data classifications performed both for the whole region and observed segments, (see section on area frame sampling). The combined estimator uses qualities of both operations:

- ground surveys: which give perfect discrimination of crops, but contain important sampling errors;

- image classification: which gives poor discrimination but full coverage, ensuring that there is no sampling error.

It is possible to associate a confidence interval with the regression estimate and also to make an economic comparison between remote sensing operations and area frame sampling surveys alone. All large programs of inventories in the USA, Canada, France, Italy, and the European Community have used such a method.

Large programs

Large programs have been initiated with Landsat MSS in North America. The philosophy has been quite different in Canada and the USA. The Canadians have studied mainly specific crops such as rape seeds and potatoes, using provincial remote sensing centres even when Statistics Canada coordinated this work (Canadian Centre for Remote Sensing, 1985). The Americans have worked on main commodities, such as wheat, corn and soy beans in the Middle-West. The area covered has been 10 states in past years. In both cases they could use their area sampling surveys to train classifications of satellite data and obtain regression estimates of crop inventories.

The NASS program

The results summarized by Allen and Hanushak (1988) show that the method was operational but results mediocre when only one satellite was in operation. The first years of the program were plagued by slow delivery of data, which prevented them from meeting the November deadline for releasing official definite surfaces. Financed by the AgRISTARS program at the beginning, it continued by itself for three years. The NASS program has now been stopped due to the decision to have only a TM sensor on the next Landsat VI. While the NASS used TM to build its area sampling frames it used only MSS for inventories through data classifications. With the actual methodology of complete coverage, TM is considered too costly. The improvement in classifications does not compensate for the increased cost. NASS is entering a phase of research in methodologies to see if SPOT or Landsat TM can be cost effective for their inventories.

Agriscat project in Italy

The Italian Ministry of Agriculture has launched the most important program of crop inventories in Europe (Consortium I.T.A., 1989). The objective is to obtain inventories of main crops (wheat and corn) both for area and yield at a national level, but also for main producing regions. Area frame sampling is adapted to each crop, as are image acquisitions of Landsat TM. The results are considered satisfactory by the Ministry of Agriculture, which has reinforced this programme each year.

Agriculture project of the EEC

Since 1982, the Joint Research Centre, Ispra, has been working on inventories in the Ardèche department of France. After work done with simulations in 1984 a complete analysis of the department was performed under contract with the SCEES for ground observations (SCEES, 1983). This has been the base of numerous studies both for stratification and multi-temporal classifications, which paved the way for regional inventories: action 1 of the agriculture project of the EEC (Hill and Megier, 1988). This pilot project was launched in 1987 (Meyer-Roux, 1987) and only the results of 1988 are known. Crop inventories have been made for 5 pilot regions. An area sampling frame was built, the statistical unit being segments of 50 ha. Both Landsat TM and SPOT data were analysed, but with one date. Windows of acquisition had been defined to discriminate either winter or spring crops in 3 regions, or summer crops in 2 regions.

Although an almost complete coverage of the regions was possible, the acquisition dates were often outside the defined windows. As a result, the quality of the classification was quite different from one crop to another and according to regions. In all cases, the results of the area frame sampling were quite good. The project is being monitored closely by the JRC team and most of the studies are carried out by national teams outside the Commission.

OTHER PROJECTS

The SCEES is conducting regional inventories with the regression method it initiated in Europe (Meyer-Roux, Fournier and Touzelet, 1983). Its main efforts have been applied to ground surveys and sample area size and shape, but there is no *a priori* stratification. SPOT images used for inventories have not always covered the selected regions for the defined window periods. The main efforts have been made to create an automated computer system since complete studies are done in-house.

As in the United States, Canada has been able to use its June enumerative survey, which has been performed for many years, in combination with satellite data. Monotemporal analysis of Landsat has generally proved adequate because the growing period is short there.

In Spain, Catalonia and Navarre have done regional inventories and are intending to enlarge their study areas. In the Netherlands, inventories of main crops, and specifically corn, are being made with Landsat TM. The regression method was not used since the results needed were localized for environmental purposes.

FEASIBILITY STUDIES

A number of other inventories have been made, but generally on small regions as feasibility studies. Large cooperatives in France have ordered inventories for specific crops, but since the mapping product is necessary for them they need two coverages during the growing season, which is sometimes difficult to get.

Conclusions

The method for establishing crop inventories over large areas is rather well established. Ground observations on a sampling base are combined with satellite data which cover the entire region. Regression estimators are generally used, unless there is a strong emphasis on cartographic products.

The programmes differ according to emphasis on area frame sampling surveys or on satellite coverage and classifications (mono- or multi-temporal). There is still a lot to be learned, but the slightly different approaches depend generally on different objectives. For example, the strategy is different if one crop is studied or a group of crops or general land cover categories. If a method is rather well established, it does not mean that the results are good and that the method is cost effective on a yearly basis. The answer of the USDA seems to have been positive with MSS, and negative with TM in the case of two existing instruments. For Italy, the Ministry of Agriculture seems to be satisfied with results, and for the European project, where conditions are quite different, it is too early to reach a definite conclusion.

The results of research have shown improvements for inventories using more bands, such as the middle-infrared, or of higher resolution, which means less mixed pixels. However, it is not at all certain that this evolution is always positive when we consider cost benefits.

The most important aspect of making an inventory is the necessity of combining ground survey with satellite data. In North America area sampling surveys are performed independently from the remote sensing analysis. In Europe they don't exist, so their cost has to be included in that of the total operation. However, they can be adapted to the methodology, which makes them more efficient for a remote sensing operation.

Monitoring and yield forecasts

Yield per unit area is often the most important variable required to assess the production of crops, the area planted being more stable than yields. Decisions have often to be taken when special events like storms or floods affect large areas. In both cases, reliable, objective and independent knowledge is very important. Historically, meteorological data have been the main source of assessment, but the LACIE programme has focused on the possible use of satellite images for such purposes.

Monitoring of vegetation; qualitative aspects

With the actual state of technology, two kinds of satellite data can be used to monitor vegetation:

- high resolution images, such as those from Landsat or SPOT with low temporal frequency (every 16 or 26 days);

- low resolution images, such as from AVHRR with high temporal frequency (once a day).

MONITORING WITH HIGH RESOLUTION IMAGES

Localized monitoring has been achieved using aerial photographs, specially in forestry where diseases are sometimes more difficult to assess from the ground than the air. Plant protection has also been monitored in specific cases by aerial remote sensing techniques but, except for the cornblight assessment in the Middle-West, investigations have generally been quite localized.

The LACIE program has initiated monitoring over large regions. Since 1977, the Foreign Agriculture Service of the USDA has set up an operational unit analyzing MSS images over different countries such as the Soviet Union, Brazil and China to monitor main crops such as wheat, corn, soybean and rice (MacDonald and Hall, 1978). The information gathered is mostly qualitative. For example, if winter wheat has frozen, lower vegetation indices in late spring will indicate this. This will mean lower yields than normal since spring wheat is less productive. Snow coverage of wheat regions during frost period will protect the seedlings. This again will be shown by Landsat images. The advantage of the technique of satellite monitoring is complete spatial observation, which is impossible by other means. It can be complementary to other sources of more precise information, like specialized articles by experts in local papers.

The limitations of satellite monitoring are numerous, e.g.,

- lack of timeliness. Complete coverage can take a long time to be completed;

- resolution constraints. The problems observed must influence large areas; for diseases, this is not the case at an early stage;

- financial constraints. In 1984 a small unit of the FAS absorbed one-quarter of the total sales of Landsat products worldwide.

For these reasons, high-resolution images tend to be used on a sample basis or to evaluate damage *a posteriori*. The method then becomes identical to the one used for inventories as discussed previously.

MONITORING WITH HIGH FREQUENCY IMAGES

Since 1980, NOAA-AVHRR images have been considered useful, especially when ground meteorological stations were scarce such as in the Sahel (Tucker, 1986; Townsend and Justice, 1986). The state of vegetation as related to water availability can be assessed from the data. Variations within the seasons or interannually can be followed for large regions which would be impossible with other means (AISC, 1985). In temperate countries, these images are generally used in combination with information from meteorological stations to expand them spatially.

Yield forecasts, quantitative method

USE OF VISIBLE AND NEAR-INFRARED BANDS

Empirical studies have shown links between yield and the level of vegetation indices (Barnett and Thompson, 1983). For example, Prevost and Vickers (1985) have found a correlation between the ratio of bands 7 and 5 of MSS and the yield of cereals. The Normalised Difference Vegetation Index (NDVI) is the most generally used with all kinds of sensors. There are, however, limits to such a method, especially when vegetation does not cover all of the land and when soil response is significant. In the case of high yields, there is also a saturation of NDVI, often characteristic throughout Europe. More fundamental or deterministic methods link NDVI to the Leaf Area Index (LAI) or to the light interception coefficient, and then use this characteristic to estimate yields (Tucker, 1979). Certain phenological stages show a stronger relationship between yield than others and the NDVI (Guerif, Delecolle, Reich, Ripoche and Seguin, 1986). The temporal evolution of NDVI with its profile can be a good indication of yield for specific crops (Baret, 1986; Boatwright and Whitehead, 1986). Research tends to use satellite data in a way similar to the results of a phenological model.

The main problem for application of remotely sensed data to yield forecasting is that frequent coverage is only possible with low resolution data, which is therefore related to a mixture of crops or other land covers (Guyot and Seguin, 1988). Even with NOAA-AVHRR, it is difficult to get images at a specified phenological stage because of clouds.

Use of thermal bands

The brightness surface temperature should be a very important indicator since it is related to the energy balance between soil and plants on the one hand and atmosphere on the other, an energy balance in which evapotranspiration plays an important role (Jackson and Reginato, 1983; Campbell; Boissard, Guyot and Jackson, these proceedings). Surface temperature could be quite complementary to vegetation indices. Water-stress, for example, should be noticed first by an increase in the brightness surface temperature and, if it affects the plant canopy, by changes in the vegetation indices. Measured brightness temperatures are affected by water vapour in the atmosphere but the method of the "Split Window", which uses two thermal bands of NOAA-AVHRR, offers accuracy in the order of 1 to 2K. Studies by Seguin (1987) have shown that the measured difference between surface and air temperatures can be related to evapotranspiration.

For monitoring purposes we still need pilot studies to assess the accuracy of such methods. We know they can be used for extreme conditions and homogenous flat zones, but we are not sure they can be used on a routine and yearly basis for most of the agricultural regions of the EEC.

Current programmes of monitoring

For temperate climates there are no large-scale programmes for monitoring crops with remote sensing techniques besides the ones of the US Department of Agriculture. There are several for developing countries (Negre, 1988; Malingreau, Bartholome and Barisano, 1988), within the FAO or the United States.

As described previously, the use of high resolution satellite data has been mainly qualitative and dependent upon the skills of the analyst. More quantitative aspects have been introduced with the comparison between vegetation indices, agrometeorological models and a large data base of conventional statistics. Because of the uncertainty of acquiring images and the required timeliness, the FAS does not want to depend upon a specific method like the LACIE one (Price, 1986). It prefers to extract as much information as possible for different crops. Their analysts have been very reluctant to introduce NOAA-AVHRR analysis because they want to be able to distinguish fields. However, since 1983-84, approximately 10 to 20% of the information has shifted from MSS to NOAA-AVHRR (Whitehead, Johnson and Boatwright, 1986).

Until now, FAS have used only MSS and not TM because of financial considerations and because the computer systems cannot handle all the data from TM. They also rely heavily on agrometeorological models which have been developed for the research programs, LACIE and AgRISTARS.

In practice, analysts tend to use high resolution satellite data as a substitute for ground observations, to check information from NOAA-AVHRR which give a general qualitative assessment and from agrometeorological models, which are more quantitative (Larrabee and Hodges, 1985). There are, however, strong limitations in combining TM and SPOT, whose resolution is more adapted to European fields than those of the Soviet Union, Argentina or Australia. The greatest need is to include remote sensing data in models of phenological development and water balance. Until now analysts have combined information from sparsely distributed meteorological stations with remote sensing data which offers complete coverage. But until a scientific method is established, the results will depend on the flair of the analyst.

The European Community pilot project of remote sensing applied to agricultural statistics

Objectives

The implementation of the Common Agricultural Policy is one of the most important objectives of the Commission of the European Communities. To discharge its duties it requires an effective system of monitoring and control so that it can provide the Commission with the facts necessary for guiding its agricultural policy. The objectives of the project are to use Remote Sensing to improve the reliability and promptness of agricultural statistics in the European Community. The system required must be capable of:

- identifying and measuring, as fast and with as much accuracy as possible, the areas under different crops. This applies particularly to those crops which enjoy a subsidy calculated by area;

- estimating regional production in near-real time;

- and estimating the production of competitors.

The pilot project should demonstrate a methodology by which remote sensing could supplement, interpret and standardise data provided by classical techniques. In the present system, information arriving at the Commission from the agricultural services of the various Member States of the European Community is collated, analysed and reported using several different methods. For this reason, Commission services sometimes find it difficult to interpret and compare data from the different countries.

The principal arguments for applying remote sensing are:

1. that the technique by its nature is insensitive to national boundaries and can thus provide impartial or objective information based on the terrain rather than on national territories.

2. that a satellite is able to provide information, using exactly defined methods, both on a continental scale and much smaller areas;

3. that, in principle, data can be collected and analysed rapidly enough to enable reports to be made available to the Commission at the moment of need, rather than months later.

The pilot project should develop the technique to a point where operational use can be made of remote sensing for agricultural monitoring. The key word for the project is **operational**: the project is not intended as an experiment, but as a programme for putting known techniques into operation. Although it is a project within the Joint Research Centre, its first priority is not research, but application. All research carried out under this project has an immediate operational aim.

Actions

The work of the project has been divided into 7 Actions, 4 of which make up the major themes, the remaining 3 supporting these main Actions. The four main Actions are:

1. The drawing up of regional inventories of crop acreage and possibly yield using a combination of ground observations and high-resolution satellite data.

2. The monitoring of the condition and development of vegetation on a continental scale using low-resolution satellite data.

3. Forecasting yield using agrometeorological models on a regional scale.

4. The calculation of rapid estimates of change in acreage and potential yield using high-resolution satellite data at sample sites scattered over Europe.

The three support Actions involve the integration of data from all sources to provide an advanced agricultural information system, the provision of data collected in the field to each of the main actions, and research into new developments in various related areas.

ACTION 1: REGIONAL INVENTORIES

Action 1 engendered the first activities of the project. By the end of 1987 a feasibility study had been carried out to evaluate the probability of sufficient remotely-sensed data being collated over agricultural areas of Europe to assess the viability of the whole concept. Cloud-cover studies suggest that one coverage a year is possible in most of the agricultural zones.

Using this feasibility study as a guide, five regions in Europe were selected for the project, the criteria being:

- reasonably flat terrain;

- high percentage of agricultural land use;

- wide differences in agricultural methods between regions;

- geographical dispersion;

- each region to correspond to a contigous area of some 20,000 square kilometres.

These regions are situated in France (covering Loir et Cher, Eure et Loir, and Loiret), Germany (Niederbayern and Oberpfalz), Greece (Kentriki, Ditiki and Makedonia), Italy (Emilia-Romagna) and Spain (Valladolid and Zamora). The results obtained at the end of 1988 have shown:

- that the methodology used both for ground surveys and the combined analysis of satellite images was rather good;

- that a complete or almost complete coverage of the 5 regions was possible with Landsat and possible with SPOT if priority was given to the action by SPOT-Image;

- that it was impossible either with SPOT or Landsat to get complete coverage during the one and a half months that correspond to the best periods for all the regions;

- that the results of ground surveys, although costly, were both timely and accurate even with the simple methodology and stratification used;

- that the ability of remote sensing to improve these results was very different from one region to the other and one crop to the other with an overall efficiency of approximately 1:7 for crops of interest, where the efficiency is defined as the variance of the conventional estimate divided by that of the estimate using remote sensing.

ACTION 2: VEGETATION CONDITION AND YIELD INDICATORS

The project supplements the detailed data that it receives from high-resolution satellites with broad information from low-resolution satellites originally designed for meteorological applications. There are two main advantages to using low-resolution satellite data. The first is that as a result of its low spatial resolution, one gets an impression of conditions over a very large area, and the second is that the data for any given site are collected frequently. For the satellites used in the project (the United States' National Oceanographic and Aeronautic Administration, or NOAA series), the radiometer on board covers a swath 3,000 kilometers wide, and can provide data on any given target once or twice a day, cloud-cover permitting.

In February 1988, an invitation to tender (ITT) was published calling for studies using low-resolution satellite data for monitoring the condition of vegetation and for providing indicators of crop yield. Among other topics, the ITT included studies on the methods to be used, including:

- the use of low-resolution satellite data for monitoring crop development and providing early warning of unusual growing conditions;

- the use of such low-resolution data in conjunction with high-resolution data or with data from external sources;

- the application to agricultural monitoring of estimates made of surface temperature through data received from meteorological satellites;

- the possibility of using low-resolution data for land-use classification.

Ten proposals from seven contractors were selected from among some 60 received. The majority of the proposals concerned the use of the visible and near infra-red channels and their combination in vegetation indices. However two teams are studying the estimation of surface temperature and land-use classification.

The data received on the ground after transmission from the satellite appears in a raw state, unsuitable for immediate analysis. In particular, the data must be corrected to take into account the altitude, position and attitude of the satellite. The data must also be corrected for the effect of the atmosphere, the degree of illumination of the ground, the presence of clouds, fog, aerosols, etc. The project team has been collaborating with the European Space Agency to develop suitable software to carry out these corrections. In October 1988, a call for tender was published for the development of sub-routines to convert the raw data into useful, corrected data. The work on this should start in April 1989.

ACTION 3: MODELS OF YIELD PREDICTION

Although the project does not contain the words "model" or "modelling" in its name, the prediction of agricultural yield using models is highly relevant to the project. Such models will be modified in such a way that remote sensing can be used to validate the results. The models to be developed under the aegis of the project are neither purely statistical nor purely deterministic, but are a third group containing some elements from both types and might be called "semi-deterministic".

Purely statistical models are based on the observed statistical relationships between yield and a variety of input data. These input data include agriculturally important information such as rainfall and temperature, but the functional relationship linking yield to these parameters is not necessarily questioned. They work best when large areas and long time periods are considered.

In purely deterministic models, the detailed functioning of the plant is modelled in all its components including root growth, biomass production, redistribution of nutrients within the plant, evapo-transpiration etc., in order to predict, with given growing conditions, what yield might be expected. These highly detailed models work best on the single plant scale, or on the scale of a single homogeneous field of a single crop. Since they depend on a quantitative understanding of the minutiae of the crop's reaction to any given input, and of precise details of the input itself, they are not suitable for modelling production over regions or nations, where it is impossible to collect all the necessary input data.

The semi-deterministic model uses a third approach, for which the level of detail is in some ways a mixture of both the other approaches. For input it takes agriculturally important information. The model then applies general, but scientifically sound, causal rules relating crop growth to environmental conditions, and finally relates the result to yield - usually, although not necessarily, through statistical relationships. The sort of inputs normally required by these models includes information on water-balance, stress-degree-days, cumulative temperatures, cumulative radiation, the relationship between day-length and flowering, etc. According to their design, semi-deterministic models can be well-suited to regional or national modelling.

The project intends to develop a comprehensive geographic information system to include all the plant, soil and meteorological data needed. The layers of data will be related mainly by phenological and water balance models. Remote sensing is seen as the main technique to obtain spatial information on plant location (Action 1 or 2). It is also considered as an objective way to control results of the models (Action 2).

In August 1988, a call for tenders was put out, intended to provide the project with a reliable data base containing historical crop, climatic and agrometeorological information, to be used in creating semi-deterministic model. These studies are under way. At the same time three other themes are being examined under contract:

- the possibility of improving estimates of evapotranspiration on a European scale;

- a method of monitoring Mediterranean grape and olive production by the analysis of atmospheric pollen loadings;

- the statistical prediction of the yield of grapes, olives and fruit for the whole of Europe.

A first review of the contracts took place in April, 1989, at Ispra.

ACTION 4: CHANGES IN ACREAGE AND POTENTIAL YIELD

One of the project's most ambitious Actions is the installation of a system in which high-resolution satellite data is received at a pre-selected station, rapidly processed and analysed by a judicious combination of computer and human expertise and the results published in a bulletin. Not only will this bulletin contain a quantitative measure of the degree of change and potential yield in an area by comparison with the situation at the same time the previous year, but also a qualitative statement concerning the condition and trend of the vegetation.

Clearly, it is not feasible to monitor the whole of Europe using high-resolution imagery. The cost would be astronomical, and the volume of data obtained so enormous that it could never conceivably be processed or analysed. The project has designed a sampling strategy which will start with 10 sites scattered over Europe, which by 1993 will have grown to 50 sites. Some ground observations will also be gathered on a sample basis in those sites. They will be used as an *a posteriori* check on accuracy, and will also enable the contractor to improve his technique as he gains experience. This action could evolve in two ways:

1. if ground observations can be completely eliminated, the method could be used to monitor foreign crops;

2. if ground observations can be integrated with satellite data, satellite images could be used for rapid estimates of crop areas and yields within Europe and the control data of ground observations will enable more accurate end of the year estimates than would be possible with satellite data alone.

SUPPORT ACTIONS

Without going into detail, project work has also been undertaken on the following:

- implementation of software to enable ground data collected in the segments to be compared with satellite data for the same site and converted rapidly into statistics;

- initiation of discussions with other projects in the JRC with the aim of cooperation in developing the application of remote sensing to agricultural statistics;

- contributing to research into the use of radar and passive microwave techniques for agricultural remote sensing.

This project tries to integrate the successful aspects of post-operational projects, but also to take into account the difficulties encountered. In particular, Action 4, dealing with a sample of sites, seems to be the only way to use high resolution images together with high frequency of coverage (3 to 5 images per year). This will oblige us to confront new problems, such as the standardisation and calibration of different sensors over time, such as TM and SPOT.

For Action 2, the use of NOAA-AVHRR over Europe, we need accurate geometric and radiometric corrections to monitor changes that are less significant than the ones observed in the Sahel or in countries with extensive farming methods and low yield.

For Action 3, agrometeorological models, the challenge will be first to set up a comprehensive geographic information system over Europe for agricultural information, then to integrate remote sensing data to control the results at different stages.

Conclusions

From the past 15 years of experience, it has already become clear that earth observation by satellite can play an important role in crop monitoring. The use of satellite images to decide upon the location of ground observations, to define homogenous units, or to stratify is quite operational, especially with the second generation satellite-sensors, SPOT HRV and Landsat TM. Thematic maps of land cover drawn by photo-interpreting techniques are also common practice, but the use of automated methods to obtain accurate inventories or to derive thematic maps has been impossible without intensive ground observations. Although the methods are rather well established, cost effectiveness has to be studied closely.

When we go from surface inventories to production and yield forecasts, we are still in a research stage and we must pursue the efforts initiated with the LACIE and AgRISTARS programmes.

However we have new means at our disposal:

- more powerful computers;

- improved earth observation satellites;

- new results from recent research.

The main problems we are facing are:

- the lack of inter-calibration between high resolution and low resolution instruments;

- the lack of means to correct the data radiometrically;

- the lack of satellite data;

- an evolution towards higher resolution in satellite data which serves cartographic purposes but not monitoring ones.

References

AGRISTARS (1986). Special issue on agriculture and resources inventory surveys through aerospace remote sensing. *IEEE Transactions on Geoscience and Remote Sensing*, **GE-24**

AISC (1985). *Climate impact assessment.* Foreign Countries Program. NOAA, Washington

ALLEN, J.D. and HANUSCHAK, G.A. (1988). *The Remote Sensing Applications Program of the National Agricultural Statistics Service 80-87.* USDA-NASS Nb SRB 88-08, USDA, Washington

BAIN, D.W., MACK, A.R. and PREVOST, C. (1986). Mapping land-use and uniform productivity areas of the prairie provinces using generalized soil landscape maps and multitemporal Landsat satellite images. Rapport CCRS

BARET, F. (1986). *Contribution au suivi radiométrique de cultures de céréales.* Université, Paris Sud, Paris

BARNETT, T.C. and THOMPSON, D.R. (1983). Large area relation of Landsat MSS and NOAA-AVHRR spectral data to wheat yields. *Remote Sensing of Environment*, **13**, 277-290

BEAUBIEN, J. (1983). Remote sensing by satellite applied in Quebec to vegetation mapping. *CRF des Laurentides Service Canadien des forets*. Canada

BOATWRIGHT, G. and WHITEHEAD, V. (1986). Early warning and crop conditions assessment research. *IEEE Transactions on Geoscience and Remote Sensing*, **GE-24**, 54-64

CANADIAN CENTRE FOR REMOTE SENSING (1985). *Crop Information System Using Remotely Sensed Data*. Canadian Centre for Remote Sensing, Ottawa

CEC (1988). Projet CORINE. *Communication de la Commission au Conseil et au Parlement Européen, Resultats du programme et orientations*. COM 88.420, CEC, Bruxelles

CONSORTIUM I.T.A. (1989). *Telerilevamento in agricoltura*. Ministero dell'Agricoltura, Italy

FECSO, R., GARDNER, W., HALE, B., JOHNSON, V. and PAVLASEK, S. (1982). *Constructions of a remotely sensed Area Sampling Frame for Southern Brazil*. USDA SRS, Staff Report NO AGES 820526, Washington D.C., U.S.A.

FERNS, D.C., ZARA, S.J. and BARBER, J. (1984). Application of high spectral resolution spectroradiometry to vegetation. *Photogrammetric Engineering and Remote Sensing*, **50**, 1725-1735

GUERIF, M., DELECOLLE, R., REICH, P., RIPOCHE, D. and SEGUIN, B. (1986). *Utilisation de la télédétection par satellite pour le suivi des cultures dans la basse vallée du Rhône*. Etude No. 85 12637000 SLA, INRA, Montfavet.

GUYOT, G. and BARET, F. (1988). Utilisation de la haute résolution spectrale pour suivre l'état des couverts végétaux. In *Proceedings 4th. International Colloquium on Spectral Signatures*. ESA, SP-287, 279-286. European Space Agency, Paris

GUYOT, G. and SEGUIN, B. (1988). Possibilité de l'utilisation de la télédétection satellitaire en agrométéorologie. *Agronomie*, **8**, 1-13

HANUSCHAK, G.A., ALLEN, R.D. and WIGTON, W.H. (1982). Integration of Landsat data into the crop estimation program of USDA's statistical reporting service. In *Machine Processing of Remotely Sensed Data Symposium*. Purdue University, Lafayette, USA

HANUSCHAK, G.A. and MORRISSEY, K. (1987). *Pilot study of the potential contribution of Landsat in the construction of Area Sampling Frames*. USDA SRS, Washington D.C., USA

HILL, J. and MEGIER, J. (1987a). Regional land cover and agricultural area statistics and mapping in the ardèche department (F) by use of thematic mapper data. In *Proceedings 7th Symposium on European Remote Sensing Needs in the 1990's*. EARSel, Noordwijkerhout, Netherlands

HILL, J. and MEGIER, J. (1987b). Cluster based segmentation of multitemporal thematic mapper data as preparation of region-based agricultural land-cover analysis. In *Proceedings IGARSS'87*. Ann Arbor, Michigan, U.S.A.

HILL, J. and MEGIER, J. (1988). Regional land cover and agricultural area statistics and mapping in the Department Ardèche, France, by use of TM data. *International Journal of Remote Sensing*, **9**, 1573-1595

JACKSON, R.D. and REGINATO, R.I. (1983). Agronomic aspects of thermal IR-measurements. In *Proceedings IIe Colloquium Internationale Signale Spectrals*. INRA Publ. No. 23. INRA, Paris

LACIE (1978). *Proceedings of Plenary Session*. The LACIE symposium. NASA JSC 14557, Houston. U.S.A.

LARRABEE, J. and HODGES, T. (1985). *NOAA-AISC user's guide for implementing CERES-MAIZE model for large area yield estimation*. AGRISTARS JSC-20237. NOAA, Washington

MacDONALD, R.B. and HALL, F.G. (1978). LACIE. An experiment in global crop forecasting. In *Proceeding of the LACIE Symposium*. Oct. 78, NASA. JSC 14551, 17-48. NASA, Washington

MALINGREAU, J., BARTHOLOME, E. and BARISANO, E. (1988). Surveillance de la production agricole en Afrique de l'Ouest. Nécessité d'une integration de différentes plates-forms satellitaires. In *Proceedings Colloque PEPS-SPOT 1*. Paris, November 1987. CNES, Paris

MEYER-ROUX, J. (1987). *The 10 Year Research and Development Plan for the Application of Remote Sensing in Agriculture Statistics*. CCR, SPI 87.39/EN. CEC, Ispra, Italy

MEYER-ROUX, J., FOURNIER, P. and TOUZELET, M. (1983). The remote sensing program of the French Agricultural Statistical Service, ERIM, Ann Arbor, Michigan, U.S.A

NEGRE, T. (1988). Sahel le saison des pluies 87 vue par les outils satellitaires d'Agrhymet. *Veille climatique et satellitaire No. 23*. 29-36

PREVOST, C. and VICKERS, H. (1985). Analyse des données Landsat MSS. Swift Current et Melfort (Saskatchewan) 1983. Index végétatif et rendement en céréales. Rapport CCRS, Janvier 85. Canadian Centre for Remote Sensing, Ottawa

PRICE, J.C. (1986). Applications of remote sensing in the U.S. Department of Agriculture. In *Proceedings of IGARSS'86 Symposium*. ESA SP-254. European Space Agency, Paris

SAINT, G., PODAIRE, A., FOURNIER, P., MEYER-ROUX, J. and CORDIER, P. (1983). Comparaisons des capteurs SPOT-HRV et Landsat-4 TM pour l'obtention de statistiques agricoles. In *Proceedings Colloquium Internationale Signatures Spectrales*. Bordeaux. INRA, Paris

SCEES (1983). *Préparation des expériences à realiser aprés le lancement des satellites de 2e génération pour l'étude des régions defavorisées*. Min. del'Agriculture, F, Contrat No. 2272-83-ID-ISP F Ispra, Italy

SCEES (1986). *Enquête sur l'utilisation du territoire effectuée en 1985 par la méthode des segments dans la région Ile-de-France et deux departements de la région Centre*. Min. de l'Agriculture, Paris. France

SEGUIN, B. (1987). Estimation de l'evapotranspiration par Télédétection satellitaire dans l'infrarouge thermique. *Academie Agriculture Francais*, **73**, 53-60

SIGMAN, R., HANUSCHAK, G., CRAIG, M., COOK, P. and CARDENAS, M. (1978). The use of regression estimation with Landsat and probability ground sample data. In *Proceedings American Statistical Society. Annual Meeting.* San Diego, California, U.S.A

TOWNSEND, J.R.G. and JUSTICE, C.O. (1986). Analysis of the dynamics of African vegetation using the normalised difference vegetation index. *International Journal of Remote Sensing*, **7**, 1435-1446

TUCKER, C.J. (1979). Red and photographic infrared linear combinations for monitoring vegetation. *Remote Sensing of Environment*, **8**, 127-150

TUCKER, C.J. (1986). Cover maximum Normalised Difference Vegetation Index images for sub Sahelian Africa for 1983-1985. *International Journal of Remote Sensing*, **7**, 1383-1384

WHITEHEAD, V.S., JOHNSON, W.R. and BOATRIGHT, J.A. (1986). Vegetation assessment using a combination of visible, near I.R. and thermal IR AVHRR data. *IEEE Transactions on Geoscience and Remote Sensing*, **GE-24**, 107-112

WIEGAND, C.L., RICHARDSON, A.J. and NIXON, P.R. (1986). Spectral component analysis: a bridge between Spectral Observations and Agrometeorological Crop Models. *IEEE Transactions on Geoscience and Remote Sensing*, **GE-24**, 83-88

23

REMOTE SENSING IN AGRICULTURE: PROGRESS AND PROSPECTS

J.L. MONTEITH

International Crops Research Institute for the Semi-Arid Tropics,[1] *Patancheru P.O., Andhra Pradesh 502 324, India*

Impressions

I believe that most participants at this Easter School will return to home base with three abiding impressions of what we have heard and seen. First, we have been left in no doubt that techniques of remote sensing are continuing to develop very rapidly, particularly in the interpretation of microwave signals and in the storage and processing of data. Second, platform speakers, along with all the enthusiasts who displayed posters, have convinced us that there are many ways in which remote sensing could, in principle, be deployed to increase the world's food supplies. Third, speakers from the floor have repeatedly pointed out that the contribution which remote sensing has so far made to agriculture lags far behind the perceived potential. In attempting to sum up conclusions from this meeting, I shall be specially concerned with the constraints which prevent that potential from being realised.

Evolution of remote sensing

Most applications of remote sensing are still in the process of evolving through stages of development familiar in the experimental sciences. After the first flash of inspiration come measurements and hypotheses, usually in that sequence but sometimes in the reverse. Hypotheses suggest how measurements should be interpreted in terms of underlying mechanisms and the number of measurements needed to support a given hypothesis often displays a broad optimum. Below the optimum, experimental support for the hypothesis is unconvincing. Above the optimum, attempts to demonstrate the validity of a hypothesis can be obscured by a fog of facts. Remote sensing often demonstrates this problem. Enormous amounts of data are generated by instrumentation on orbiting satellites but usually only a small fraction is subsequently used. Data banks, however comprehensive, cannot generate hypotheses spontaneously. This process is always limited by the human "eye-brain" system that Allan talked about, and in many remote sensing laboratories the ratio of minds to megabytes seems very small!

---

[1] Submitted as Conference paper No. 1485 by the International Crops Research Institute for the Semi-Arid Tropics (ICRISAT)

The next step is to use hypotheses, singly or in groups, to generate what used to be called a "theory" but which is now usually referred to as a "model". A model is a quantitative description of a system derived from a limited set of data but capable of predicting how it will behave in response to changes either in the external environment or within the system itself. We have considered several types of agricultural models, but have concentrated on dry matter production and light interception by canopies (papers by Kanemasu and Prince), and on the relation between transpiration rate and surface temperature (papers by Campbell and Guyot). I shall return to both these areas. Models

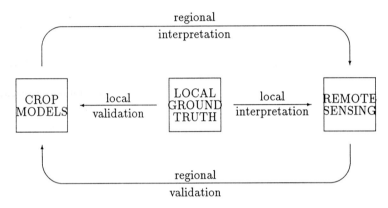

FIGURE 23.1. Interaction of remote sensing and crop models.

and remote sensing can usefully reinforce each other, as *Figure 23.1* suggests. Neither can progress far without "ground truth" on the scale of a single field (for models) or of a few pixels (for remote sensing). Models which have been validated by appeal to ground truth (which is usually limited in scope and sometimes not entirely truthful), can be used on a regional scale to interpret imagery from aircraft or satellites in terms of the state of soil and/or vegetation. Similarly, the application of models on a regional scale can be validated by using remote sensing, in the form of satellite images for example, to provide independent estimates of crop yields.

Eventually, a good model should be extracted from the perfectionist clutches of its creators and used to improve the management of an agricultural system. The sophisticated scheme for managing water resources in part of the Netherlands, described by Nieuwenhuis, was one of the few cases we heard of where this last step had been successfully reached.

I have suggested that a disproportionate amount of effort seems to have been expended on the data-collecting end of remote sensing, but it is only fair to recognize that this is typical of a discipline in the Natural History phase of its development (so it was unkind of Rutherford to describe biologists as "stamp collectors"!). The historical background to this state of affairs is that most satellite technology was developed for military surveillance. Applications to earth resources in general and to agriculture in particular are therefore a bonus; but Evans and other speakers reminded us that simpler techniques, like aerial photography, may often provide quicker and more precise answers to agricultural problems.

Models of agricultural systems have lagged behind the availability of measurements from aircraft and satellites because most have relied for their validation on sets of observations made at a very limited number of sites. Management has lagged further behind still because farmers and their advisers remain unconvinced that satellite technology can beat inherited skills. This point of view is justified at present: there is no evidence that the invention of the telescope in 1608 encouraged farmers to direct field operations from their bedroom windows!

Evaporation

To derive his formula for estimating evaporation, Penman (1948) deliberately eliminated surface temperature as a variable because, 40 years ago, it was hard to observe in the field and even harder to record. For the measurements needed to establish a wind function, he was forced to use cumbersome mercury-in-steel thermometers to record the surface temperature of water in evaporation pans.

In the 1960s, the development of compact and fairly stable radiation thermometers that could be operated in the field made it possible to measure the Radiative Temperature (T_r) of soils and crops to about 1K and encouraged a reassessment of Penman's algebra to see what additional information could be obtained about responses of crops to their water supply. The USDA group in Phoenix, Arizona, became the main exponents of techniques for using measurements of T_r to assess rates of evaporation from crops, shortage of water in terms of "stress degree days" and implications for yield (Idso, Jackson and Reginato, 1977). However, other groups working at cloudier sites with random fluctuations of solar radiation found it more difficult to interpret T_r in terms of crop water supply.

More problems arise when T_r is measured from a satellite, because of calibration drift and because of the need to correct for atmospheric absorption and emission. These topics have barely been touched on at this meeting but cannot be ignored in any realistic assessment of the potential for remote sensing in agriculture. Errors can be large when latent heat loss is estimated from the difference between net radiation and sensible heat loss and when the sensible heat calculation involves the difference between T_r and the temperature at screen height measured with a completely different system. Error can be minimized by using a reference surface (Nieuwenhuis) but even then it is desirable to have additional information about aerodynamic surface characteristics and about weather at screen height (Gash, 1986).

I therefore believe that agronomists, like hydrologists and ecologists, are likely to have to wait for some years before remote sensing can provide them with estimates of evaporation better than what is now available from formulae that incorporate an informed guess about the magnitude of a surface resistance. Microwave radiometry, as described at this meeting by Luzi, Paris and Holmes, has a number of potential advantages, including the ability to penetrate clouds, to measure soil water (but only close to the surface), and to monitor changes of plant water content. Another type of remote sensing (referred to by Kanemasu) is the use of eddy correlation equipment on low-flying aircraft to measure fluxes of both water vapour and CO_2, so that water-use efficiencies can also be found, as described by Schuepp, Austin, Desjardins, MacPherson and Boisvert (1987). This is a powerful way of exploring transfer processes in the atmospheric boundary layer, but costs are likely to remain prohibitive for most agricultural applications.

Dry matter production

Several papers and posters have demonstrated how remote sensing can be used to estimate rates of dry matter accumulation by crop stands, given daily mean values of incoming solar radiation and the fraction of that radiation intercepted by foliage as obtained from its spectral characteristics. It is customary to work with wavebands just below and just above a wavelength of 700 nm to give maximum discrimination between foliage and underlying soil, but this is a small fraction of the spectral information available from most satellites. I agree with Steven that more effort should now be spent on looking at derivative spectra and at other wavebands to obtain indices that can be correlated with stress. It may also be possible to obtain more precise information about seasonal and secular changes of ground cover by combining measurements of the Normalised Difference Vegetation Index, (NDVI) with microwave polarization differences (Becker and Choudhury, 1988).

To estimate rates of dry matter production during the growing season and final yield, it is necessary to know the appropriate value of e – the mass of dry matter accumulated per unit of radiation intercepted (This quantity is often referred to as a "radiation use efficiency" but it is not a true efficiency until it is multiplied by the energy equivalent of biomass). Many systematic measurements of e for field crops have been reported recently (e.g. Kiniry, Jones, O'Toole, Blanchot, Cabelguenno and Spaniel, 1989) and have prompted me to examine the question of whether e and the corresponding water use efficiency can both be conservative at the same time (Monteith, 1989). The answer appears to be "Yes, to a good approximation when water supply is unrestricted; no, when there is a shortage of water". Nutrient shortage appears to operate in the same way but on a restricted scale (Green, 1987).

The first response of most plants to drought (or nutrient deficiency) is to slow the expansion of leaves and to allocate a larger fraction of current assimilate to extending roots. This helps to keep water supply and demand in balance and to stabilize the value of e.

In contrast, if stress builds up rapidly when a plant has a substantial amount of foliage to support, closing stomata is the only way in which the demand for water can be reduced to the level of supply. Because stomatal resistance has to be more or less proportional to demand, it is often found to be a strong function of saturation vapour pressure deficit in these circumstances. Stomatal closure reduces the photosynthesis rate per unit of intercepted radiation and therefore the value of e. At present we have no reliable way of estimating the non-potential value of e because so little is known about how the size and activity of the root system determines the maximum rate of water supply. However, it may be possible to obtain useful guesses of non-potential e by correlating measurements of e at the ground with the rate of change of vegetative cover determined spectrally. In this case, differences of population would need to be taken into account.

Prince gave an impressive demonstration of how values of e characteristic of vegetation in a semi-arid part of West Africa (and clearly sub-potential) were obtained from careful sampling on the ground and from satellite records. Without more ground studies of this type, enormous amounts of valuable information will be archived and eventually destroyed unused - a tragic waste of resources.

Time and space

As Allan pointed out at the beginning of this meeting, the agricultural potential of remote sensing is circumscribed by limitations of space and time. On the scale of hours to days, farmers in many parts of the world already obtain weather forecasts developed with the help of satellite images of cloud systems and interpreted in terms of the timing of rainfall, extremes of temperature, risk of high winds, etc. In relatively unpopulated regions, where observing stations are sparse, remote sensing should play a much more central role in forecasting weather. The technique for estimating rainfall from cloud-top temperature as described by Milford is a most encouraging example of this process and it is reassuring to know that FAO is now involved through its ARTEMIS project.

Still within a relatively short time-scale, but on a larger scale spatially, satellite images, released internationally and in "real" time could provide governments and extension services with the information they need to monitor and control fungal epidemics and pest invasions, to assess surface water resource for irrigation schemes, to monitor the extent of floods, etc.

On a medium time-scale (weeks to months), the main potential for remote sensing appears to be in the assessment of crop growth rates as a basis for predicting yield and as an index of the need for irrigation, application of fertilisers, control of disease, frost protection, etc. As an example of a successful regional survey, the contribution from Brooms Barn Experimental Station by Jaggard and ClarK described how annual sugar production was estimated from the spectral properties of radiation reflected from representative fields of beet and regularly monitored from a light aircraft. Observations from a satellite are now being used at the WMO regional centre in Niger (AGRHYMET) to assess rainfall distribution and the seasonal progress of crop production in the western Sahel. Both China (Zheng Dawei) and India (Sahai and Navalgund, 1988) are using satellites in this way, despite problems created by small farm sizes, the prevalence of mixed cropping and extensive cloud cover during rainy seasons.

On a much longer time-scale (years) the main potential of remote sensing appears to be the development of detailed inventories for soils and crops, a process likely to become much faster and more efficient with the development of the Geographical Information Systems referred to by several speakers; the accumulation of crop histories over several consecutive years as a basis for improving assessments of stress and yield; and the monitoring of land degradation as a consequence of pollution, erosion, poorly designed irrigation schemes, etc.

Postscript

In a rapidly developing field of research and technology, it is dangerous to pontificate about what may or may not be possible in the foreseeable future. I have felt bound to speculate a little but find it salutary to recall how ignorant I was of the potential for remote sensing when the first Sputnik started bleeping its way round the world in October, 1957. The Annual Report of Rothamsted Experimental Station for 1957, which appeared a few months later, contained these words: "Measurements of reflection coefficient may give useful estimates of leaf growth without destructive sampling". When I wrote that, it never occurred to me that my crude, home-made solarimeters would be replaced within my lifetime by satellite-borne radiometers, scanning the continents to estimate net primary production from a Normalised Difference Vegetation Index.

The continuing development of satellite and space station technology makes it impossible to predict what remote sensing may be able to do for agriculture in another 30 years or even in 10. As we consider the surpluses of food which now embarrass many western countries, it is clear that two items should appear high on the agenda for agricultural remote sensing: (a) increasing food production and distribution in countries with chronic malnutrition and widespread poverty; and (b) in all parts of the world, endeavouring to identify and then to minimize damage to the environment caused by agricultural practices. As several participants have pointed out, the achievement of these goals calls for the political will to stimulate both national and international action and to remove scientists from "the bottom of the pecking order" when priorities are assigned for access to data.

References

BECKER, F. and CHOUDHURY, B. J. (1988). Relative sensitivity of normalised difference vegetation index and microwave polarization index for vegetation and desertification monitoring. *Remote Sensing of Environment.* **24**, 297-311

GASH, J. (1986). A note on estimating the effect of a limited fetch on micrometeorological evaporation measurements. *Boundary Layer Meteorology.* **35**, 409-413

GREEN, C. F. (1987). Nitrogen nutrition and wheat growth in relation to absorbed solar radiation. *Agricultural Meteorology.* **41**, 207-248

IDSO, S. B., JACKSON, R. D. and REGINATO, R. J. (1977). Remote Sensing of crop yields. *Science.* **196**, 19-25

KINIRY, J. R., JONES, C. A., O'TOOLE, J. C., BLANCHOT, R., CABELGUENNO, M. and SPANIEL, D. A. (1989). Radiation use efficiency and biomass accumulation prior to grain filling for five grain legume species. *Field Crops Research.* **20**, 51-64

MONTEITH, J. L. (1989). Conservative behaviour in the response of crops to water and light. In *Theoretical Production Ecology: Hindsights and Perspectives.* Pudoc, Wageningen. (in press)

PENMAN, H. L. (1948). Natural evaporation from open water, bare soil and grass. *Proceedings of the Royal Society of London.* **A193**, 120-146

SAHAI, B. and NAVALGUND, R. R. (1988). Indian remote sensing utilization programme for agriculture. In *Remote Sensing in Agriculture* Ed. by B. Sahai. Indian Society of Remote Sensing, Ahmedabad, India

SCHUEPP, P. H., AUSTIN, L. B., DESJARDINS, R. L., MacPHERSON, J. I. and BOISVERT, J. (1987). Airborne determinations of water use efficiency and evapotranspiration. *Agricultural and Forest Meteorology.* **41**, 1-19

LIST OF POSTER PRESENTATIONS

Optimal sampling for agricultural remote sensing
P.M. Atkinson
Department of Geography, University of Sheffield, Western Bank, Sheffield S10 2TN, U.K. and Rothamsted Experimental Station, Harpenden, Hertfordshire AL5 2JQ, U.K.
Remote sensing of photosynthetic activity by fluorescent lidar
Giovanni Cecchi and Luca Pantani
Consiglio Nazionale della Ricerche, Instituto di Ricerca sulle Onde Elettromagnetiche, Via Panciatichi 64, I-50127 Firenze, Italy
The derivation of a weighted infrared - red vegetation index for the estimation of LAI
J.G.P.W. Clevers
Wageningen Agricultural University, Department of Land Surveying and Remote Sensing, P.O. Box 339, 6700 AH Wageningen, Netherlands
High spectral resolution derivatives in remote sensing
T.H. Demetriades-Shah
Evapotranspiration Laboratory, Waters Annex, Department of Agronomy, Kansas State University, Manhattan, Kansas 66506, U.S.A.
Soil ground data collection for agricultural applications of microwave remotely sensed data
G.M. Foody
School of Geography, Kingston Polytechnic, Penrhyn Road, Kingston-upon-Thames KT1 2EE, U.K.
Mapping agricultural crops from synthetic aperture radar data
G.M.Foody
School of Geography, Kingston Polytechnic, Penrhyn Road, Kingston-upon-Thames KT1 2EE, U.K.
Calibration and performance of a C-band ground-based scatterometer
M.R. Gorman[+], M.G. Holmes[*], P.L. Mitchell[*], Z.T. Whishaw[*] and A.D. Diament[*],
[*] Department of Botany, University of Cambridge, Downing Street, Cambridge CB2 3EA, U.K.
[+] Scott Polar Research Institute, University of Cambridge, Lensfield Road, Cambridge CB2 1ER, U.K.
Crop monitoring in Sweden
K. Hall-Könyves
Department of Physical Geography, Remote Sensing Laboratory, University of Lund, Sölveg 13, S-223 62 Lund, Sweden
Remote sensing in the Pacific Islands
M. Kumar
School of Pure and Applied Sciences, University of the South Pacific, Suva, Fiji, and D. van R. Classen
Institute of Natural Resources, University of the South Pacific, Suva, Fiji

Crop canopy spectral reflectance
M. Kumar
School of Pure and Applied Sciences, University of the South Pacific, Suva, Fiji

Phytophenological mapping
D. Lloyd
Bristol University Remote Sensing Unit, University Road, Bristol BS8 1SS, U.K.

Reflectance of layered bean leaves over different soil backgrounds: measured and simulated spectra
J.R. Miller*, M.D. Steven[+] and T.H. Demetriades-Shah[+#]
* Physics Department, and Centre for Research for Experimental Space Science, York University, Toronto, Ontario M3J 1P3, Canada
[+] Department of Geography, University of Nottingham, Nottingham NG7 2RD, U.K.
[#] Current address: Waters Annex, Kansas State University, Manhattan, Kansas 66506, U.S.A.

Soil variability for microwave remote sensing
P.L. Mitchell*, Z.T. Whishaw*, A.D. Diament*, M.R. Gorman[+] and M.G. Holmes*
* Botany School, University of Cambridge, Cambridge CB2 3EA, U.K.
[+] Scott Polar Research Institute, University of Cambridge, Cambridge CB2 1ER, U.K.

Developing an optimum reflectance index for use in crop yield models
E.O. Mongain[+], J. Walsh[+], J. Burke[°] and M. MacSuirtain*
[+] Physics Department, University College, Dublin, Eire
* Forestry Department, University College, Dublin, Eire
[°] Agriculture and Food Research Authority (Teagasc), Oakpark Research Centre, Carlow, Eire

Hand-held radiometry and IR-thermography of plant diseases in field plot experiments
H-E. Nillson,
Swedish University of Agricultural Sciences, Uppsala, Sweden

Satellite identification of armyworm sites in East Africa
T. Robinson,
Department of Geography, University of Reading, Reading RG6 2AB, U.K.

Study of estimation of area of land irrigated by underground water
P. Ruiz Zanon,
Ibersat S.A., Madrid, Spain

Spectral estimates of intercepted photosynthetically active radiation and biomass production of a durum winter wheat affected by nitrogen and water deficiences
S. Steinmetz*, M. Guerif, R. Delecolle and F. Baret
INRA/Station de Bioclimatologie, B.P. 91 – 84140 Montfavet, France
* On leave from EMBRAPA/CNPAF, CP 179, 74000 Goiânia –GO, Brazil

Middle-infrared vegetation indices and their relationship with final grain yield in cereals
A.D. Zmuda and J.C. Taylor,
Silsoe College, Silsoe, Bedfordshire, U.K.

LIST OF PARTICIPANTS

Abdulla, H.	Department of Meteorology, University of Reading, 2 Earley Gate, Reading, RG2 2AU, U.K.
Abreu, F.	Escola Superior Agraria, Praceta Rainha D., Leonor, 7800 - Beja, Portugal
Albar, O.	Department of Physiology and Environmental Science, University of Nottingham, School of Agriculture, Sutton Bonington, Loughborough, LE12 5RD, U.K.
Allan, J.A.	School of Oriental and African Studies, Thornhaugh Street, Russell Square, London, WC1H 0XG, U.K.
Allen, S.J.	Institute of Hydrology, Wallingford, Oxfordshire, OX10 8BB, U.K.
Apponi, G.	Telespazio, Via Alberto Bergamini, 50, 00159 Roma, Italy
Aran, M.	Remote Sensing Department, Trabajos Catastrales SA, Carretera Del Sadar, Edificio El Sario, 31006 Pamplona, Spain
Atherton, J.G.	Department of Agriculture and Horticulture, University of Nottingham, School of Agriculture, Sutton Bonington, Loughborough, LE12 5RD, U.K.
Atkinson, P.	Geography Department, University of Sheffield, Sheffield, S10 2TN, U.K.
Azam Ali, S.	Department of Agriculture and Horticulture, University of Nottingham, School of Agriculture, Sutton Bonington, Loughborough, LE12 5RD, U.K.
Baban, S.	School of Environmental Sciences, University of East Anglia, Norwich, NR4 7TJ, U.K.
Backhaus, R.	DLR, Linder Hohe, 5000 Koln 90, Federal Republic of Germany
Bates, S.	Department of Physiology and Environmental Science, University of Nottingham, School of Agriculture, Sutton Bonington, Loughborough, LE12 5RD, U.K.
Batts, G.R.	Department of Agriculture and Horticulture, University of Nottingham, School of Agriculture, Sutton Bonington, Loughborough, LE12 5RD, U.K.
Bennett, J.	Central Avenue, Chatham Maritime, Chatham, Kent, ME4 4TB, U.K.
Bensaal, D.	Department of Meteorology, University of Reading, 2 Earley Gate, Reading, RG2 2AU, U.K.

Birnie, R.V.	Macaulay Land Use Research Institute, Craigiebuckler, Aberdeen, AB9 2Q3, Scotland, U.K.
Blakeman, R.H.	Aerial Photography Unit, ADAS, Block B, Government Buildings, Brooklands Avenue, Cambridge, CB2 2DR, U.K.
Bosac, C.	Department of Physiology and Environmental Science, University of Nottingham, School of Agriculture, Sutton Bonington, Loughborough, LE12 5RD, U.K.
Bouazie, A.	Department of Meteorology, University of Reading, 2 Earley Gate, Reading, RG6 2AU, U.K.
Buker, C.	Institut fur Pflanzenbau, University of Bonn, Katzenburgweg 5, 5300 Bonn 1, Federal Republic of Germany
Campbell, G.S.	Department of Agronomy and Soils, Washington State University, Pullman, Washington 99164, U.S.A.
Clark, J.A.	Department of Physiology and Environmental Science, University of Nottingham, School of Agriculture, Sutton Bonington, Loughborough, LE12 5RD, U.K.
Clark, C.N.A.	Broom's Barn Experimental Station, Higham, Bury St. Edmunds, Suffolk, IP286NP, U.K.
Clark, A.	Department of Physiology and Environmental Science, University of Nottingham, School of Agriculture, Sutton Bonington, Loughborough, LE12 5RD, U.K.
Clevers, J.G.P.W.	Department of Land Surveying and Remote Sensing, Wageningen Agricultural University, Postbus 339, 6700 AH Wageningen, Netherlands
Colls, J.J.	Department of Physiology and Environmental Science, University of Nottingham, School of Agriculture, Sutton Bonington, Loughborough, LE12 5RD, U.K.
Corlett, J.	Department of Physiology and Environmental Science, University of Nottingham, School of Agriculture, Sutton Bonington, Loughborough, LE12 5RD, U.K.
Craig, M.	Room 4168, South Building, Washington D.C., 20250-2000, U.S.A.
Craigon, J.	Department of Physiology and Environmental Science, University of Nottingham, School of Agriculture, Sutton Bonington, Loughborough, LE12 5RD, U.K.
Crane, A.J.	74 The Street, Bramford, Ipswich, Suffolk, U.K.
Csornai, G.	FOMI Remote Sensing Centre, H-1149 Budapest, Bosnyak ter 5, Hungary
Cutler, J.	Nigel Press Associates, Edenbridge, Kent, TN8 6HS, U.K.

List of Participants

Dambe, D.	Department of Meteorology, University of Reading, 2 Earley Gate, Reading, RG6 2AU, U.K.
Danson, M.	Department of Physiology and Environmental Science, University of Nottingham, School of Agriculture, Sutton Bonington, Loughborough, LE12 5RD, U.K.
Davies, I.O.G.	Smith Associates Ltd., Chancellor Court, Surrey Research Park, Guildford, Surrey, GU2 5YP, U.K.
De Sousa Otto, A.M.	Universidade De Algarve, R. Francisco Horta 9-2.C, 8000 Faro, Portugal
Deane, G.C.	Hunting Technical Services Ltd., Thamesfield House, Boundary Way, Hemel Hempstead, HP2 7SR, U.K.
Decelle, Y.	Unite de Phytotechnie des Regions Temperate, Esa Gembloux, Passage des Deportes 2, B-5800 Gembloux, Belgium
Demetriades-Shah, T.H.	Evapotranspiration Laboratory, Waters Annexe, Department of Agronomy, Kansas State University, Manhattan, Kansas 66506, U.S.A.
Deng, N.	Department of Physiology and Environmental Science, University of Nottingham, School of Agriculture, Sutton Bonington, Loughborough, LE12 5RD, U.K.
Dennis, R.	Inchdryne, Nethy Bridge, Inverness-shire, Scotland, U.K.
Diaz, M.E.	Manuel Perez, 2 P-2 4 D, 41005 Sevilla, Spain
Di Marco, I.	Italeco, Via Carlo Pesenti 108, Roma 00156, Italy
Disfani, M.N.	Department of Geography and Topographic Science, University of Glasgow, Glasgow, G12 8QQ, U.K.
Dockter, K.	Institut fur Pflanzenbau, University of Bonn, Katzenburgweg 5, D-5300 Bonn 1, Federal Republic of Germany
Dumur, D.	Department of Agriculture and Horticulture, University of Nottingham, School of Agriculture, Sutton Bonington, Loughborough, LE12 5RD, U.K.
Evans, R.	Department of Geography, University of Cambridge, Cambridge, U.K.
Faulkner, R.	Civil Engineering Department, Loughborough University, Loughborough, LE11 3TU, U.K.
Figueiredo, A.	Department of Geography, University College London, 26 Bedford Way, London, WC1H 0AP, U.K.
Foody, G.M.	School of Geography, Kingston Polytechnic, Penrhyn Road, Kingston-upon-Thames, Surrey, KT1 2EE, U.K.
Fuchs, M.	Institute of Soil and Water, Volcani Centre, PO Box 6, Bet Dagan 50-250, Israel

Gale, T.	10 Tyneside Road, Loughborough, LE12 5RD, U.K.
Gharres, S.	Department of Physiology and Environmental Science, University of Nottingham, School of Agriculture, Sutton Bonington, Loughborough, LE12 5RD, U.K.
Gilabert, A.	Departement of Termodynamica, Facultat Fisica, Universitat de Valencia, Dr Moliner, 50, 46100 Burjassot, Valencia, Spain
Gomes, I.	I.N.M.G., Rva C, Aeroporto Lisboa, 1700 Lisboa, Portugal
Gonzales-Torralba, F.	Department of Geography, University of Nottingham, Nottingham, NG7 2RD, U.K.
Gorman, M.R.	Scott Polar Research Institute, University of Cambridge, Lensfield Road, Cambridge, CB2 1ER, U.K.
Gray, D.	Department of Geography, University of Edinburgh, Drummond Street, Edinburgh, U.K.
Greenwood, N.	Hunting Technical Services Ltd., Thamesfield House, Boundary Way, Hemel Hempstead, Herts, HP2 7SR, U.K.
Gregson, K.	Department of Physiology and Environmental Science, University of Nottingham, School of Agriculture, Sutton Bonington, Loughborough, LE12 5RD, U.K.
Gunaratne, S.P.	Department of Agriculture and Horticulture, University of Nottingham, School of Agriculture, Sutton Bonington, Loughborough, LE12 5RD, U.K.
Guyot, G.	INRA - Station de Bioclimatologie, B.P. 91, 84 140 Montfavet, France
Guzman, O.	Department of Meteorology, University of Reading, 2 Earley Gate, Reading, RG6 2AU, U.K.
Hall-Konyves, K.	Department of Physical Geography, University of Lund, Solvegatan 13, S-22362 Lund, Sweden
Hanan, N.	School of Biology, Queen Mary College, Mile End Road, London E1, U.K.
Hassan, B.	School of Plant Biology, University College of North Wales, Bangor, Gwynedd, U.K.
Hebblethwaite, P.D.	Department of Agriculture and Horticulture, University of Nottingham, School of Agriculture, Sutton Bonington, Loughborough, LE12 5RD, U.K.
Hervas De Diego, F.J.	ESA/Estec Code OP, Postbus 299, NL2200 AG Noordwijk, Netherlands
Hinde, B.	NERC Scientific Services, Polaris House, North Star Avenue, Swindon, SN2 1EU, U.K.

List of Participants

Holmes, M.G.	Department of Botany, University of Cambridge, Downing Street, Cambridge, CB2 3EA, U.K.
Hollwill, P.	Remote Sensing, Institute of Hydrology, Crowmarsh Gifford, Wallingford, Oxon, OX10 8BB, U.K.
Horwood, P.	Butterworths, PO Box 63, Westbury House, Bury Street, Guildford, GU2 5BH, U.K.
Hutchinson, P.	Department of Meteorology, University of Reading, 2 Earley Gate, Reading, RG6 2AU, U.K.
Igualada, F.J.	Trabajos Catastrales SA, Carretera Del Sadar, Edificio El Sario, 31006 Pamplona, Spain
Istasse, A.	Station de Phytopathologie, Chemin de Liroux, B5800 Gembloux, Belgium
Jaggard, K.W.	Broom's Barn Experimental Station, Higham, Bury St. Edmunds, Suffolk, IP28 6NP, U.K.
Jensen, A.	Botanical Institute, Nordlandsvej 68, 8240 Risskov, Denmark
Kanemasu, E.	Evapotranspiration Laboratory, Waters Annexe, Kansas State University, Manhattan, Kansas 66506, U.S.A.
Kantar, F.	Department of Agriculture and Horticulture, University of Nottingham, School of Agriculture, Sutton Bonington, Loughborough, LE12 5RD, U.K.
Kershaw, C.D.	Statistics Department, Rothamsted Experimental Station, Harpenden, Herts, AL5 2J4, U.K.
Kumar, M.	Physics Department, School of Pure and Applied Sciences, University of the South Pacific, Suva, Fiji
Lamontagne, S.	Department of Meteorology, University of Reading, 2 Earley Gate, Reading, RG6 2AU, U.K.
Legg, B.J.	AFRC Institute of Engineering Research, Wrest Park, Silsoe, Bedford, MK45 4HS, U.K.
Mat Lela, M.S.	Department of Geography, University of Nottingham, Nottingham, NG7 2RD, U.K.
Li Yuzhu,	Institute of Agrometeorology, Academy of Meteorological Science, State Meteorological Administration, Baishiqiaolu 46, Beijing 100081, People's Republic of China
Lichtenthaler, H.K.	Botanisches Institut II, Universitat Karlsruhe, Kaiserstrasse 12, D-7500 Karlsruhe, Federal Republic of Germany
Lloyd, D.	Remote Sensing Unit, Department of Geography, University of Bristol, Bristol, BS8 1SS, Avon, U.K.

Luo Cheng,	Department of Agriculture and Horticulture, University of Nottingham, School of Agriculture, Sutton Bonington, Loughborough, LE12 5RD, U.K.
Luzi, G.	Centre of Microwave Remote Sensing, Viale Galileo 32, Florence, Italy
McArthur, A.J.	Department of Physiology and Environmental Science, University of Nottingham, School of Agriculture, Sutton Bonington, Loughborough, LE12 5RD, U.K.
Madeira, A.C.	Instituto Superior de Agronomia, (Agrometeorologia), Universitaat de Lisboa, Tapada da Ajuda, Lisboa, Portugal
Madouh, M.	Department of Meteorology, University of Reading, 2 Earley Gate, Reading, RG2 2AU, U.K.
Makhanya, E.M.	University of Zululand, Private Bag X10, Isipingo, 4110, Zululand, Republic of South Africa
Malthus, T.J.	Department of Physiology and Environmental Science, University of Nottingham, School of Agriculture, Sutton Bonington, Loughborough, LE12 5RD, U.K.
Mather, P.	Department of Geography, University of Nottingham, Nottingham, NG7 2RD, U.K.
Meyer-Roux, J.	Institute of Applications of Remote Sensing, CEC Joint Research Centre, 21020 Ispra (Varese), Italy
Milford, J.R.	Department of Meteorology, University of Reading, 2 Earley Gate, Reading, RG6 2AU, U.K.
Miller, J.	Complex Systems, Institute for Earth, Oceans and Space, University of New Hampshire, Durham, NH 03824, Canada
Mitchell. P.L.	Botany School, University of Cambridge, Downing Street, Cambridge, CB2 3EA, U.K.
Moggridge, H.T.	Colvin and Moggridge, Landscape Consultants, Filkins, Lechlade, Gloucestershire, GL7 3JQ, U.K.
Monteith, J.L.	Resource Management Programme, ICRISAT, Patancheru PO, AP 502324, India
Munikil, A.	Department of Agriculture and Horticulture, University of Nottingham, School of Agriculture, Sutton Bonington, Loughborough, LE12 5RD, U.K.
Munthali, G.	Department of Meteorology, University of Reading, 2 Earley Gate, Reading, RG6 2AU, U.K.
Musa, M.R.	Department of Agriculture and Horticulture, University of Nottingham, School of Agriculture, Sutton Bonington, Loughborough, LE12 5RD, U.K.

List of Participants

Nichol, C.	Delta-T Devices, 128 Low Road, Burwell, Cambridge, CB5 0EJ, U.K.
Nieuwenhuis, G.J.A.	Staring Centre, Postbus 35, 6700 AA Wageningen, Netherlands
Nillson, H-E	Department of Plant and Forest Protection, Swedish University of Agricultural Sciences, Box 7044, S-75007, Uppsala, Sweden
Nmiri, A.	Department of Meteorology, University of Reading, 2 Earley Gate, Reading, RG6 2AU, U.K.
O'Rourke, E.	1 York Road, Rathmines, Dublin 6, Ireland
Oxley, E.R.B.	School of Biological Sciences, University College of North Wales, Bangor, Gwynedd, U.K.
Paris, J.	Department of Geography, California State University, Fresno, California 93740-0069, U.S.A.
Partington, K.C.	GEC-Marconi Research Centre, West Hanningfield Road, Great Baddow, Chelmsford, Essex, CM3 4QX, U.K.
Pillinger, R.N.	Shell Research Ltd., Sittingbourne Research Centre, Sittingbourne, Kent, ME9 8AG, U.K.
Posselt, W.	Messerschmitt-Bolkow-Blohm GMBH, PO Box 80 11 69, 8000 Munchen 80, Federal Republic of Germany
Prince, S.D.	Code 623, Earth Resources Branch, NASA/Goddard Space Flight Centre, Greenbelt, Maryland 20771, U.S.A.
Quarmby, N.A.	Department of Geography, University of Reading, Whiteknights, Reading, RG6 2AB, U.K.
Radhi, M.I.	School of Environmental Sciences, University of East Anglia, Norwich, NR4 7TJ, U.K.
Read, G.	47 Brookfield Avenue, Loughborough, Leics, LE11 3LN, U.K.
Remotti, D.	Italeco, Via Carlo Pesenti 108, Roma 00156, Italy
Robinson, T.	Department of Geography, University of Reading, Whiteknights, Reading, RG6 2AB, U.K.
Rohr, W.	Angstorf, CH-3186 Dudingen, Switzerland
Rollin, E.	Department of Geography, University of Southampton, Southampton, SO9 5NH, U.K.
Rossini, P.	Telespazio, Via Alberto Bergamini, 50, 00159 Roma, Italy
Rothfus, H.	DLR-Oberpfaffenhofen, Munchener Strasse, 8031 Wessling, Federal Republic of Germany
Ruiz-Zanon, P.	IBERSAT S.A., Velazquez 24, 28001 Madrid, Spain

Ruwwe, T.	Institut fur Photogrammetrie, University Bonn, Nu Ballee 15, 5300 Bonn, Federal Republic of Germany
Sanders, G.	Department of Physiology and Environmental Science, University of Nottingham, School of Agriculture, Sutton Bonington, Loughborough, LE12 5RD, U.K.
Scott, R.K.	Department of Agriculture and Horticulture, University of Nottingham, School of Agriculture, Sutton Bonington, Loughborough, LE12 5RD, U.K.
Scurlock, J.	Department of Biology, Cambridge University, King's College, Cambridge, U.K.
Sedgley, R.H.	School of Agriculture, University of Western Australia, Nedlands, Western Australia 6009, Australia
Shueb, S.	Geography Department, University of Durham, Durham, U.K.
Sousa, L.C.	Escola Superior Agraria de Beja, Praceta Rainha D., Leonor, 7800 - Beja, Portugal
Stafford, J.V.	AFRC Institute of Engineering Research, Wrest Park, Silsoe, Bedford, MK45 4HS, U.K.
Steinmetz, S.	INRA - Station de Bioclimatologie, B.P. 91, 84140 Montfavet, France
Steven, M.D.	Department of Geography, University of Nottingham, Nottingham, NG7 2RD, U.K.
Sweet, N.	Department of Physiology and Environmental Science, University of Nottingham, School of Agriculture, Sutton Bonington, Loughborough, LE12 5RD, U.K.
Tal, A.	Space and Remote Sensing Division, Centre for Technology, Analysis and Forecasting, Tel-Aviv University, 69978 Ramat-Aviv, Israel
Tempany, K.	St. Mary's, Avoca Avenue, Blackrock, Co. Dublin, Ireland, U.K.
Tobin, A.	Logica Space and Defence Systems, Ltd., 68, Newman Street, London, W1A 4SE, U.K.
Tsivion, Y.	Givat Ada, 37808, Israel
Unsworth, M.H.	Department of Physiology and Environmental Science, University of Nottingham, School of Agriculture, Sutton Bonington, Loughborough, LE12 5RD, U.K.
Vossen, P.	CEC Joint Research Centre, 21020 Ispra (Varese), Italy
Walsh, J.	Physics Department, University College Dublin, Belfield, Dublin 4, Ireland,
Webster, R.	Rothamsted Experimental Station, Harpenden, Herts, AL5 2JQ, U.K.

List of Participants

Wegmuller, U.P.	Institute of Applied Physics, University of Bern, Sidlerstrasse 5, CH-3012 Bern, Switzerland
Wheldon, A.	10 Enfield Street, Beeston, Notts, NG9 1AL, U.K.
Whishaw, Z.	Botany School, University of Cambridge, Downing Street, Cambridge, CB2 3EA, U.K.
Wilton, B.	Department of Agriculture and Horticulture, University of Nottingham, School of Agriculture, Sutton Bonington, Loughborough, LE12 5RD, U.K.
Winings, S.	USDA-NASS, Room 4168 South, Washington, D.C., 20250-2000, U.S.A.
Wood, J.	Delta-T Devices, 128 Low Street, Burwell, Cambridge, CB5 0EJ, U.K.
Worthington, A.	Department of Physiology and Environmental Science, University of Nottingham, School of Agriculture, Sutton Bonington, Loughborough, LE12 5RD, U.K.
Wright, C.J.	Department of Physiology and Environmental Science, University of Nottingham, School of Agriculture, Sutton Bonington, Loughborough, LE12 5RD, U.K.
Wyatt, B.K.	Monk's Wood Experimental Station, Abbots Ripton, Huntingdon, PE17 2LS, U.K.
Xavier, S.	Unite de Phytotechnie des Regions Temperate, Telsat/03/2, Passage des Deportes 2, 5800 Gembloux, Belgium
Xie Jingrong,	Department of Agriculture, University of Aberdeen, 581 King Street, Aberdeen, AB9 1UD, U.K.
Zheng Dawei,	Institute of Comprehensive Development of Agriculture, Beijing Academy of Agricultural and Forestry Sciences, PO 2449, Beijing, China

Index

above-ground biomass, 187
accumulated intercepted PAR, 190
accuracy of predictions, 204
acquisition dates, 383
acquisition windows, 383
Action
    CEC Remote Sensing programme, 387
active-microwave, *see* microwave and radar
active-microwave and optical data, 355–373
Advanced Very High Resolution Radiometer, 150
Advanced Very High Resolution Radiometer, *see* AVHRR
aerial photographs
    crop husbandry
        mistakes, 245
aerial photography, 229–250, 398
    advantages, 229
    barley yellow dwarf virus, 237
    barley yellow mosaic virus, 241
    crop reflectance, 229
    crop trials, 247
    film type
        black and white infra-red, 233
        colour infrared, 233
        panchromatic film, 232
        true colour film, 232
    image analysis, 247
    interpretation, 231
    potato blight, 236
    resolution, 230, 231
    rhizomania, 244
    soil structure, 245
    striping, 245, 246
    sugar beet virus yellows, 236
    take-all, 238
    yellow rust, 236
aerodynamic resistance, 46, 47
    stability corrected, 48
aerosol, 31
aerospace data, 138
age classification
    crop, 322
Agricultural Development and Advisory Service (ADAS), 229
agricultural information needs, 7
agricultural space organisation, 8
agricultural statistics
    pilot project, 387
Agricultural Vegetation Index, 142
Agriculture project
    EEC, 382
Agriscat project, 382
AgRISTARS, 359, 381, 382
agrometeorological models, 386
agronomic data, 139, 150

air photos, 76, 85
    best acquisition time, 82
    oblique, 82
air temperature, 46
airborne imaging radar, 361
airborne reflectance measurements, 204
airborne scanner
    instability, 117
airborne scanner imagery
    geometric correction, 117
    radiometric correction, 117
airborne sensors, 127
airborne systems
    chlorophyll fluorescence, 300
aircraft, 398
aircraft measurement, 336
albedo
    single scattering, 347
alfalfa, 338, 366
algorithms
    edge detection, 133
all-weather systems, 9
angular dependence
    radar backscatter, 316
annual leaf production
    woody vegetation, 172
area estimation, 154, 162
area prediction, 12
area sampling frame, 138, 379, 383
areal green cover, 356, 364, 368, 372
arid environment, 269
array devices, 4
ARTEMIS project, 12, 401
atmospheric
    water vapour, 386
atmospheric absorption, 277, 399
atmospheric aerosols, 31
atmospheric correction, 277, 279
    NDVI, 176
atmospheric effects, 151
    greenness index, 152
atmospheric optical thickness, 177
atmospheric scattering, 277
atmospheric stability, 113, 260
atmospheric water vapour
    surface temperature, 67
available water, 107
AVHRR, 127, 154, 355, 384
    multidate composites, 173
    nominal resolution, 170

backscatter, 313
    instrument characteristics, 309
    polarization
        microwave, 366

row direction, 318
soil moisture, 316
backscattering coefficients
   diurnal variations, 373
   woody vegetation, 371
bare soil, 341
barium sulphate
   standard panel, 189
barley yellow dwarf virus
   aerial photography, 237
   distribution of, 239
   effect of drilling date, 240
   focal pattern, 239
barley yellow mosaic virus
   aerial photography, 241
   varietal susceptibility, 241
Bayesian classifier, 160
beet crops
   productivity, 202
beet yield
   forecasts, 201
biomass
   rainfall, 196
biomass per unit ground area, 185
biomass production, 107
biomes, 170
Boltzmann's constant, 332
boundary layer resistance, 257
brightness, 332
brightness temperature, 332, 336, 339, 347, 385
   crop type, 339
British Sugar, 201
broad-band optical data, 356

C-band backscatter, 366
C-band microwave scatterometer, 371
calibration localisation, 103
calibration procedure
   NOAA-9 AVHRR, 176
calibration zones, 106
Canada, 382, 383
candidate index, 222
candidate indices
   for stress, 218
canopy, 398
   air temperature, 255
   development, 204
   diffusion, 256
   emissivity, 65
   geometry, 34, 55, 185, 186
   illumination, 29
   infrared temperature, 268
   layer
      microwave propogation, 334
   leaf angle distribution, 190
   microwave penetration, 311
   pruned, 222

radiance, 53
red-edge, 217
reflectance, 33, 275
   saturation, 27
   viewed area, 29
reflectance dependence, 31
resistance, 47, 64, 255, 257–259
roughness, 48
size distribution, 310
temperature, 45, 255–270
   distribution, 45
   effect of stomatal closure, 61
   inversion of measurements, 256
   response to environment, 262
   sensitivity, 262
   spatial variability, 54
   uncertainty, 262
   view angle, 267
   water deficit, 60
canopy height, 260
canopy structure, 61
canopy temperature, 255
   response to water stress, 265
canopy-air temperature difference, 64, 255, 262
carbon
   assimilation, 287
   respiratory loss, 176
cartography, 7
cell content, 21
cereal production
   estimation, 137
China, 137
chlorophyll, 19, 23, 24, 220
   content, 279, 294
   fluorescence, 287
      in vivo, 287
      induction kinetics, 288
   fluorometer, 291
   reflectance, 359
chlorophyll fluorescence, 302
   light-induced, 289
   remote sensing, 300
chlorophyll fluorescence spectra, 292
chlorosis, 24, 205, 210, 213, 219, 220
choice
   film, 233
citrus, 273
citrus-canopy reflectance, 281
class membership
   probability, 130, 132
classification, 118, 160, 300, 381, 383
   accuracy, 249, 323
      contextual information, 131
      sample size, 131
   correctness, 164
   decision rules, 127
   map, 163

Index 417

method, 162
multi-temporal, 382
multispectral images, 128
of soil using radar, 320
per pixel approach, 128
problems, 127
rules, 127
success rate, 323
training, 382
classifier
    calibration, 130
    contextual, 131
    per-field, 163
cloud cover
    irradiance, 31
    problems, 9
cloud duration, 99
cloud free views, 176
clustering
    analysis, 139
clustering method, 160
cold cloud
    duration and flood forecasting, 108
    statistics, 98
cold damage, 274
cold damage algorithm, 281
colour composite
    Landsat, 161
colour film, 79
colour infrared, 79
Columbus Project, 3, 14
combined optical and active-microwave data, 323, 355
commodities
    main, 382
Common Agricultural Policy
    CEC, 387
complementarity of data sources, 4
constraints
    financial, 385
    resolution, 385
conversion coefficient (conversion efficiency), 202, 209
corrections
    geometric, 391
correlation spectrum, 221
cost effectiveness, 392
cost per square kilometer, 15
coverage
    high frequency, 391
    Landsat, 388
    SPOT, 388
crop
    age classification, 322
    characteristics, 114
    classification, 248
    identification, 230, 248, 360

confidence, 164
crop acreage, 388
crop assessment, 149
crop canopies, 218
crop chlorophyll content, 154
crop classification, 111, 118, 320
    accuracy, 321
crop condition, 229
crop cover, 275
crop damage
    identification, 322
    identification on aerial photographs, 231
crop development, 159
crop discrimination by radar, 310
crop disease, 233, 236
    and stress, 229–250
crop growth rate, 401
crop height, 117
crop husbandry, 244
    mistakes
        aerial photographs, 245
crop inventory, 159, 160, 377, 379, 380
    objectives, 383
    pilot regions, 382
crop mapping, 159, 160
crop models, 398
crop monitoring, 159, 377, 386
crop patterns, 77, 83
    appearance, 83
crop phenology, 8, 10
crop production, 401
crop reflectance, 231
    spectral, 144
crop stress, 233
    and disease
        identification, 249
crop survey, 161
crop temperature, 111, 112, 118
crop transpiration, 112, 118
    determination, 113
    regional, 111, 121
crop trials
    aerial photography, 247
crop type, 118
    brightness temperature, 339
crop vigour and yield, 231
crop water status, 112
crop water stress, 255, see water stress, 269
    prediction from canopy temperature, 256
Crop Water Stress Index (CWSI), 255
crop yield, 83, 385, 389
    forecasting, 159
cross-polarization, 312
    scattering coefficient, 311
curve fitting, 218

Daedalus airborne scanner, 117, 119

dark-object subtraction technique, 277
data
    processing, 159, 217
    processing resolution, 10
    reduction, 11, 217
        algorithms, 224
    timeliness, 249
data banks, 397
data integration, 12, 13
density slicing, 154
derivative indices, 218
derivative spectra, 400
derivative techniques, 218
dielectric behaviour
    soil-water mixtures, 314
dielectric constant, 333
    mixture, 334
    of vegetation, 319
dielectric layers, 344
diffuse radiation, 190
diffusion resistance, 257
digital field database, 162
digitised maps, 111
directional properties of leaves, 22
discrimination, 381
    visual, 378
diseases
    colour change, 235
divergence analysis, 129
drainage class, 118
drought, 400
    and light intercepted, 212
drought detection, 322
dry matter accumulation, *see* dry matter production
dry matter per unit of radiation intercepted, 400
dry matter production, 185, 194, 257, 400
    above ground, 192
    standard deviation, 195

Earth Observing System (EOS), 308, 324, 355
    capability, 310
    Synergism Study, 372
Earth resources satellite, 308
economics, 3
eddy correlation, 399
EEC, 382
Effective Aerial Film Speed (EAFS)
    black and white infrared, 233
    colour infra-red film, 233
    panchromatic film, 232
    true colour film, 232
efficiency of conversion, 179
    dry matter production, 196
electromagnetic radiation
    interaction with plants, 19
emission, 399

emission spectra
    fluorescence, 287
emissivity, 46, 332, 351
    plant surfaces, 65
    soil, 65
energy balance, 46, 385
erosion, 75, 85
    rate, 85, 86
erosion and overgrazing, 85
error analysis, 161
error of predicted yield, 155
errors
    radiative temperature, 267
estimating production of winter wheat, 137–146, 149–156
estimation
    of grain production, 187
    yield, 142
estimation of total production, 143
European Remote Sensing Satellite I (ERS I), 14
evaporation, 107, 399
evaporative cooling of leaves, 47
evapotranspiration, 46, 386, 390
    flux, 112
eye-brain in remote sensing, 7

facet transformation method, 117
farm maps, 162
farmers yields, 204
feasibility studies, 160
feature analysis, 76
field boundaries, 130, 163, 339, 362
field homogeneity, 163
field measurements
    Mali, 175
field sampling method, 172
field size, 380
fields surveyed, 204
film
    choice of, 233
    panchromatic, 232
    type, 79
film/filter combinations, 229
filtering, 281
flight direction, 117
flood forecasting, 108
fluorescence
    life-time, 301
    blue-light induced, 296
    emission spectra, 287
    ground level, 288
    kinetics
        spruce, 296
    laser-induced, 291
    parameters, 288
    physiological state, 295

Index

pulse kinetics, 288
ratio, 296
red-light induced, 298
steady-state, 289
fluorescence emission spectra
chlorophyll, 297
fluorescence kinetics, 291
fluorometer
Laser-Induced Computer Aided, 296
portable, 291
pulse-modulation, 298
foliage, see canopy
footprint size, 9
forecasting system, 204
forecasts
yield per unit area, 201
forest, 35, 121, 357
chlorophyll fluorescence, 299
reflectance, 27
forest decline research, 290
forest inventories, 378
forest survey, 9
fraction of light intercepted, 203
fraction of PAR intercepted, 188
fraction of radiation intercepted, 400
frequency histograms
radiative temperatures, 57
Fresnel reflectivity, 344
Fresno County experiment, 361
friction velocity, 260
frost damage, 273, 279
identification and evaluation, 279
frosts
evaluating damage, 281
future platforms, 3
future sensors, 3

geographical correction, 151
Geographical Information System (GIS), 12, 13, 111–122, 159, 162, 390, 401
use in water management, 115
geographical location, 336
geometric correction, 278
geostationary satellites, 97
Germany, 388
global climate, 13
global systems, 12
global vegetation
stratification, 170
Global Vegetation Index, 170
goals of remote sensing, 355
grain number, 188
grassland, 118
canopy interception, 190
biomass, 185–197
reflectance, 189
grassland savannas, 169

Green Leaf Area Index, 356, 368, 372
Green Range, 142
greenness, 12, 356, 360
Kauth-Thomas, 356
greenness index
atmospheric effects, 152
greenness indices, 150
gross primary production, 170
ground control points, 117
ground cover, 33
radiative temperature, 57
ground data, 250
ground measurements, 191
ground monitoring, 150
area classification, 152
ground monitoring network, 137, 146, 150
ground network, 149
ground observations, 378
ground radar images, 84
ground radiance, 128
ground roughness, 84
ground survey, 381, 383
ground truth, 398
ground-based measurements, 189
ground-truth by chlorophyll fluorescence, 301
groundwater
regime, 114
groundwater extraction, 119
effects on transpiration, 120
groundwater table, 119
growing season
duration, 178
Sahel, 173
growth efficiency, 170
seasonal, 171
growth vigour monitoring, 152

harvest, 172
harvest method, 175
haze radiance, 278
herbaceous production, 173
rangeland, 176
Sahel, 177
HIgh Resolution Imaging Spectrometer (HIRIS), 14, 308, 324
High Resolution Multi-frequency Microwave Radiometer, HMMR, 15
high spectral resolution, 215
high temporal frequency remote sensing, 169–179
Hungary
soils, 160
hydraulic conductivity, 120
hydrological model, 112

image
analysis

aerial photography, 247
classification, 127–134, 381
  cost, 149
  digital analysis, 80
  processing, 115, 159
  segmentation, 133
  type, 79
image distortion, 308
image superposition, 278
image texture, 130
images
  thermal infrared, 115
imaging radar, 307
imaging spectrometers, 215, 216, 223
imaging system
  geometry, 127
imaging systems, 307
incidence angle
  influence, 312
  leaf reflectance, 22
incident radiation, 171
information transmission, 155
infrared aerial photographs, 141
infrared channels, 14
infrared film, 145
infrared radiometer, 336
  imaging, 49
infrared reflectance, 20
  fungal infection, 235
  vigour, 235
infrared temperature, 339
  canopy, 268
Inter-Tropical Convergence Zone, 173
intercepted PAR, 193
intercepted radiation, 400
  estimated from NDVI, 190
interpretation
  stereoscopic, 78
inventories, 401
inventory
  automated methods, 392
  crop, 380
inverse problem analysis, 255–270
inversion technique, 223
irrigation, 11
irrigation of grassland, 119
isothermal net radiation, 46

Kauth-Thomas Greenness, 356
Kauth-Thomas yellowness, 358

L-band, 360
Lambert's law, 51
Lambertian diffusers, 30
Lambertian reference, 357
land class, 381
land cover, 378–380

land use, 378, 379
  patterns, 81
land use map, 118
land-use classification, 389
Landsat, 11, 15, 78, 141, 146, 149, 150, 159–164,
    273, 276, 355, 356, 361, 377, 382, 384
  cloud free images, 248
Landsat MultiSpectral Scanner (MSS), 128, 130,
    189, 382
Landsat Thematic Mapper (TM), 115, 117, 128,
    215, 359
Landsat with radar, 323
Large Area Crop Inventory Experiment (LA-
    CIE), 12, 149, 356, 381, 384, 386
laser-induced fluorescence, 291
latent heat flux, 46, 112, 260
leaf
  age, 23
  colour, 219
  condition, 231
  inclination distribution function, 34
  layering, 219
  optical properties, 22, 23
    disease attacks, 25
    pest attacks, 25
  orientation, 231
  pigments, 20
  reflectance, 19, 24, 190
  senescence, 23
  structure, 273
  thickness, 221
  transmittance, 19
leaf angle, 34
leaf angle distribution, 186–188
leaf area, 231
leaf area index, 34, 139, 185, 186, 190, 209, 220,
    335, 336, 341, 347, 385
  (corn), 349
  alfalfa, 349
  canopy reflectance, 27
leaf cover, 219
leaf optical properties, 24
leaf reflectance spectrum, 215
leaf structure, 22
leaf turgor, 258
leaf water
  content, 23, 221
  potential
    critical, 258
leaf water content and reflectance, 24
leaves
  evaporative cooling, 47
  horizontal, 339
light diffusion by leaves, 21
light interception, 185, 187, 203, 385, 398
  nitrogen application, 210

Index

light use efficiency, *see* efficiency of conversion, 192
  rainfall, 196
line of soils, 362
local area coverage
  AVHRR, 173
location, 377, 378
locusts
  breeding conditions, 105
lodging, 231

Mahalanobis distance, 130
Mali, 169, 171, 174
map
  production, 118
  scales, 75
mapping soils, 87
masking, 281
Maximum Likelihood, 131
  algorithm, 118
  rule, 128
Maximum Likelihood classifier, 128
measured radiance, 278
measurement conditions, 51
meteorological data, 114
meteorological satellite, 169
meteorological variables, 255
Meteosat, 8, 12, 98, 102, 141, 146
micro wave bands, 9
microwave
  active, 307
  angle of incidence, 312, 361
  channels, 14
  passive, 307
  resolution, 10
  vegetation penetration, 361
microwave backscatter, 84, 309
  crop morphology, 318
  soil, 313
microwave data, 9, 84
microwave emission
  dielectric constant, 331
  soil and vegetation
    scattering, 343
  soil moisture, 334
microwave emissivity, 84, 331
microwave penetration
  canopy, 313
microwave polarization, 311, 336
microwave propogation
  air-vegetation mixture, 335
microwave radiometer, 332, 399
  K-band, 336
  L-band, 336
  X-band, 336
microwave remote sensing, 308
microwave scatterometer

  C-band, 366
microwave sensors
  active, 331
  passive, 331
microwave signatures of canopies, 10
microwave sounder, 4
microwaves
  penetration depth, 317
middle-infrared, 115
  reflectance, 21
mineral deficiencies, 24
mis-classification, 160, 163
model, 171, 398
  agronomic, 143
  atmospheric radiative transfer, 177
  light transmission and reflection, 219
  plant growth and yield, 185
  radar backscatter, 318
  radiation canopy, 186
  semi-deterministic, 390
  statistical, 390
  thermal radiance, 268
  water stress, 256
  yield prediction, 389
model statistic, 98
models
  deterministic, 390
Moderate Resolution Imaging Spectrometer (MODIS), 15, 324
moisture availability, 119
moisture-stress index, 359
monitoring
  vegetation, 384
  crop growth and development, 144
  crops, 386
  high frequency, 385
multi-spectral radar, 10
multi-spectral scanner, 3, 4, 378
multi-temporal studies
  radar, 320

nationwide crop estimation in China, 149
NDVI, 179, 385
  atmospherically corrected data, 175
  average, 179
  composites, 177
  drought, 179
  scene, 179
near-infrared, 169
near-infrared bands, 385
necroses, 25
net climatic radiation, 46
net radiation, 45, 46, 112, 260, 265
neutral conditions, 47
Newton-Raphson procedure, 261
nitrogen, 219
  deficiency, 24

NOAA AVHRR, 8, 32, 102, 149, 153, 169–179, 385
Normalised Difference Vegetation Index (NDVI), 33, 107, 118, 142, 171, 186, 213, 275, 277, 358, 385, 400, 402
normalised temperature, 332
Normalised-Difference Stem-Angle Index, 370
nutrient deficiency, 400

observation
　interval, 169
observation angle, 338
observing programmes, 13
observing technologies, 4
off-nadir reflectance, 31
operational application
　satellite images, 380
optical
　data, 355–373
　properties, 34
　　canopy, 275
　　of bark, 25
　　of cones, 25
　　of flowers, 25
　　of leaves, 19
　　of soil, 19, 25
　　of vegetation, 19–37
optimal monitoring phases, 151
orchard, 357, 362
　data, 363
orientated rows, 57
output services, 155
overgrazing, 85

panchromatic film, 79
parcel boundaries, 9
parcel size, 9
parcels
　citrus, 275
partitioning, 379
pattern recognition, 248
penetration depth
　polarization differences, 317
　wavelength-dependence, 317
Penman transform, 260
per-pixel approach
　difficulties, 130
percent correct classification, 160
percent crop cover, 276
periglacial features, 77
Perpendicular Transformation, 356
Perpendicular Vegetation Index, 142, 356
phenological development, 386
phenological stage, 150
photo-interpretation, see aerial photography, 231
photochemical quenching coefficient, 298
photographs
　digitised, 248
photography
　multi-temporal, 250
photosynthesis, 185, 187, 287
photosynthetic capacity, 288
photosynthetic induction, 288
photosynthetic pigments, 287
photosynthetic rates
　fluorescence, 290
photosynthetically active radiation
　absorbed, 177
photosynthetically active radiation (PAR), 170, 186
physiographic maps, 78
pixel area, 9, 128
pixel context, 130, 131
pixel size, 127
　radiative temperature, 55
pixels
　percentage correctly classified, 381
plant height, 336
plant stress
　microwave radiometry, 332
plant water
　content, 335, 336, 347, 399
　status, 255–270
platform stability, 82
platforms, 3–15, 186
　large, 14
polar-orbiting satellite, 169
polarization, 309, 337
　microwave, 372
Polarization Index, 342
portable fluorometer, 291
portable radiometer, 144
potato blight, 229, 236
potential soil moisture deficit (PSMD), 83
potential transpiration, 114
prairie grasslands, 189
precipitation monitoring
　accuracy, 97
　costs, 97
predicting yield of sugar beet, 201–205
prediction
　experiments, 155
　models, 153
　quasi-operational, 155
primary production, 169–179
　global, 170
　gross, 176
　net, 175
production
　below ground, 172
　confidence limits, 174
　forecasts, 201
　models, 175
　regulation, 11

Index

productivity, 210
Progressive Transformation, 362
psychrometer constant, 257
    apparent, 257
'push-broom' devices, 3

quasi-operational prediction, 155

radar, 9, 307–324, 331
    advantages, 308
    attenuation
        seasonal variation, 321
    frequency, 310
    incidence angle effects, 312
    multi-temporal, 323
    real aperture, 307
    row direction, 318
    synthetic aperture, 307, 360
radar and weather, 308
radar backscatter, 84
    angular dependence, 316
    canopy structure, 319
    crop geometry, 318
radar imagery, 307
radar remote sensing, 308
radar return
    crop species, 319
radar return signal
    row direction, 319
radiance
    canopy, 53
radiation
    use efficiency, 400
radiation absorbed, 263
radiation thermometry, 255
radiative temperature, 45, 51, 111, 399
    angular variation, 268
    apparent, 51
    atmospheric effects, 67
    canopy geometry, 55
    emissivity, 65
    errors, 267
    evapotranspiration rates, 60
    ground cover, 57
    of leaves, 25
    pixel size, 55
    row orientation, 59
    sun angle, 49
    variability, 49
    view angle, 52, 53, 56
radiative temperatures
    canopy height, 59
    frequency histograms, 57
radiative transfer theory, 344
radiometer
    4-band, 189
    infrared, 333

rain-fed
    agriculture, 107
rain-fed crops, 105
raindays, 102
rainfall, 139, 197
    actual, 103
    calibration, 100
    data skewness, 100
    dekads, 103
    distribution, 401
    estimation, 97
        accuracy, 102, 104
        calibration, 99
        uncertainty, 98
        use, 105
        validation, 99
    Ethiopia, 105
    prediction, 12
    ten-day, 100
rainfall amounts, 100
rainfall estimation, 102, 106, 109
raingauge measurements, 98
rangeland herb production, 176
    Sahel, 172
raster data, 14
Rayleigh-Jeans formula, 332
Real Aperture Radar (RAR), 307
reconnaissance stage, 76
red-edge, 216
reflectance
    angular variation, 30
    citrus, 276
    coniferous forests, 29
    nadir viewing, 30
    soil contribution, 34
    variation coefficient, 27
    wheat leaves, 23
    wind speed, 33
reflectance ratio
    canopy IR/R, 186
    near-infrared to red, 185, 203
reflectance spectra
    during senescence, 24
    plant canopies, 26
    plant leaves, 19
refractive index, 22
regional inventories, 382, 383, 388
regional production, 387
regional transpiration, 111
regression estimator, 381
regression model, 98
Relative Vegetation Index, 142
remote sensing
    applications, 377
        ground based, 255
        operational use, 387
        techniques, 87

tractor mounted, 224
resistance
    canopy, 255
resolution
    MSS, 378
    nominal, 170
    SPOT, 378
    TM, 378
resolving power of observing system, 9
rhizomania
    aerial photography, 244
    soil movement, 244
rice production, 137
Richardson number, 48
rill and gully erosion, 85
root distribution, 257
root/shoot ratio, 175
roughness, 113
row crops
    temperature variability, 51
row orientation
    reflectance, 34
    temperature distribution, 57
runoff, 103, 108

Sahel, 169, 385
    rainfall, 97
    rangelend production, 172
    West Africa, 170
Sahel grasslands, 177
Sahel growing season, 173
Sahel rainfall, 103
sample areas, 172
sample size, 99
sampling, 10, 54
    error, 152
    multi-temporal, 139
    strategy, 379
        Europe, 391
sampling error, 140
sampling interval, 103
satellite, 397
    data
        atmospheric correction, 171
        high resolution, 386
        low resolution, 388, 389
        uncorrected, 171
    measurements, 169, 191
    meteorological, 169
    monitoring
        limitations, 385
    polar-orbiting, 169
    sensors, 248
saturation vapour density, 257
saturation vapour pressure deficit, 400
scale, 11
scattering function, 345, 346

scatterometer measurements, 316
scatterometers, 307, 331
seasonal change, 360
seasonal NDVI, 177
seasonal production, 173
Senegal, 169, 173, 174
senescent leaves
    stress index, 294
sensible heat flux, 46, 112
sensing frost damage, 273–282
sensor degradation, 176
sensors, 3–15
    calibration, 336, 391
Shuttle Imaging Radar, 360, 365
Side-Looking Airborne Radar (SLAR), 307
simple-ratio vegetation index, 358
simulated transpiration, 121
simulation map, 112
simulation model
    agro-hydrological, 112
simulation models, 111
single scattering albedo, 347
soil
    background, 33, 179, 219, 222
        reflectance, 189
    bare, 341
    boundaries, 81
    brightness, 356
    characteristics, 114
    classification criteria, 76
    colour, 84
    cover, 33
    density, 336
    dielectric constant, 84
    erosion, 75
    heat, 112
    iron content, 25
    map, 115, 244
        fieldwork, 76
        large scale, 76
        medium scale, 77
        scale, 76
    mapping, 75, 76
    microwave backscatter, 313
    moisture
        content, 341
    moisture content
        dielectric constant, 333
    organic matter, 25
    permittivity, 314
    roughness, 336
    salinisation, 75, 85
    temperature, 84
    temperature differences, 83
    transpiration, 121
soil colour
    soil moisture, 84

soil degradation, 75, 85
soil dielectric properties, 309
soil losses, 86
soil moisture, 7, 25, 84, 314
    content, 336
        radar, 313
soil monitoring, 75–87
soil organic matter
    radar signal, 315
soil pattern, 77
soil reflectance, 25, 80, 185, 190, 358
    and mineral composition, 25
soil reflectivity, 346
soil roughness, 26
    and colour, 84
    microwave backscatter, 316
soil signatures, 80
soil structure
    aerial photography, 244, 245
soil surface characteristics, 80
soil temperature, 52, 346
soil water, 257, 399
    potential, 257, 259
soils, 163
    dielectric properties, 314
solar elevation, 28
    reflectance, 28
solar energy intercepted
    leaf area, 197
solar irradiance, 202
solar radiation, 46, 179, 211, 262, 400
    canopy, 56
    incident, 171
    total, 171
solar zenith angle, 277
sources of error
    temperature, 269
Spain, 276, 383, 388
spatial data, 14
spatial resolution, 7, 8, 55, 97, 98
    of meteorological data, 10
spatial variability, 54
    of surface temperatures, 54
spectral characteristics of crops, 400
spectral classes
    statistically separable, 129
spectral indices, 144, 188
spectral radiance, 30
spectral reflectance, 222
    leaf area index, 222
    leaf cover, 222
spectral response
    citrus, 273
    frost, 273
spectral sensitivity, 150
    film, 233
spectral signature, 210, 220, 248, 360, 381

spectral transmittance
    leaf, 273
spectral vegetation index, *see* vegetation index
spectro-radiometer
    IRIS, 215
spectro-radiometric measurements, 150
spectrometer, 4
spectrophotometer, 203
specular reflection
    of leaves, 22
Split Window method, 386
spongy mesophyll, 22
SPOT, 15, 78, 115, 130, 154, 191, 249, 281, 355, 358, 377, 382, 384, 386
SPOT High Resolution Visible sensor, (HRV), 128
sprinkling irrigation, 120
spruce
    damaged, 299
stability correction, 47
stability parameter, 260
standing biomass, 210
    woody, 366, 367, 372
standing crop
    above-ground, 172
statistic
    rainfall estimate, 98
Stefan-Boltzmann constant, 46
stomatal control
    evapotranspiration, 61
stomatal opening, 258
stomatal resistance, 61, 257, 400
strain, 212
strata boundaries, 161
stratification, 377, 379, 382
    applications, 379
    automated, 380
stratified map, 138
streamflow forecasts, 108
stress, 212
stress detection, 288
stress effects, 213
    in vegetation, 210
stress index, 292
    of evaporative demand, 257
stress indicator, 296
stress physiology, 287–302
stress-adaptation index A_p, 293
stress-degree-days, 262, 399
stressed vegetation monitoring, 212
striping
    aerial photography, 245, 246
Sudan
    dry dekads, 107
sugar beet, 201, 219
    virus yellows, 210, 219, 236
        aerial photography, 236

sugar production, 401
sugar yield, 202
   national forecasts of, 204
sun azimuth, 117
sun azimuth angle, 267
sun zenith angle, 267
sunflower, 338
surface energy balance, 111
surface roughness, 260
surface temperature, 46, 54, 389, 398
   measurement, 267
      accuracy, 269
      precision, 269
      true, 66
surface-air temperature difference, 111
Synthetic Aperture Radar (SAR), 15, 307, 360

take-all
   aerial photography, 238
   drainage, 241
target scale, 131
Tasseled Cap Transformation, 356, 362
technological development, 4
temperature
   low, 273
   minimum, 139
   radiative, 45
   spatial variability, 54
temperature distribution
   row orientation, 57
   within canopy, 49
temperature histograms, 57
temperature threshold, 98
temperatures
   aerodynamic, 268
temporal coverage, 97
temporal resolution, 7, 10
terrain variations, 130
test areas, 161
thematic data, 162
thematic map, 162
Thematic Mapper, 21, 81, 129, 273, 359, 369
   band reflectance, 363
thematic maps, 378
thermal
   infrared radiation, 45
thermal bands, 385
thermal images, 112
thermal imaging devices, 45
thermal infrared, 98, 338
   resolution, 359
   sensor, 332
thermal infrared images, 115
thermal sensors, 4
thermographs
   histograms, 60
thermography, 115

thermometer
   infrared radiation (IR), 45, 255, 267
threshold temperature, 99
time resolution, 102
topographic maps, 162
tractor mounted remote sensing, 224
training area, 275, 279
training data, 130
Transformation
   Perpendicular, 356
   Progressive, 362, 364
   Tasseled Cap, 356, 362
Transformed Divergence, 129
Transformed Vegetation Index, 356
transmittance
   infrared, 273
transmittance spectra of wheat leaves, 19
transpiration, 112, 185, 257
   map, 115, 119
   measured, 257
   potential, 257
   rate, 259, 398
   reduction, 120
   simulated, 121
tree leaf production, 174

uncorrected satellite data, 171
use of computer, 7

validation, 379
vapour pressure deficit, 47, 61, 255, 260, 262
variability of pixel size, 28
variation coefficient, 33
   of viewed area, 27, 29
vector mapping, 14
vegetation
   amount, 358
   condition, 389
   index
      cover, 222
      radiation intercepted, 213
   microwave emission, 334
   responses to stress, 214
   semi-arid, 169
   type, 170, 355
Vegetation Index
   Perpendicular, 356
   Transformed, 356
vegetation index, 113, 140, 144, 146, 150, 154, 160, 169, 209–224, 277, 281, 360, 385
   high-spectral resolution, 209
   interpretation, 209
   limitations, 210
   sun angle effects, 151
   testing, 218
   winter wheat
      frozen, 384

Index

view angle, 30, 77, 117, 176
    and shadows, 50
view area, 27
view azimuth, 268
view zenith, 267
virus yellows of sugar beet, 205
visible bands, 385
visible channels, 14
visible reflectance, 19
visual discrimination, 378
visual interpretation, 118
vitality index, 289
von Karman's constant, 47, 260

walnuts, 369
water
    bound, 314
    crop
        use of, 114
    dielectric constant, 333
    free, 314
    uptake model, 114
water availability, 385
water available, 83, 113
water budget, 97
water deficit
    canopy temperature, 60
    evaporation, 61
water erosion, 86
water flow
    soil to leaf, 257
water management, 111–122
water management research, 121
water potential
    critical, 258
    leaf, 258
    soil, 257
water resources, 398, 401
water shortage
    microwave radiometry, 332
water stress, 24, 220, 257, 386
    measurement, 256
    fluorescence, 293
water table, 119
water thickness
    in leaves, 222
water use estimates, 11
water vapour flux, 399
water-use efficiency, 399, 400
weather forecasts, 401
wetness, 359
wheat, 338
    light interception
        effects of nitrogen, 187
    yields, 149
wheat-soil correlation model, 154
wheat-soil ratio, 154

wilting, 61
wind erosion, 86
wind speed, 260
wind velocity, 46, 113
    effect of, 47
windy conditions, 33
winter wheat
    area evaluation, 154
    chlorophyll content, 154
    estimation of growing area, 141, 142
    growth assessment, 153
    production of North China, 137
    vigour, 151
    yield estimation, 153
woody vegetation
    annual leaf production, 172
woody vegetation amount, 363

X-band attenuation, 321
X-band backscattering
    by plants, 369

yellow rust
    aerial photography, 236
yellowness, 356
    Kauth-Thomas, 358
yield, 399, 401
    cereals, 385
    estimation, 142, 153
    forecasting, 385
    greenness, 153
    microwave radiometry, 332
    winter wheat, 153
yield fluctuation
    meteorological conditions, 143
yield forecasting, 204, 384, 388
yield indicators, 389
yield prediction, 377
yield prediction model, 389
yield reporting, 137

zenith angle, 267
zero plane displacement, 260
zero plane height, 260
zoning, 377, 378
    methods, 379